Cyber-Physical Systems and Supporting Technologies for Industrial Automation

R. Thanigaivelan
A.K.T. Memorial College of Engineering and Technology, India

S. Kaliappan
Velammal Institute of Technology, India

C. Jegadheesan
Kongu Engineering College, India

A volume in the Advances in Systems Analysis, Software Engineering, and High Performance Computing (ASASEHPC) Book Series

IGI Global
PUBLISHER of TIMELY KNOWLEDGE

Published in the United States of America by
IGI Global
Engineering Science Reference (an imprint of IGI Global)
701 E. Chocolate Avenue
Hershey PA, USA 17033
Tel: 717-533-8845
Fax: 717-533-8661
E-mail: cust@igi-global.com
Web site: http://www.igi-global.com

Library of Congress Cataloging-in-Publication Data

Names: Thanigaivelan, Rajasekaran, 1977- editor. | Kaliappan, S., 1975- editor.
Title: Cyber-physical systems and supporting technologies for industrial automation / edited by: R. Thanigaivelan and S. Kaliappan.
Description: Hershey PA : Engineering Science Reference, [2023] | Includes bibliographical references. | Summary: "This edited book attracts readers including engineers, researchers, technologists, students at both the graduate level and undergraduate level, and even entrepreneurs. It unfolds the latest trends of CPS in health care, manufacturing processes, energy and mobility industry. It also focuses on advanced subsystems required for communication of real time data. Many publishers are publishing books in this area and received good response. This theme will attract my scholars and students who are working in this field to identify the research gap. This book will foresee more citations"-- Provided by publisher.
Identifiers: LCCN 2023008220 (print) | LCCN 2023008221 (ebook) | ISBN 9781668492673 (hardcover) | ISBN 9781668492710 (paperback) | ISBN 9781668492680 (ebook)
Subjects: LCSH: Automatic control. | Cooperating objects (Computer systems) | Automation. | Robots, Industrial.
Classification: LCC TJ225 .C93 2023 (print) | LCC TJ225 (ebook) | DDC 629.8/9--dc23/eng/20230328
LC record available at https://lccn.loc.gov/2023008220
LC ebook record available at https://lccn.loc.gov/2023008221

This book is published in the IGI Global book series Advances in Systems Analysis, Software Engineering, and High Performance Computing (ASASEHPC) (ISSN: 2327-3453; eISSN: 2327-3461)

British Cataloguing in Publication Data
A Cataloguing in Publication record for this book is available from the British Library.

All work contributed to this book is new, previously-unpublished material.
The views expressed in this book are those of the authors, but not necessarily of the publisher.

For electronic access to this publication, please contact: eresources@igi-global.com.

Advances in Systems Analysis, Software Engineering, and High Performance Computing (ASASEHPC) Book Series

Vijayan Sugumaran
Oakland University, USA

ISSN:2327-3453
EISSN:2327-3461

MISSION

The theory and practice of computing applications and distributed systems has emerged as one of the key areas of research driving innovations in business, engineering, and science. The fields of software engineering, systems analysis, and high performance computing offer a wide range of applications and solutions in solving computational problems for any modern organization.

The **Advances in Systems Analysis, Software Engineering, and High Performance Computing (ASASEHPC) Book Series** brings together research in the areas of distributed computing, systems and software engineering, high performance computing, and service science. This collection of publications is useful for academics, researchers, and practitioners seeking the latest practices and knowledge in this field.

COVERAGE

- Metadata and Semantic Web
- Storage Systems
- Computer Graphics
- Computer System Analysis
- Engineering Environments
- Virtual Data Systems
- Human-Computer Interaction
- Performance Modelling
- Distributed Cloud Computing
- Computer Networking

The Advances in Systems Analysis, Software Engineering, and High Performance Computing (ASASEHPC) Book Series (ISSN 2327-3453) is published by IGI Global, 701 E. Chocolate Avenue, Hershey, PA 17033-1240, USA, www.igi-global.com. This series is composed of titles available for purchase individually; each title is edited to be contextually exclusive from any other title within the series. For pricing and ordering information please visit http://www.igi-global.com/book-series/advances-systems-analysis-software-engineering/73689. Postmaster: Send all address changes to above address.

Titles in this Series

For a list of additional titles in this series, please visit: http://www.igi-global.com/book-series/

Perspectives and Considerations on the Evolution of Smart Systems
Maki K. Habib (American University in Cairo, Egypt)
Engineering Science Reference • © 2023 • 419pp • H/C (ISBN: 9781668476840) • US $325.00

Neuromorphic Computing Systems for Industry 4.0
S. Dhanasekar (Department of ECE, Sri Eshwar College of Engineering, India) K. Martin Sagayam (Karunya Institute of Technology and Sciences, India) Surbhi Vijh (ASET, Amity University, Noida, India) Vipin Tyagi (Jaypee University of Engineering and Technology, India) and Alex Norta (Tallinn University, Estonia)
Engineering Science Reference • © 2023 • 377pp • H/C (ISBN: 9781668465967) • US $270.00

Business Models and Strategies for Open Source Projects
Francisco José Monaco (Universidade de São Paulo, Brazil)
Business Science Reference • © 2023 • 339pp • H/C (ISBN: 9781668447857) • US $270.00

Advanced Applications of Python Data Structures and Algorithms
Mohammad Gouse Galety (Department of Computer Science, Samarkand International University of Technology, Uzbekistan) Arul Kumar Natarajan (CHRIST University (Deemed), India) and A. V. Sriharsha (MB University, India)
Engineering Science Reference • © 2023 • 298pp • H/C (ISBN: 9781668471005) • US $270.00

Handbook of Research on Machine Learning-Enabled IoT for Smart Applications Across Industries
Neha Goel (Raj Kumar Goel Institute of Technology, India) and Ravindra Kumar Yadav (Raj Kumar Goel Institute of Technology, India)
Engineering Science Reference • © 2023 • 542pp • H/C (ISBN: 9781668487853) • US $345.00

IGI Global
PUBLISHER of TIMELY KNOWLEDGE

701 East Chocolate Avenue, Hershey, PA 17033, USA
Tel: 717-533-8845 x100 • Fax: 717-533-8661
E-Mail: cust@igi-global.com • www.igi-global.com

Table of Contents

Detailed Table of Contents

The adoption of new technologies can pose challenges for organizations, such as significant investments in hardware, software, human capital, and the need to overcome technical and cultural barriers. The impact of Industry 4.0 on enterprises will depend on their ability to effectively adopt and integrate these new technologies into their operations. Semi structured interviews were conducted with leaders driving in the country from academia, companies, government, and society, with the purpose of gathering their opinions about the challenges and opportunities to adoption of Industry 4.0. Leaders in adoption mentioned that the lack of qualification of human capital in the manufacturing industry hinders the adoption of Industry 4.0. Mexican leaders describe important advances in the manufacturing sector, and they consider as having the greatest potential for growth. Mexico still should be increases initiatives to promote the adoption of Industry 4.0 technologies to drive growth and innovation across multiple sectors.

There are a number of diseases that are attacking human beings every day and everywhere. Chronic disease patients should be monitored continuously, and actions should be taken if it goes beyond the level of tolerability before one has to pay some large amount of money to health care centre. Financially struggling people cannot afford that amount, or even the equipment for monitoring purposes only. Generally chronic diseases are affected only by adults; they can be controlled but not cured. The authors propose chapter to monitor chronic disease patients instead of admitting them in hospital—a wearable bio sensor will monitor the patient efficiently and continuously, changes in their reading will be seen sooner by the caretaker or medical advisor. This chapter analyse wearables applied for chronic disease like cardiovascular disease and respiratory disease. Analysis on these devices is based on the parameters like convenience, cost, and efficiency of the device.

Hash functions map data of arbitrary size to fixed-size values, though there are some that support variable length output. The values returned by a hash function are called hash values. Hash functions are used in many fields such as generation of passwords and encryption which are the building blocks of cryptology, digital signatures, data integrity control, crypto storage, and random number generation. However, some hash functions like MD5 and SHA1 are now broken. Using a weak or broken hash function can leave data vulnerable, and thus should not be used. This chapter presents definitions, structures, performances, as well as security vulnerabilities, of hash functions.

The integration of additive manufacturing (AM) technology, commonly known as 3D printing, into industrial processes has led to the adoption of Industry 4.0 principles. This includes the integration of dynamic changes in the user experience, data-driven smart systems, and simpler production processes. In this research, a consumer-driven cyber-physical systems (CPS) framework is proposed, utilizing customer demand data to automate the entire process. The use of cloud computing, specifically Microsoft Azure, is utilized to optimize the process parameters and improve the efficiency of the 3D printing process using fused deposition modeling (FDM) technology. Additionally, machine learning techniques, specifically the multi-layer perceptron (MLP) neural network model, are implemented to further optimize the process parameters. A raspberry pi is used to connect to the Azure IoT hub and machine learning studio, where the algorithm is automatically reviewed, and the best value is selected.

Digital technology has transformed the whole world. Starting from the day-to-day activities of almost all of transactions, communication is happening in the digital format. The data that is communicated is also getting stored in a number of devices, starting from sensors to mobile and other computing devices. These devices can be digital or analog in nature. All the devices communicate within them and exchange the needed data among them. A cyber physical system (CPS) combines the digital and analog devices that are present in the physical and cyber world. The CPS integrates sensors and networks of the physical process to the computational components. These are done so as to control the physical process. There have been a number of applications where CPS have been deployed. Though there are vast applications, the CPS faces a lot of issues like the network security. This chapter aims at providing the current scenario the network security plays and its applications.

Cyber physical systems (CPS) are a combination of physical and cyber systems that interact with each other to perform a specific task. Industry 4.0, also known as the fourth industrial revolution, relies heavily on CPS to improve efficiency and automation in various industries such as mobility, healthcare, energy, and manufacturing. Design challenges in CPS include ensuring control and communication between the machine and the physical systems, modeling the complex interactions between the cyber and physical systems, integrating sensors and wireless networks, and ensuring high-speed data transfer. Additionally, interfacing actuators and sensors, and ensuring network security are also important considerations. CPS is a critical component of Industry 4.0 that enables machines, devices, and systems to communicate and collaborate with each other, resulting in more efficient and effective operations. The future of CPS in Industry 4.0 and the impact that it may have on the manufacturing industry as a whole is discussed in this Chapter.

Autonomous physical systems like unmanned air vehicles are creating a large market in the current technological scenario. This rapid technology is implemented in military, civil, and other complex applications. This new technology inspired many organizations and other individuals to create a UAV. In recent years with the availability of UAVs and autonomous vehicles familiarly called drones, there must be a threat to the security of autonomous vehicle operating domains or countries. It is often used to connect wireless networks, the cloud, or mobile applications. This physical system has to be monitored properly by the user to restrict unauthorized user interfaces in drones. This chapter deals with the integration of UAVs and cloud computing flying ad-hoc network (FANET), mobile application interface in IoT. The authors will discuss the components of drones used for computing. The drone

The integration of cyber-physical systems (CPS) with energy infrastructure has the potential to revolutionise the way we generate, distribute and consume energy. CPS enables the monitoring and control of energy systems in real-time, and allows for the integration of renewable energy sources and energy storage management. This research aims to explore the use of CPS in the design and development of a housing microgrid, with a focus on optimising interconnected energy systems and controlling and sensing parameters. To accomplish this, the authors propose a model-based design (MBD) approach, which uses simulation tools to create and test numerous control procedures for a housing microgrid. The imitation tool, GridMat, is a MATLAB/Simulink toolkit that can co-simulate control techniques written in Simulink and power systems modelled in GridLAB-D. This research demonstrates how GridMat can create, simulate, debug, and evaluate multiple layers of control algorithms for a residential microgrid.

With the invention of Quick Response (QR) code technology by the Japanese automotive company Denso Wave, the ease of use and the development of image processing receivers are making its use widespread in all sectors day by day. The widespread use of QR code, especially in applications containing personal data such as banking, finance, and messaging applications, has made it the focus of fraudsters and cyber attackers. QR code supported phishing attacks are among the new methods of fraudsters because awareness of the safe use of QR code has not yet been formed. In this context, in this chapter, the authors try to draw attention to the safe use of QR code, and present cyber security risks that QR codes may contain, common usage areas of QR codes, QR code generators, and issues related to QR code awareness.

Preface

The advancement of operations in various industries, including architecture, engineering, and construction (AEC), depends on the development of cyber-physical systems (CPSs). Technological developments that support the CPS's objectives of automation and intelligent interaction between cyber and physical areas have aided in the development of the system. The evaluation of CPS supporting technologies in AEC projects is the primary objective of this book. This book chapter provides the complete analysis of the academic and industrial literature, the technologies, their degree of implementation and application across the project lifecycle for horizontal and vertical project scenarios. A classification was created to show the relationships between the technologies and offer fresh perspective on further research into CPS applications in the AEC.

The emergence of the Internet of Things (IoT) is one of the most rapid and fascinating advances in information and communications technology. Despite the fact that networking technologies have proliferated over the past 20 years, until recently they were mostly used to link together conventional end-user devices like mainframes, desktop and laptop computers, and, more recently, smartphones and tablets.In recent years, a significantly wider variety of devices have been connected to the network. Vehicles, home appliances, medical equipment, electric metres and controllers, street lighting, traffic lights, smart TVs, and virtual assistants like Amazon Alexa and Google Home have all been included. Currently connected to the network are more than eight billion such gadgets, according to industry analysts, who predict that this number will rise.

Various universities and Institutes are teaching advanced materials and manufacturing and many students, industrialist, professors and research scholars will be interested in it.

The cyber physical system finds application in healthcare, transportation, aerospace and defense, food and beverages, and electronics and semiconductors. The cyber physical system market, by geography, is segmented into four regions, namely, North America, Europe, Asia-Pacific (APAC), and the Rest of the world (RoW). The employment demand for highly skilled labour relevant to cyber physical system is increasing. The training for developing a skilled labour pool in cyber physical system will be improved through various program such as vocation training and education programmes, as well as continuing professional development schemes. Various universities and Institutes are teaching advanced materials and manufacturing and many students, industrialist, professors and research scholars will be interested in it.

ORGANIZATION OF THE BOOK

The book is organized into sixteen chapters. A brief description of each of the chapters follows:

Ch. 1: A Review on Artificial Intelligence based Cyber Physical System for Industry 4.0 focus on cutting-edge features ie the Internet of Things (IoT) and Web of Things (WoT), which finds application in industrial systems.

Ch. 2: Advances and Challenges to Adoption of Industry 4.0 in the Manufacturing Sector: Case Mexico discussed the impact of Industry 4.0 on enterprises and their dependance on their ability to effectively adopt and integrate these new technologies into their operations.

Ch. 3: Analysis on IoT Wearable Sensors to Monitor Chronic Diseases focuse on monitor chronic disease patients instead of admitting them in hospital and paying too much of money, wearable bio sensor will monitor the patient efficiently and continuously, changes in their reading will be intimated in prior to the care taker or medical advisor. This chapter analyse wearable's applied for chronic disease like cardio vascular disease and respiratory disease. Analysis on these devices based on the parameters like convenient, cost and efficiency of the device.

Ch. 4: Analysis and Comparison Of Hash Functions in Terms of Security and Performance presents definitions, structures, performances as well as security vulnerabilities of hash functions.

Ch. 5: Analysis and monitoring of design challenges in the cyber physical systems in manufacturing sectors using cloud computing dicuss the integration of additive manufacturing (AM) technology, commonly known as 3D printing, into industrial processes has led to the adoption of Industry 4.0 principles. In this chapter, a consumer-driven cyber-physical systems (CPS) framework is proposed, utilizing customer demand data to automate the entire process.

Ch. 6: Cyber Physical Systems and Network Security: The Present Scenarios and Its Applications aims at providing the current scenario the network security plays and its applications

Ch. 7: Cyber Physical Systems Design challenges in the areas of mobility, healthcare, energy and manufacturing discuss the design challenges in CPS include ensuring control and communication between the machine and the physical systems, modeling the complex interactions between the cyber and physical systems, integrating sensors and wireless networks, and ensuring high-speed data transfer.

Ch. 8: Cyber Security in Drone Technology deals with the integration of UAVs and cloud computing Flying Ad-hoc NETwork (FANET), mobile application interface in IoT. We will discuss the components of drones used for computing. The drone integration of computing system architecture and types has an important role in the physical system. This chapter discusses the technology development in drones and the novel ideas may incorporate in future.

Ch. 9: Design and implementation of production line through cyber physical system in manufacturing sector in Industry 4.0 using machine learning explores the critical steps for mixing three CPS columns (manufacturing line, sequence, and conveniences) into modern smart industrial settings, in instruction to approve an manufacturing Cyber-Physical Systems model. The method focuses on incorporating various digital capabilities into a cloud platform, to facilitate real-time multi-device communication, information analytics/distribution, and deep learning-based worldwide regeneration.

Ch. 10: Detection of fault analysis in sensor network system through embedded cyber physical system in manufacturing sector explores the issue of sensor reliability in the adoption of Internet of Things (IoT) data in critical industries such as motorized and industrial sectors. It summarizes significant European research and innovation programs, as well as recent investigations on sensor dependability in cyber-physical systems (CPS).

Ch. 11: Face Biometric Authentication System for ATM using Deep Learning discuss an automatic teller machine security model that would combine a physical access card and electronic facial recognition using Deep Convolutional Neural Network. If this technology becomes widely used, faces would be protected as well as their accounts. Face Verification Link will be generated and sent to user to verify the identity of unauthorized user through some dedicated artificial intelligent agents, for remote certification.

Ch. 12: Immediate fault detection in production and communication line in industrial sector through artificial intelligence presents a deep learning-based model for network fault analysis and investigates its effectiveness through experiments and simulations. The proposed technique is aimed to prevent errors produced by the created model and enhance the resilience, universality, and accuracy of the network fault diagnostic model. The results of the study suggest that the proposed technique is capable of identifying and locating faults in a timely and efficient manner, thus reducing network interruption and increasing network reliability.

Ch. 13: Implementation of cyber physical system in smart cities through augmented reality networks in the mobility decade discuss the implementation of CPS in smart cities through the use of AR networks to enhance mobility. AR technology can provide real-time information and guidance to citizens, improving the navigation and transportation experience. Additionally, the integration of CPS and AR can lead to the optimization of city operations such as traffic management and maintenance of infrastructure.

Ch. 14: Implementation of human machine interface module and control system based on controller area network focus on universal CAN bus with an IoT focus, which aims to establish a system that can connect and communicate with different devices using the CAN protocol, while also considering the unique requirements of IoT applications.

Ch. 15: Interfacing of Actuators and Sensors In this chapter provides a thorough review of the sensor and actuators and where these models are implemented in various real contexts of mechanical parts of the industry, electrical and electronic circuits and other essential areas.

Ch. 16: Modelling and development of energy system through cyber physical system with optimising interconnected with control and sensing parameter explores the use of CPS in the design and development of a housing microgrid, with a focus on optimising interconnected energy systems and controlling and sensing parameters. To accomplish this, we propose a model-based design (MBD) approach, which uses simulation tools to create and test numerous control procedures for a housing microgrid

Ch. 17: New Generation Mobile Cyber Security Threats: Qr Codes And Social Engineering Threats in this chapter, the safe use of QR code, and present cyber security risks that QR codes may contain, common usage areas of QR codes, QR code generators, and issues related to QR code awareness.

Ch. 18: Prediction and development of health care system through cyber physical system using cloud computing and big data This chapter aims to demonstrate the potential of big data and cloud technologies to enhance healthcare system efficiency and provide patients with access to a wide range of intelligent healthcare applications and services.

Ch. 19: Recognition and implementation of cyber physical system in the development of smart factories in Industry 4.0 through optimization techniques this article investigates the recognition and implementation of cyber physical systems (CPS) in the development of smart factories in Industry 4.0. With the integration of physical and digital components, CPS allows for the seamless integration of control systems, communication networks, and sensor systems leading to the optimization of factory processes.

Ch. 20: Securing Decentralized Finance through Blockchain Technologies focuses on DeFi oracles, starting with introducing DeFi oracle application scenarios, then moves onto characterizing them into several types based on their design factors. Finally, it analyzes and address the security aspects of oracles in DeFi and explores the risks associated with oracles, such as data inaccuracy, centralization, manipulation, and oracle failure, and proposes solutions to mitigate these risks by focusing on the Chainlink network.

Ch. 21: Recognition of cyber physical system in smart manufacturing system with wireless connectivity through deep learning techniques in industrial process in this chapter a concept of a smart factory that integrates deep learning algorithms into a defect detection system and outlines the network architecture for the proposed smart factory is discussed.

Ch. 22: Security for Data Communication in Cyber Physical System Limitation and Issues to Analyse the Performance Level of Networks: A Secure mode of Data Transmission in Networks Using Different Types of Component Layers focus on the elements of CPS such as networked systems with processors, sensors, cloud service provider, these platform can also used in car, medical device, and IoT sensors for creating platforms and the network performance also increased.

This book acts as a bridge between this high-level research and the large number of academics and practitioners looking to cyber physical systems for innovative solutions, providing them with practical and approachable information. Applications in aerospace, automotive, medical, and food industries are addressed, featuring technical details that will help successful implementation. This unique book also provides broad coverage of the theory behind this CPS,IOT,Sensors andNetworking,as well as the technical details required for readers to investigate the novel applications of the involved methods for themselves.

First and foremost, the Editors would like to thank the Almighty for His blessings for completing the work to their satisfaction. The Editors would also like to thank all the Chapter Contributors, the Reviewers, Book Development Editor, and the team of IGI global Personnel for their availability for work on this editorial project. Last, but positively not least, the editors would like to thank all the well-wishers, colleagues, students, and all who were directly or indirectly involved in this project.

R. Thanigaivelan
A.K.T. Memorial College of Engineering and Technology, India

S. Kaliappan
Velammal Institute of Technology, India

C. Jegadheesan
Kongu Engineering College, India

Acknowledgment

The editors would like to recognize the help of all the people concerned in this project and, more particularly, to the authors and reviewers that took part in the review process. Without their support, this book would not have become a reality.

First, the editors would like to thank each one of the authors for their contributions. Our sincere gratitude goes to the authors of chapter's who contributed their time and expertise to this book.

Second and Third editors wish to acknowledge the valuable contributions of the reviewers regarding the improvement of quality, coherence, and content presentation of chapters. Most of the authors also served as referees; we highly appreciate their double task.

R. Thanigaivelan
A.K.T. Memorial College of Engineering and Technology, India

S. Kaliappan
Velammal Institute of Technology, India

C. Jegadheesan
Kongu Engineering College, India

Chapter 1
A Review on Artificial Intelligence-Based Cyber Physical Systems for Industry 4.0

G. Shanmugaraj
Velammal Institute of Technology, India

G. Sangeetha
RMK College of Engineering and Technology

G. Sethuram Rao
https://orcid.org/0000-0003-1878-8599
Velammal Institute of Technology, India

K. Mohanambal
Velammal Engineering College, India

ABSTRACT

The cyber physical system (CPS), which combines cutting-edge features via the internet of things (IoT) and web of things (WoT), has produced an excellent foundation for creating innovative industrial systems and applications. This has made it possible to link computing and communication infrastructures with the processes of the physical world. The challenges that the industrial industry is experiencing are changing quickly. It's important to gather, comprehend, and utilize the enormous amount of data the Industrial internet of things generates (IIoT). There is an urgent need for methodical artificial intelligence (AI) development and application in order to observe its true influence in the next generation of industrial systems, namely Industry 4.0. Industry 4.0 is transforming the businesses production, enhancement, and distribution of their goods. The internet of things, cloud computing, analytics, AI, and machine learning are among the cutting-edge technology that manufacturers are incorporating into their manufacturing processes.

DOI: 10.4018/978-1-6684-9267-3.ch001

1. INTRODUCTION

Artificial intelligence (AI) is a branch of cognitive science that has active research programmes in a variety of fields, including robotics, machine learning, natural language processing, and image processing. Machine learning and artificial intelligence (AI) have historically been seen as black-art approaches, and there is frequently a dearth of strong data to persuade industry that these techniques will deliver a return on investment time and time again. At the same time, a developer's choices and experience have a big impact on how well machine learning algorithms function. As a result, AI has had mixed success in industrial applications. Industrial AI, on the other hand, is a methodical field that focuses on creating, testing, and implementing various machine learning algorithms for applications in the industrial sector with long-term performance. It serves as a bridge between academic research findings in AI to industry practitioners and serves as a systematic technique and discipline to offer solutions for industrial applications. The relevance of artificial intelligence (AI) and the need for a well-defined roadmap and strategic investment in this field are highlighted in the most recent White House report on the subject (Lee, 2016).

This study gives an insight into the present status of AI technologies and the eco-system needed to harness the potential of AI in industrial applications within the 5C architecture originally suggested in Lee et al. (2015). AI and machine learning, manufacturing companies can fully benefit from the wealth of information generated not just on the factory floor but also across all of their business units, as well as from partners and outside sources. With the use of AI and machine learning, operations and business processes can become more visible, predictable, and automated .Smart use of data is made possible by integrated cutting-edge technology in the Digital Enterprise portfolio. It is also possible to develop cognitive architectures that precisely replicate the motivations, emotions, personalities, and other essential traits of truly intelligent humans (Sun 2020).

The rise of productivity has not yet been significantly affected quantitatively by AI-driven automation (Lee. K, 2016). In addition, modern enterprises face fresh difficulties due to market demand and competitiveness. They require the drastic transformation known as Industry 4.0.Industry operations will be more flexible, effective, and environmentally friendly by the integration of AI with current developing technologies including the Industrial Internet of Things (Da Xu L, 2014), big data analytics (Ardakani. HD, 2015), cloud computing (Greer. MJ, 2013), and cyber physical systems (Baheti. R, 2011). Industrial AI is still in its infancy, hence it is crucial to distinctly define its framework, methodology, and obstacles in order to deploy it in industry. In order to do this, we created an industrial AI ecosystem that encompasses all relevant components and serves as a roadmap for improved comprehension and application. The enabling technologies that may be used to build an Industrial AI system are also outlined.

2. WORKING OF ARTIFICIAL INTELLIGENCE

Using a range of technologies, artificial intelligence enables software and computers to sense, comprehend, act, and learn independently or to support human tasks. In comparison to human workers, artificial intelligence can increase the efficiency of industrial production. Robots may also be equipped with AI to carry out activities that a human would find difficult or impossible to complete, such as handling hazardous raw materials or microscopic components. It is crucial to understand that many of these industrial robots are currently not very intelligent. They are frequently capable of carrying out a variety of tasks effectively, although their programming is often constrained. You must reconfigure it if you require more. The fact that this technology improves with Industry 4.0 is what fuels the region's ongoing expansion. The cost of artificial intelligences falling as it develops. Using more sophisticated AI algorithms has also made it possible for industries to assess the purchase of new technologies that help them make decisions in a more complex and secure manner. Figure. 1. shows how does Artificial Intelligence works.

Figure 1. Working of artificial intelligence

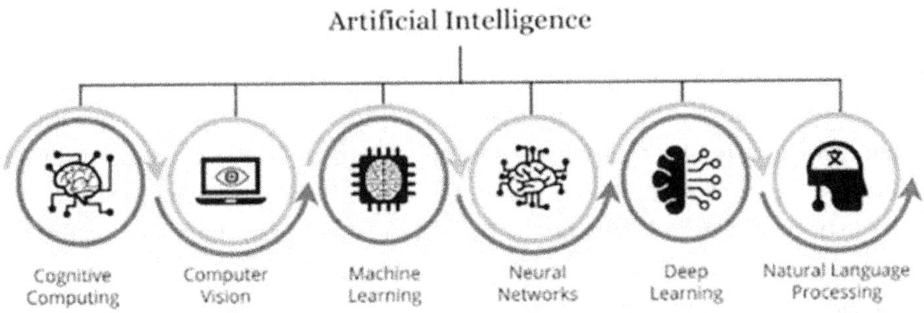

3. SIGNIFICANT ADVANCEMENTS IN INDUSTRY 4.0 THROUGH ARTIFICIAL INTELLIGENCE

The fact that this technology improves with Industry 4.0 is what fuels the region's ongoing expansion. The cost of artificial intelligence is falling as it develops. Using more sophisticated AI algorithms has also made it possible for industries to assess the purchase of new technologies that help them make decisions in a more complex and secure manner.

Error Reduction: Intelligent algorithms can do activities that are prone to faults in processes carried out by humans very successfully after being educated. Algorithms shouldn't likely experience the effects of external stimuli as they are not vulnerable to them.

Revenue Growth: Decision-makers will have more time to consider the core business and abandon other AI tasks as there will be less errors and workers will be concentrated on more crucial operations.

Cost Reduction: Many banks or online retailers use robots to start customer support interactions. Only in cases when the issue is more complicated is a human attendant contacted. With this strategy, businesses can cut personnel costs or assign workers to other, more important duties that can boost revenue and help them concentrate on their core operations.

4. APPLIED ARTIFICIAL INTELLIGENCE IN MANUFACTURING SYSTEM

Artificial intelligence has been incorporated into industrial lines during the past few years. Nevertheless, businesses from various sizes and industries around the world are already implementing intelligent technologies that enable them to automate and enhance processes.

Intelligent Manufacturing (Industrial IoT And AI): The following efficiencies are made possible by intelligent manufacturing, which applies real-time data analysis, artificial intelligence (AI), and machine learning to the production process. According to Trend Force, there will be a noticeable increase in this industry, which predicted that the market for smart manufacturing would reach more than $ 320 billion by 2020 and more than 2.6 million industrial robots would be placed in companies.

AI in Transportation: Smart technology can support the route's projected transit time or offer recommendations for more effective modes of transportation, other routes, sign recognition, self-driving cars, and sensors.

AI in Agriculture: There are several autonomous machine prototypes that can harvest without human assistance, and there are even automated systems for spotting illnesses and pests in certain crops.

5. ARTIFICIAL INTELLIGENCE VS. MACHINE LEARNING IN CYBER PHYSICAL SYSTEM

It would be beneficial to evaluate the domains of artificial intelligence and machine learning jointly before we start looking at how artificial intelligence is influencing Industry 4.0. Because the word "AI" is also frequently used to refer to machine learning algorithms in the subsequent sections of this paper, for the comfort and experience, dear readers. Even while AI and machine learning are closely related, they are not the same. An example of AI is machine learning. The need familiarity with system sociology (Dombrowski and Wagner 2014), multi-discipline testing (Balaji et al. 2015), and social networks (Wan et al. 2015).

Artificial intelligence (AI) refers to a computer system's capacity to imitate human cognitive processes like learning and making decisions. A computer system simulates human learning processes by using arithmetic and logic to make decisions and learn from new information.A machine can learn without being explicitly instructed by using machine learning (ML), which uses data-driven mathematical models. This enables a computer system to learn on its own and get better with practise.

The nine pillars include the internet of things, cloud computing, autonomous and robotics systems, big data analytics, augmented reality, cyber security, simulation, system integration, and additive manufacturing. Figure 2 shows Pillars of artificial Intelligence.

6. ARTIFICIAL INTELLIGENCE IN INDUSTRY 4.0

To properly leverage the enormous amounts of data created on the factories, business units, partners, and third-party sources, manufacturers use AI and machine learning. Artificial intelligence (AI) in Industry 4.0 gives production processes and business procedures visibility, predictability, and automation through the generation of insights.

For manufacturing operations, artificial intelligence in Industry 4.0 offers a number of practical advantages. Error reduction is primary and foremost. On tasks that are prone to human mistake, well-trained machine learning algorithms surpass humans in terms of accuracy and efficiency. As a result, less time and money are wasted on ineffective processes and subpar goods, which lead to lower costs and higher income. And just a small number of AI applications in industry today provide these significant advantages. Figure 3 shows the graph of Industry 4.0 Market Size.

Figure 2. Pillars of artificial intelligence

Figure 3. Industry 4.0 market size

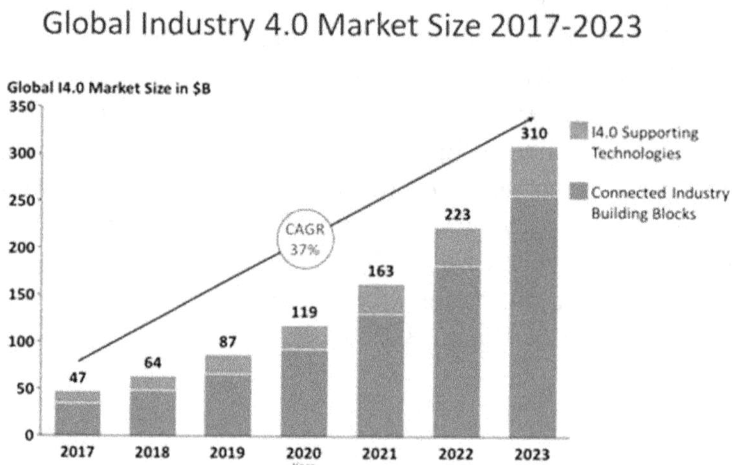

Predictive Analytics: In predictive analytics, predictions about future events are made using historical data, statistical modelling, data mining techniques, and machine learning, a branch of artificial intelligence. In order to identify trends in their data and spot errors and anomalies in their processes, manufacturers use predictive analytics. The economy and culture are already being changed by artificial intelligence (AI), and the rise of AI decision-making has generated debate about potential risks and the need for more openness in AI decision-making (de Fine Licht et al. 2020).

Predictive Maintenance: The condition, status, and performance of an asset are tracked using predictive maintenance, an AI-based proactive maintenance technique. By enabling manufacturers to schedule maintenance around their manufacturing schedule, predictive maintenance attempts to lower expensive and unplanned failures. In a sense, machine learning algorithms forecast the mistakes that will happen on the assembly line, ensuring that the problems are fixed before they happen.

Industrial Robotics: Industrial robots have been in use for more than 50 years, but recently they have lost their reputation as cold competitors who replace human workers with merciless efficiency and have gained a reputation as friendly helpers that may ease line workers' lives rather than snatching their jobs. Robots with artificial intelligence can move with greater sophistication and with greater accuracy and dependability. The work that goes into developing and implementing industrial automation may be greatly reduced thanks to artificial intelligence in Industry 4.0, which is perhaps the most important development . Figure 4 shows the image of Industrial robotics.

Figure 4. Industrial robotics

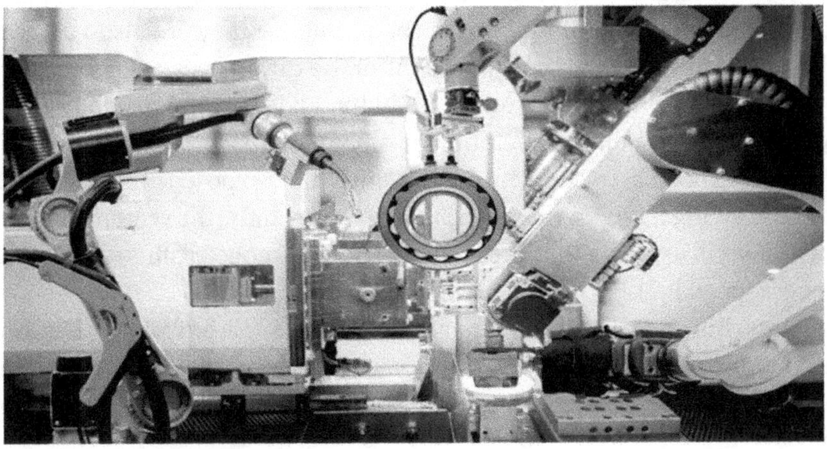

Computer Vision: Visual inspections are the main focus of industrial Computer Vision applications. Swiftness and accuracy are two areas where artificial intelligence clearly outperforms humans in visual assessment. An AI-enhanced Computer Vision system with cameras which are more sensitive than the human eye can find minor problems right away that human inspectors would miss.

Inventory Management: Industry 4.0's use of artificial intelligence lowers inventory-related costs in a variety of ways, from maximising the amount of inventory maintained on hand to foreseeing shortfalls before they materialise. AI is a perfect tool for inventory management because of its capacity to analyse enormous volumes of data and uncover hidden patterns. The most well-known corporation to employ AI in inventory management is Amazon. Industry 4.0 also uses artificial intelligence in a number of other areas, such as supply chain management, sustainability, and waste reduction. This tendency was sparked by the enormous economic possibilities of a hyper connected economy. According to literature, the digital economy's networking potential will offer promising economic opportunities in the future (Nicolescu et al. 2018).The number of AI industrial applications will surely increase as computer resources become more affordable and domain expertise multiplies.

7. CHALLENGES OF INDUSTRIAL ARTIFICIAL INTELLIGENCE

Industrial AI is expected to provide a wide range of diverse and significant benefits, and even a partial fulfilment of these benefits would bring unique and substantial hurdles for applying AI to industries. The following issues are more crucial and top priority among the current difficulties and complexity.

Machine-To-Machine Interaction: Although AI algorithms are capable of properly mapping a collection of inputs to a set of outputs, they are also vulnerable to slight deviations in the inputs brought on by differences across machines. It must make sure that, in the long run, individual AI solutions do not interfere with or conflict with the operation of other systems.

Cyber security: The smart manufacturing system is more exposed to cyber hazards due to the growing usage of linked technology. Currently, the sector is unprepared for the security risks that are present and the scope of this vulnerability is underappreciated.

Data Quality: Big, bias-free data sets are necessary for AI systems. The findings farther down the line may be erroneous as a result of learning from faulty or insufficient data sets.

8. CONCLUSION

Industry 4.0 completely automates the administration of the various production process phases through the application of AI. Based on the product specifications, any stage of the production process can be improved in real time. It is possible to integrate the entire development process, and different departments can split the workload of the data procedures. Manufacturing processes can incorporate data feedback and collection systems using AI. Production procedures and assembly lines may work together more efficiently with the use of this technology. Intelligent AI systems are used to forecast repairs and asset failure. For many businesses throughout the world, Industry 4.0 is now a reality. This is a continuing, constantly-improving process, similar to digital transformation and automation.

REFERENCES

Baheti, R., & Gill, H. (2011). *Cyber-physical systems*. Impact Control Technology.

Balaji, B., Al Faruque, M. A., Dutt, N., Gupta, R., & Agarwal, Y. (2015) Models, abstractions, and architectures. In: *Proceedings of the 52nd annual design automation conference on—DAC'15*, (pp. 1–6). ACM. 10.1145/2744769.2747936

Da Xu, L., He, W., Li, S. (2014). *Internet of things in industries: A survey*. IEEE Trans. Ind. Inf.

de Fine Licht K. (2020). *Artifcial intelligence, transparency, and public decision-making*. Springer.

Dombrowski, U., & Wagner, T. (2014). Mental strain as feld of action in the 4th industrial revolution. *Procedia CIRP*, *17*, 100–105. doi:10.1016/j.procir.2014.01.077

Lee, J., Ardakani, H. D., Yang, S., & Bagheri, B. (2015). Industrial big data analytics and cyberphysical systems for future maintenance & service innovation. *Procedia CIRP*, *38*, 3–7. doi:10.1016/j.procir.2015.08.026

Lee, K. (2016). *Artificial intelligence, automation, and the economy*. The White. House Blog.

Nicolescu, R., Huth, M., Radanliev, P., & De Roure, D. (2018). Mapping the values of IoT. J Inf. *Technoloy*, *33*(4), 345–360.

Sun, R. (2020). Potential of full human–machine symbiosis through truly intelligent cognitive systems. *AI & Society*, *35*(1), 17–28. doi:10.100700146-017-0775-7

Wan, J., Cai, H., & Zhou, K. (2015). Industrie 4.0: enabling technologies. In: *Proceedings of 2015 international conference on intelligent computing and internet of things,* (pp. 135–140). IEEE.

Wan, J., Yan, H., Li, D., Zhou, K., & Zeng, L. (2013). Cyber-physical systems for optimal energy management scheme of autonomous electric vehicle. *The Computer Journal, 56*(8), 947–956. doi:10.1093/comjnl/bxt043

Wu, D., Greer, M. J., Rosen, D. W., & Schaefer, D. (2013). Cloud manufacturing: Strategic vision and state-of-the-art. *Journal of Manufacturing Systems, 32*(4), 564–579. doi:10.1016/j.jmsy.2013.04.008

Chapter 2
Advances and Challenges to Adoption of Industry 4.0 in the Manufacturing Sector:
Case of Mexico

Eduardo Cornejo-Velazquez
https://orcid.org/0000-0002-0653-9459
Universidad Autónoma del Estado de Hidalgo, Mexico

Clavel-Maqueda Mireya
https://orcid.org/0000-0002-5487-9888
Universidad Autónoma del Estado de Hidalgo, Mexico

ABSTRACT

The adoption of new technologies can pose challenges for organizations, such as significant investments in hardware, software, human capital, and the need to overcome technical and cultural barriers. The impact of Industry 4.0 on enterprises will depend on their ability to effectively adopt and integrate these new technologies into their operations. Semi structured interviews were conducted with leaders driving in the country from academia, companies, government, and society, with the purpose of gathering their opinions about the challenges and opportunities to adoption of Industry 4.0. Leaders in adoption mentioned that the lack of qualification of human capital in the manufacturing industry hinders the adoption of Industry 4.0. Mexican leaders describe important advances in the manufacturing sector, and they consider as having the greatest potential for growth. Mexico still should be increases initiatives to promote the adoption of Industry 4.0 technologies to drive growth and innovation across multiple sectors.

DOI: 10.4018/978-1-6684-9267-3.ch002

INTRODUCTION

Industry 4.0 represents a major transformation in the way goods and services are produced and delivered. It is characterized by the integration of advances into the manufacturing and supply chain processes. Also, is reshaping organizations and disrupting business models, traditional companies will have to transform themselves to adopt these new processes or accept their obsolescence. The term Industry 4.0 was first used in a study published in Germany in April 2011 at the Hannover Fair, as one of the ten projects for 2020, it was officially presented under a subtitle of high-tech strategy (Grabowska & Saniuk, 2022; Kagermann et al., 2022). With this step, Germany has marked the leadership in the automotive, mechanical engineering, and Internet of Things (IoT) sectors in industrial environments; according to the prospects Germany was posed to be a pioneer in this new market (Kagermann et al., 2011).

Industry 4.0 encompasses the integration of digital technologies, automation, and data exchange within industrial processes, aiming to create more efficient and intelligent systems. Several studies (Inkulu et al., 2022; Saniuk et al., 2022; Caiado et al., 2022; Javaid et al., 2022a) provide additional insights and empirical evidence that support the opportunities and challenges associated with adopting Industry 4.0 technologies in enterprises.

For instance, research by Caiado et al. (2022) and Javaid et al. (2022a) offers empirical evidence highlighting the positive impact of Industry 4.0 technologies on efficiency, productivity, and profitability within organizations. These studies demonstrate that the integration of Internet of Things (IoT) and Artificial Intelligence (AI) enables predictive maintenance, improved supply chain management, and smarter decision-making (Javaid et al., 2022b; Radanliev et al., 2022). By leveraging data-driven insights, organizations can streamline processes, reduce costs, identify new revenue streams, make informed decisions, optimize supply chains, and ultimately increase profitability.

Industry 4.0 also provides opportunities for innovative producers and system suppliers. Implementing smart manufacturing processes, harnessing data analytics, and integrating various systems in the value chain can enhance efficiency, yield cost savings, and improve customer satisfaction. Moreover, regions that actively embrace and promote Industry 4.0 technologies can attract investments, foster innovation, and develop robust industrial ecosystems. This can result in job creation, heightened productivity, and increased competitiveness at the regional level.

However, failing to adapt to the changing technological landscape poses a serious risk for organizations and regions. Those that lag behind in embracing Industry 4.0 technologies may experience decreased productivity, loss of market share, and diminished relevance in the industry. The rapid advancement of Industry 4.0, if not effectively implemented, can lead to a widening productivity gap between leaders and laggards (Lorenz et al., 2015).

The adoption of new technologies can require significant investments in hardware, software, and human capital. In addition, organizations may also need to overtechnical and cultural barriers to effectively adopt and implement these technologies (Dieste et al., 2022; Senna et al., 2022). The impact of Industry 4.0 on enterprises will depend on their ability to effectively adopt and integrate these new technologies into their operations (Sigov et al., 2022; Ammar et al., 2022). Those organizations that can do so will be better positioned to compete in the rapidly changing business landscape, while those that fail to keep pace may fall behind.

In this chapter, interviews were used to identify advances, challenges, and opportunities in the process of adoption of Industry 4.0 in the manufacturing sector in Mexico, from the point of view of those who lead projects related to these innovations. Qualitative data obtained through the interviews also provide an account of the skills required to favor the adoption of Industry 4.0 in the manufacturing sector in Mexico.

BACKGROUND

The concept of Industry 4.0 advanced the idea of Cyber Physical Systems (CPS) into Cyber Physical Production Systems (CPPS) aim to empower manufacturing companies to strong collaboration between departments, making the right information available to the right people on a real-time basis (Kagermann et al., 2013; Klitou et al., 2017; Yang & Gu, 2021; Xu et al. 2021).

To achieve success in Industry 4.0, it is important to standardize the communication architecture for hardware and manage complex systems effectively to facilitate appropriate decision-making at the right time, thereby increasing efficiency and productivity (Marnewick & Marnewick 2019; Sony et al., 2021). Standardization of communication architecture is crucial for effective communication between companies involved in the production process. Managing complex systems is important as production processes become more complex with technological advancements. Critical success factors for Industry 4.0 include leadership, change management, continuous improvement, and quality and flexibility-based competition (Pozzi et el., 2023).

For the implementation of Industry 4.0, it is necessary to establish reliable, comprehensive, high-quality, and high-speed internet networks, as well as to ensure that all parties involved in the processes exploit them properly. In addition, it is essential to ensure the security of information (Kagermann et al., 2013; Rose et al., 2016).

Also, the organization and work design must include a system of continuous education to improve employees' professional competencies and thus be able to adapt to the constant changes in the labor market, especially in the era of Industry 4.0 (Mikhailov et al., 2018).

Legislation needs to adapt to the new industrial era and consider new agents involved. The efficient use of resources is increasingly important due to the potential problems and costs that may arise from emerging technology. Legislation and regulations need to be prepared for new ways of working, and laws must adapt to new operations. Therefore, it is important to consider the strengths and weaknesses of emerging technologies and their implications for society and the environment (Ponce-Romero et al., 2017).

The *Future of Work* (2020) highlights the importance of understanding the existing skills base in certain countries and industries, as well as how disruptive change will dictate new skill requirements in each case.

The report notes that discrepancies can arise not only between the current supply and demand for contemporary skills, but also between those skills and those that will be needed in the future.

To eliminate these gaps, a solid knowledge base is required to identify the skills that are currently in demand and those that will be needed in the future. This will help to ensure that the workforce is equipped with the necessary skills to meet the demands of the future.

By staying ahead of the curve and anticipating future skill requirements, individuals and organizations can position themselves for success in the rapidly changing world of work (WEF, 2020).

The main challenges of adopting Industry 4.0 include the need to define a strategy for Industry 4.0, rethink the organization and processes to maximize results, understand the business model, carry out successful pilots, make the organization realize that action is needed, manage change, foster an appropriate business culture, achieve true interconnection of departments, and have the appropriate human talent (Rose et al., 2016; Arboleda et al., 2020). In addition, the lack of tools remains an obstacle to exploiting the great potential of Industry 4.0, and formal methods and systems need to be implemented to enter Industry 4.0. Higher education institutions also face challenges in their articulation in Industry 4.0 (Pérez Rojas, 2020).

Given the need to take advantage of the possibilities offered by Industry 4.0, it is required to adopt and adapt the qualifications and competencies of human capital to be able to perform the tasks produced in a factory of the future, which will naturally be different (Habraken & Bondarouk, 2017; Bissola & Imperatori, 2019; Liboni et al., 2019; Silva et al., 2020; da Silva et al., 2022).

Industry 4.0 requires a workforce with high skill levels and, if this is not the case, companies must find ways to replace or retrain it.

To make this process possible, it is necessary to develop and establish training courses for employees, focusing on new key tasks such as the management and control of digitized systems.

It should be noted in this context that the biggest burden for companies undertaking the adoption of Industry 4.0 is employee expertise in the area of process and control know-how related to digitization (Baum & Kabst, 2014; Gospel, 2015).

Therefore training, and continuous professional development of employees is of great importance to succeed in the early stages of the transition to digitalization (Kagermann et al., 2013). The transformation will considerably require specific employee competencies, especially in the field of Information Technology (Okkonen et al., 2019; Schneider & Sting, 2020).

Typical programs to promote learning can include simulations of the digitized production process of the digitized production process to generate an understanding of Industry 4.0 procedures.

Industry 4.0 procedures. Other practices proposed by the literature are digital learning techniques (Kohnke, 2017; Zangiacomi et al., 2020).

Whether the company chooses to retrain employees or to look for new professionals depends largely on the internal capabilities of the organization.

While it is beneficial for companies to leverage the expertise of existing employees, it may also make sense to look for new professionals who are fit for transformation, and the strategy that each organization will depend on its own circumstances, but also on the context in which it operates. But also, on the context in which it develops (Agostini et al., 2020).

Human talent is one of the main values of the organization and this should be translated into the quality of its services and products, the entry into new markets, sales, innovation and profitability, with which the company has a sustainable competitive position over time. Strategy is the basis for the development and accumulation of organizational skills and competencies; in fact, it is through strategy that the company makes decisions that determine the resources that the organization will allocate to its key activities (Terán, 2013).

The strategies formulated by companies must be consistent and take advantage of the talent of the people at its disposal. Therefore, the company needs to know the strengths and weaknesses of its personnel, so that based on this information the company can decide on the possible actions to be taken.

Industry 4.0 Technologies

Role of Industry 4.0 technologies, such as Artificial Intelligence (AI), Internet of Things (IoT), Big Data, and Machine Learning (ML), are creating a sustainable environment in manufacturing and other industries. Javaid et al. (2022b) studied the benefits of Industry 4.0 for sustainable manufacturing and the ways in which it can improve environmental sustainability. Research was based on a literature review and identifies twenty major applications of Industry 4.0 in creating a sustainable environment. The authors concluded that Industry 4.0 technology has the potential to improve environmental sustainability through more efficient manufacturing processes and reduced resource consumption.

Khan and Javaid (2022) discuss the role of Internet of Things (IoT) in Industry 4.0 and its applications in monitoring production systems in manufacturing and services.

IoT allows for data collection and sharing through internet-connected devices, providing opportunities for improvement in performance. Researchers studied the potential of IoT to improve Industry 4.0 by analyzing its major drivers and how it can create a smart factory.

With proper implementation of IoT, industries can expect improved efficiency and lower costs in manufacturing, although there is still more to be achieved.

Adoption of Industry 4.0

Ancarani et al., (2019) analyzes the relationship between back shoring and the adoption of Industry 4.0 technologies. The study was based on secondary data from relocation initiatives to Europe and examines the competitive priorities of back shoring companies that may drive their adoption of new technologies. Their findings suggest that back shoring companies that prioritize high quality, and the reduction of non-conformance costs were more likely to adopt Industry 4.0 technologies. Back shoring initiatives that focus on reducing direct costs or increasing responsiveness were not significantly linked to the adoption of Industry 4.0. Results also highlight that companies that compete on quality and are involved in product innovation will be more likely to adopt new technology.

About the barriers to the implementation of Industry 4.0 technologies in the manufacturing sector, Raj et al. (2022) through a literature review and discussions with industry experts, 15 barriers were identified and analyzed using a Grey Decision-Making Trial and Evaluation Laboratory (DEMATEL) approach. The most significant barrier found was the lack of a digital strategy and resource scarcity in both developed and developing economies.

Study suggests that improvements in standards and government regulation could promote the adoption of Industry 4.0 in developing countries, while technological infrastructure is needed in developed countries. Highlight the difficulties in diffusing technological innovation due to a lack of coordinated national policies on Industry 4.0 in developing countries, which may limit companies from fully experiencing the Industry 4.0 revolution.

The results of the study can assist decision-makers and practitioners in overcoming the barriers to successful implementation of Industry 4.0 across the manufacturing sector.

Senna et al. (2022), summarizes the challenges that companies face in adopting Industry 4.0 and the lack of understanding of the interrelationships of these barriers. Authors used interpretive structural modeling and matrix impact of cross multiplication analysis to categorize these barriers using the Technology-Organization-Environment framework in the Portuguese manufacturing industry. Results indicate that standardization and lack of off-the-shelf solutions were the root barriers and that organizational barriers have the highest dependency and lowest driving power. Authors provide recommendations for managers and policymakers in the areas of standardization dissemination, infrastructure 93 development, and digital strategy.

Agostini and Nosella (2020), investigated the impact of financial resources invested in advanced manufacturing technologies and social capital on the adoption of Industry 4.0 technologies. It also examines if the organizational context enhances these relationships.

Nguyen and Luu (2020) aims to explore the factors that influence the adoption of Industry 4.0 by Small and Medium Enterprises (SME) in Ho Chi Minh City, Vietnam. Researcher used a mixed-method approach, including in-depth interviews and a quantitative survey of 396 SME representatives. Results of the study showed that factors such as perceived development of human resources, perceived cost savings, perceived product quality improvement, perceived ease of use, business resources, business environment conditions, perceived usefulness, and perceived enhanced customer relationships all had a positive and significant effect on the adoption of Industry 4.0. Authors suggest that efforts to increase the perceived importance of Industry 4.0, and the personal relevance of the technology to the adopters can contribute to successful implementation.

Finally, Corò and Volpe (2020) discusses the challenges and opportunities of Industry 4.0 and digitization in the world of technology. They mention that the biggest challenge will be job substitution and social inequality, while the benefits include substantial gains in every industry. Difficulties in defining Industry 4.0 were due to different perceptions by companies and stakeholders. Main barriers to creating value from data management and analytic were managerial and cultural rather than technological. Overcoming these barriers can lead to an improvement in a firm's human capital and enable the creation of new data-driven business models.

METHODOLOGY

Exploratory research based in semi-structured interviews were conducted in 2021, period of COVID-19 pandemic, with leaders driving Industry 4.0 in Mexico, with the purpose of gathering their opinions to understand the concept, problems faced in the adoption of the associated technologies, human capital needs in the sector, their perception of the skills required, and their projections regarding its adoption.

Determination of the Sample of Industry 4.0 Leaders

Industry 4.0 leaders were selected based on their specialization in the field using personal experience, professional contacts, and complementary data.

A non-probabilistic sample (Battaglia et al., 2008; Johnson and Christensen, 2013; Hernandez-Sampieri and Torres, 2018) of experts was formed according to the following two criteria:

- ***CR1.*** They are experts in the field, recognized by the role that they play in academia (research centers), leaders of civil, governmental organizations, or leading entrepreneurs in Industry 4.0 adoption identified through their public curriculum.
- ***CR2.*** They voluntarily agreed to participate by invitation (Battaglia et al., 2008).

To obtain a list of experts that met the first criterion (*CR1*) the following procedure was carried out:

1. For each region of country identified in the literature as relevant in the technology adoption process, research centers related to manufacturing and Industry 4.0, civil and governmental associations, and companies with activities to adoption through scientific and dissemination publications (conferences or industry magazines).
2. To identify candidates for interview, the approach of Ivlev et al. (2015) was considered, which defines the following objective criteria to identify expert agents in a subject matter:
    - ***CR3***. Job position.
    - ***CR4***. Level of education.
    - ***CR5***. Years of total experience.
    - ***CR6***. Years of experience in the issue.
    - ***CR7***. Level of participation in the problem.

However, there was not database in the country that publicly reports this information, there is not an organization that systematizes such data. Therefore, the LinkedIn network was used as a resource to obtain information because it is the most widely used professional social network (Ollington et al., 2013; Kluemper et al., 2016); moreover, it has been specifically designed for professional networking (Van-Dijck and Poell, 2013).

According to Tang et al. (2008), it is possible to identify in an academic social network, connections between people through an association search, and to identify experts in the network by considering not only the person-specific information, but also that of their relationships with other people by means of two steps: initialization and propagation (Zhang et al., 2007).

Initialization has to do with a node in the network that has certain relevance, from this and having as a basis the propagation theory, relationships with other people were identified, the basic idea is that, if a person knows many experts or if his name coincides many times with that of another expert, it is likely that person is an expert on the subject.

So, with that approach:

1. Individuals related to Industry 4.0 proposals (see specifics *CR1*), who gave conferences on I4.0 in recognized universities or events, were considered.
2. LinkedIn profiles were identified, and a table was systematically generated with a list of possible candidates.
3. Highest profiles resulting from the weightings of the five inclusion criteria (*CR3-CR7*) were selected to generate a list of 50 candidates.

We were sent an invitation to participate in this process that was answered by 13 leading experts in the adoption of Industry 4.0 in Mexico (*CR2*). Which are described in Table 1 based on their geographic location (5 from northern, 5 from bajio, and 3 from center zones) and the sector to which they belong (3 from academia, 4 from business, and 6 from government/society).

To carry out the interviews, an appointment was arranged with each one of them and personal contact was made after a few exchanges of follow-up messages. In accordance with the ethical approval obtained, the interviewees remain anonymous. Follow-up meetings and calls were organized at meetings and follow-up calls were arranged to address any unclear issues and to avoid any misinterpretation.

Using a semi structured interview script, the experts were contacted through Google document forms survey, telephone conversation, and email communication. The means of communication synchronous and asynchronous employed, were widely used, and recognized as regular and functional means of communication.

Table 1. Sample of leading experts in Industry 4.0

Sector / Region	North	Bajio	Center	Total
Academy	1	2	-	**3**
Company	1	2	1	**4**
Government/Society	3	1	2	**6**
Total	**5**	**5**	**3**	**13**

Source: Authors

Instrument to Obtain Information

The instrument used to obtain information from Industry 4.0 driving leaders was based on the study of Acosta (2018) titled *Skills needed in Industry 4.0 and related challenges, a response to Industry 4.0 challenges*, that has been adapted to manufacturing sector in Mexico.

The opinions of specialists were collected to identify key elements related to the current and future needs of the manufacturing sector in the framework of Industry 4.0. Table 2 presents the dimensions of the instrument.

Table 2. Dimensions of the instrument applied to experts

Dimension	Questions
Expert data	4
Industry 4.0 Knowledge	2
Drivers for Industry 4.0 adoption in Mexico	2
Decision factors for joining Industry 4.0 in Mexico	4
Industry 4.0 technologies being implemented	2
Challenges to incorporate Industry 4.0 in Mexico	2
National strategies for the promotion of Industry 4.0 in Mexico	1
Skills required to implement Industry 4.0	1
Availability of skills	1
Skills training initiatives for Industry 4.0	1
Future of Industry 4.0 in Mexico	3

Source: Authors based on Acosta (2018)

RESULTS AND DISCUSSIONS

This section presents the results of the analysis of the data obtained from the semi structured interviews conducted with the leaders in the adoption of Industry 4.0. The sample of persons interviewed consisted of 77% male and 23% female; 61% of the leaders interviewed were under 50 years of age: 8% were under 30 years old, 38% were between 31 and 40 years old, 15% between 41 and 50 years old, 31% between 51 and 60 years old, and 8% between 61 and 70 years old.

In each of the analysis phases, coding categories and subcategories were generated in accordance with the objectives of the study. To carry out the process systematically, MAXQDA 20 software (MaxQDA, 2023) was used for the organization and analysis of qualitative data.

The qualitative data analysis strategy was based on the four steps according to Tutty et al. (1996) described below:

Step 1. Identify codes units of meaning, from the participants' responses to the interview questionnaires. Codes were organized by grouping similar responses into a category or theme and separating dissimilar ones to create distinct categories (Karasar, 2012).
Step 2. Sort the codes by units of meaning and emergent categories were generated, the categories were analyzed for themes or patterns.
Step 3. Examine the categories for meaning and interpretation.
Step 4. Construct a diagram to illustrate the codes and themes found in the data (Miles and Huberman, 1994).

As a result of the four-step analysis, a graphic representation of the units of meaning and categories identified in the interviewees' responses was constructed, presented as a word cloud in Figure 1.

Together they describe the interest of the leaders promoting Industry 4.0 in Mexico in the relationship between the available technologies and the possibilities it offers to companies and industry when integrating them into their work and processes. As shown in Figure 1, the size of words represents the relevance of them. Highlights topics were training, knowledge, and education of the workforce to strengthen integration and innovation, improve productivity, and competitiveness of companies.

The analysis allowed us to describe some patterns in the data. Figure 2 shows a conceptual map of the categories and subcategories identified. The nodes represent the themes and sub-themes, while the thickness of the lines represents the terms that appeared most frequently in the responses.

Figure 1. Words cloud used in the interviews
Source: Authors

Figure 2. Categories and subcategories in Industry 4.0 adoption
Source: Authors

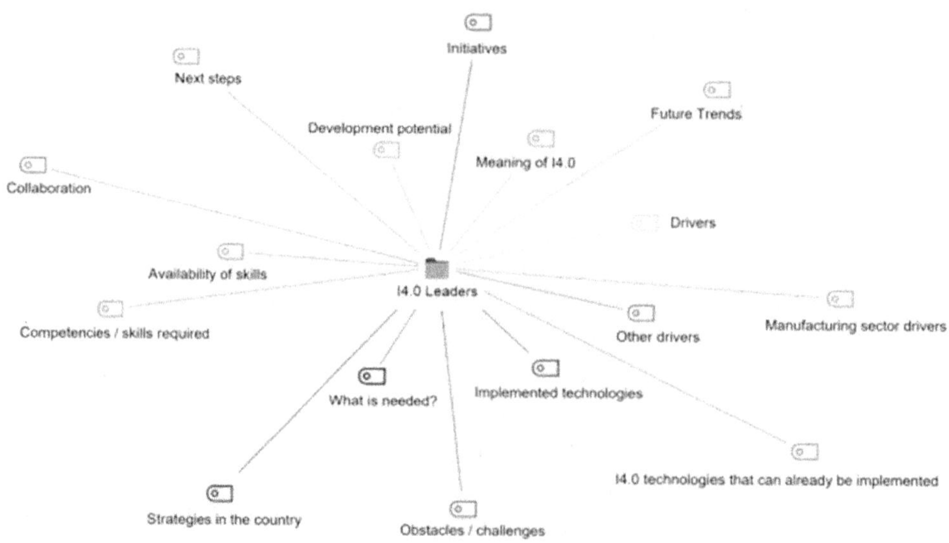

As a result of our analysis, the following sections describe our findings.

Industry 4.0 Knowledge

The most important units of meaning identified in the analysis are shown in Figure 3. Derived from this and from the information obtained in the interviews, it was found that Industry 4.0 was defined as the integration of digital technologies in the industry, primarily the manufacturing sector. The experts emphasized the importance of combining technological and human elements, such as information systems, cyber-physical systems, and tools. Leaders also acknowledged that Industry 4.0 brings about new ways of working and production.

Figure 3. Meaning of Industry 4.0
NOTE: Thickness of the lines represents the terms that appeared most frequently in the responses.
Source: Authors

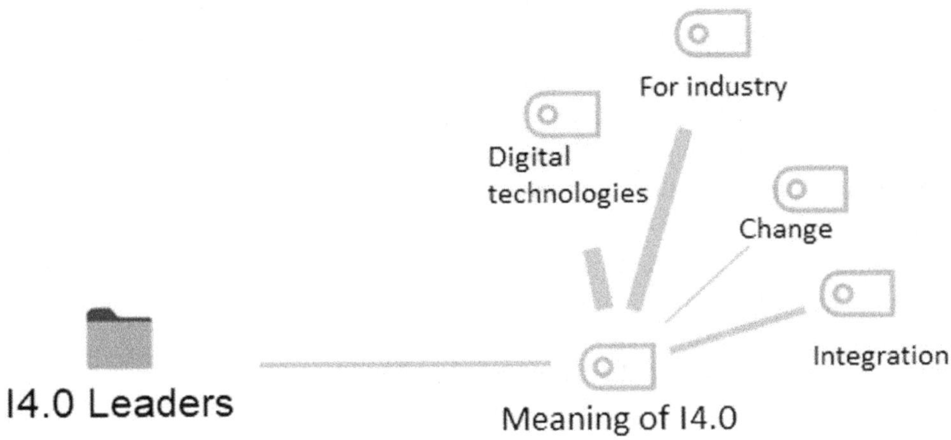

Also, they describe the meaning of Industry 4.0 from different professional contexts. Leaders conceived an ecosystem of digital technologies that enables their integration into production systems to impact the productivity, competitiveness, and efficiency.

In relation to the future trends of Industry 4.0 shown in Figure 4, it is stated that it will promote the growth and evolution of companies through the optimization of processes and the implementation of intelligent systems such as Smart Factory, Smart Production, and Artificial Intelligence.

Figure 4. Future trends of Industry 4.0.
NOTE: Thickness of the lines represents the terms that appeared most frequently in the responses.
Source: Authors

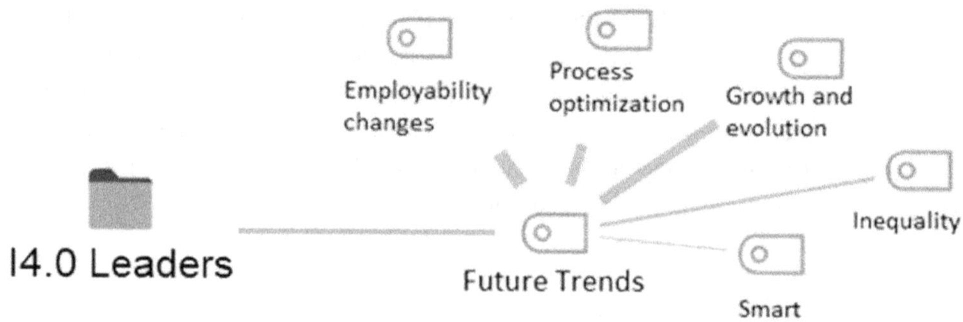

It is stated that the changes brought about by Industry 4.0 will have important consequences on employability and inequality among individuals, companies, and nations. Promoter leaders interviewed were asked to give their opinion on future trends in the themes: society, industry, education, and workforce. They describe a scenario in which the effects of Industry 4.0 will extend to the processes of companies to improve their efficiency and drive improvements in productivity and competitiveness.

In addition, they considered the effects on labor markets that will require new knowledge and skills to achieve effective integration of the associated technologies.

Drivers for Industry 4.0 Adoption in Mexico

Interviewees consider that, within the companies, an important promoter is the drive to improve productivity, increase investment, and integration into regional, national, and international value chains. Factors external that were mentioned: public policies and the effects of the COVID-19 pandemic, as is shown at Figure 5.

It is noteworthy that research and development carried out in companies, as well as in research centers, were mentioned as promoting factors. In addition, training provided to workers to fulfill with international quality standards.

Figure 6 shows the factors promoting the manufacturing sector in Mexico, where the following were reported, productivity, which is what companies are actively seeking, and Industry 4.0 can have a direct impact in the short and medium term.

Figure 5. Drivers for adoption of Industry 4.0
NOTE: Thickness of the lines represents the terms that appeared most frequently in the responses.
Source: Authors

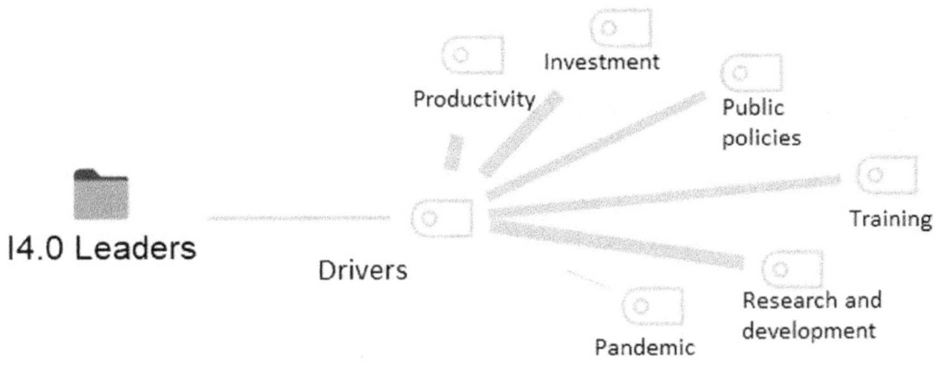

Figure 6. Drivers for adoption of Industry 4.0 in manufacturing sector
NOTE: Thickness of the lines represents the terms that appeared most frequently in the responses.
Source: Authors

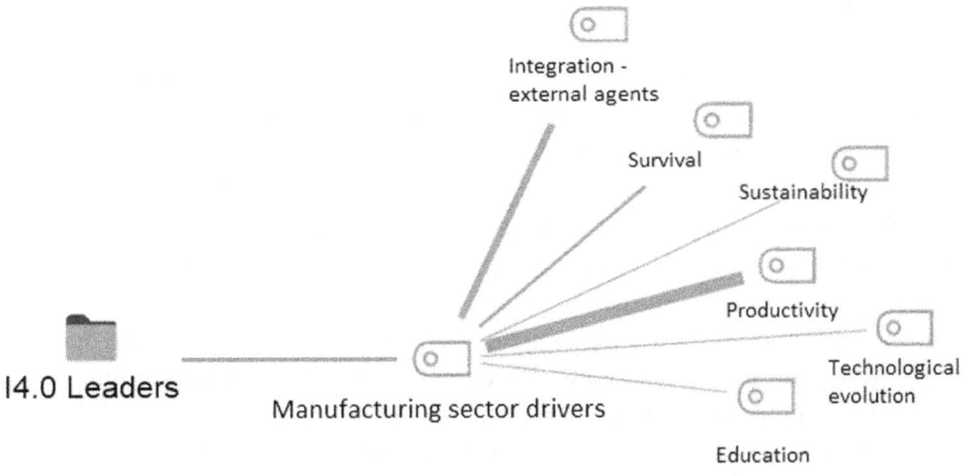

Integration or external agents, due to the linkage with companies in other geographic regions, the productive chaining, as well as the requirements and standards to which Mexican companies must respond. Survival, in the sense of adapting to the current evolution of the forms of production, which is a determining aspect in the growth and continuity of the companies.

The interviewees remarked that education and knowledge are important agents to promote the adoption of Industry 4.0. In general terms, the importance of factors such as technological evolution involving the integration of technology, use of high-end technology, technological transformation, and digital transformation were also taken up as external pressures that promote the adoption of this new paradigm, the identification and grouping of the responses related to this topic is shown in the Figure 7.

Figure 7. Other drivers promoting adoption of Industry 4.0
NOTE: Thickness of the lines represents the terms that appeared most frequently in the responses.
Source: Authors

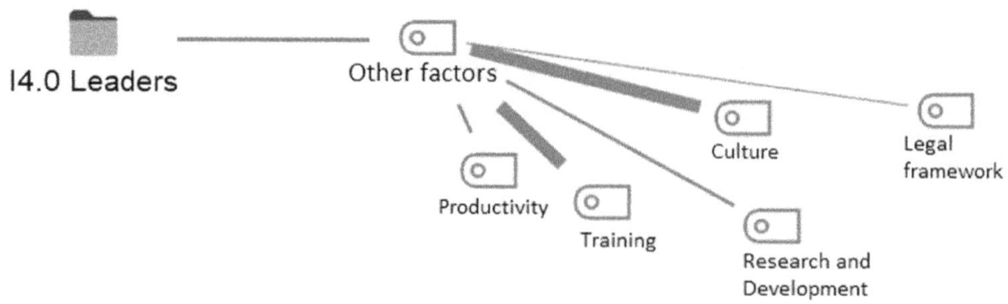

Their responses refer to the terms related to sustainability as a promoter that pressures the manufacturing sector to adopt an environmentally friendly and efficient Industry 4.0 to generate social welfare.

Promoting factors of adoption, included in the group context and external pressures, were related to aspects of participation in global value chains, chambers, associations, and specialized business delegations that promote technological transformation and the adoption of new technologies.

Group of factors linked to education includes knowledge, education, and training both at trade fairs and corporate events, research centers and formal institutions of technical and higher education.

Factors considered in the group of governments with investment were linked to public policies that promote investment in initiatives for the development of labor skills and entrepreneurial capabilities for the development of regions of the country linked to Industry 4.0.

Group of resilience factors includes the continuity of business operations and its survival despite the effects of the global COVID-19 pandemic.

Responses specific to the manufacturing sector, refer to context, market, productivity, and competitiveness, with no mention of governance.

Performance data and integration with other operators were highlighted as the main drivers, followed by education and resilience.

Decision Factors for Joining Industry 4.0

Regarding the relevance of the promoting factors described in the scientific literature: human capital or labor force; technologies and infrastructure requirements; company, region, state, and country strategies; and benefits.

In order to express their opinion, the people interviewed assigned weighting on a scale of three levels (Little, Regular, and Great) to promoters factors: human capital/workforce, technology and infrastructure requirements, company, region, state, and country strategies, and profits, as shown in Figure 8.

Figure 8. Promoter factors for joining Industry 4.0. Axis Y represents the frequency of responses.
Source: Authors

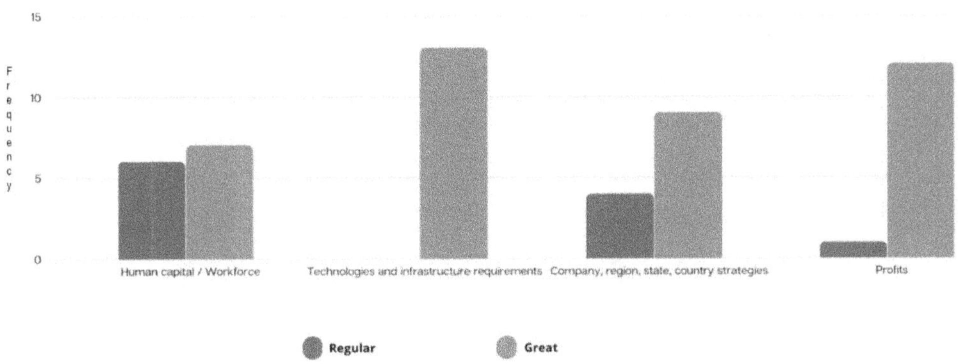

Leaders interviewed agree on the greater relevance of having the required technologies and infrastructures in the company. The favorable opinion of the benefits that the company will be able to build from the adoption of Industry 4.0 stands out.

Strategies of the company, region, state, and country were considered by the interviewees to be of great or regular importance to promote the adoption of Industry 4.0.

In the case of human capital/workforce as a promoter factor, the responses describe two opinions: one group of leaders considered this factor to be of great importance, while a second group considered it to be of regular importance.

Other factors that they suggested were education (*training, culture, research, and development*), economics (*return on investment and costs*), cultural factors, and legal framework regulating data protection were highlighted.

Industry 4.0 Technologies Being Implemented

Leaders interviewed, 33%, consider that all Industry 4.0 technologies can already be implemented in Mexico. This argument is supported by the feasibility of connectivity.

As shown in Figure 9, leaders answered that other technologies that were pointed out as feasible were IoT, mobility, cloud computing, autonomous robots, smart manufacturing, 3D (manufacturing), Big Data, and analytics.

Figure 9. Technologies of I4.0 that could be implemented in Mexico
NOTE: Thickness of the lines represents the terms that appeared most frequently in the responses.
Source: Authors

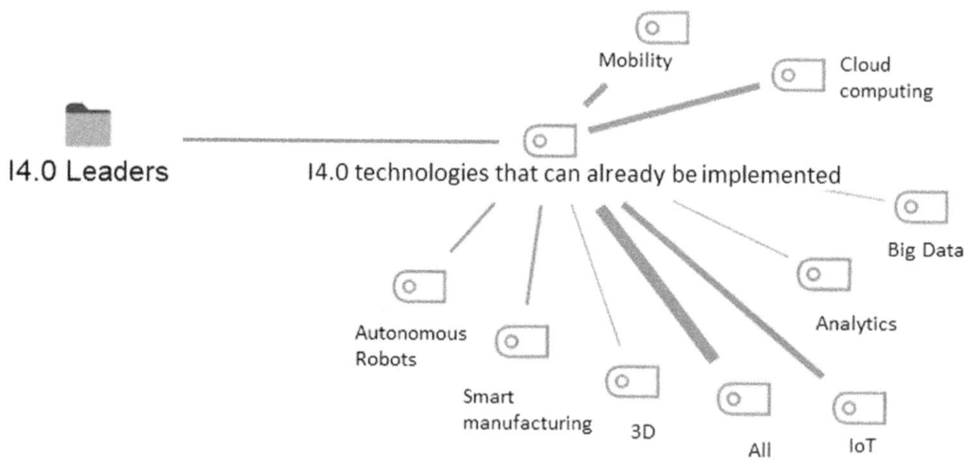

Their responses included real-time location systems technologies, mobile technologies, Big Data and data analytics, autonomous robots, Internet of Things (IoT), cloud computing, and Artificial Intelligence (AI). In addition, additive manufacturing, augmented reality, and 3D manufacturing were mentioned.

Industry 4.0 enabling technologies, which were not mentioned in the analyzed responses were horizontal and vertical systems integration, as well as cybersecurity.

According to the opinion of the interviewees, the following technologies are already being implemented in the Mexican manufacturing sector: IoT, Big Data, additive manufacturing, digital twins, virtual reality, autonomous robots, AI, machine learning, and data analytics. In addition, the answers include real-time location technologies and mobile technologies. Classification, coding, and analysis of their responses is presented in Figure 10.

Figure 10. Technologies of I4.0 that already are implemented in Mexico
NOTE: Thickness of the lines represents the terms that appeared most frequently in the responses.
Source: Authors

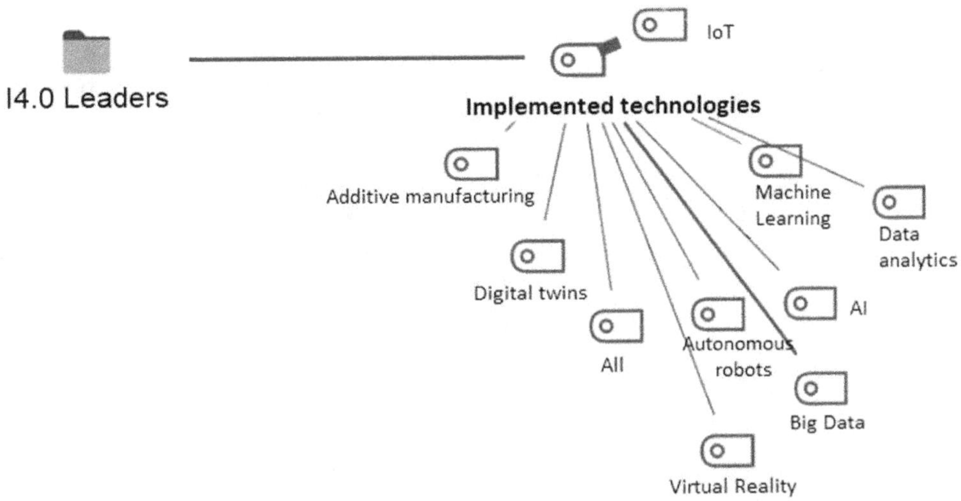

Challenges to Incorporate Industry 4.0 in Mexico

The obstacles or challenges faced by companies and individuals for the adoption of Industry 4.0 in Mexico. Leaders' opinions were organized as show Figure 11, and after classified into: *Culture*, linked to resistance to change and disarticulation between industries (lack of horizontal integration and heterogeneity between sectors and regions), as some of the main challenges.

- ***Knowledge or digital competencies***, related to the cognitive and technical capabilities needed to efficiently exploit the available tools and technologies to stop being just technology operators.
- ***Lack of leadership***, linked to a lack of knowledge and investment (including government incentives), deficient support for research and development, lack of impetus for the development of the infrastructure required for the efficient use of Industry 4.0 technologies, and the lack of an organized and coordinated long-term strategy.

Figure 11. Challenges to adoption of Industry 4.0
NOTE: Thickness of the lines represents the terms that appeared most frequently in the responses.
Source: Authors

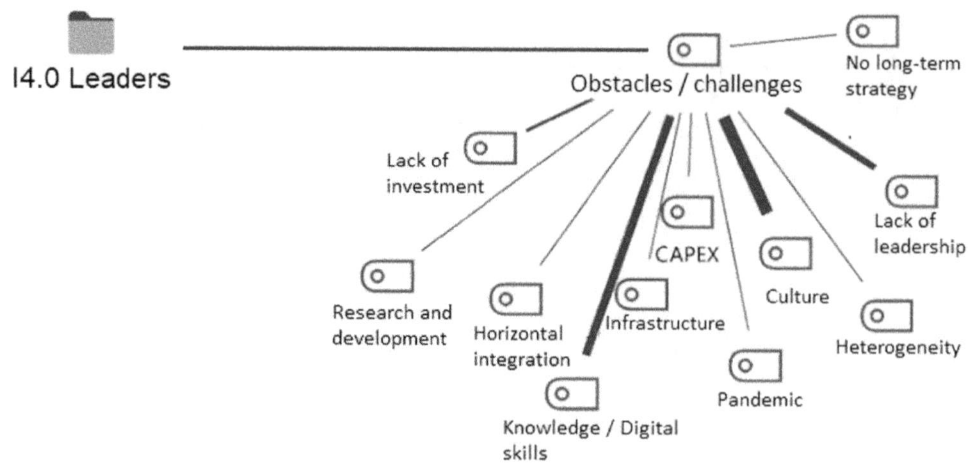

Table 3 presents a classification of the obstacles or challenges that slow down the adoption of Industry 4.0 in Mexico: *education* (culture, knowledge or digital skills, research and development, lack of leadership), *internal conditions* (poor horizontal integration, pandemic, heterogeneity in the country, lack of long-term strategy), and *infrastructure and investment* (infrastructure, capital expenditure, lack of investment, government incentives).

Regarding the needs that are essential for the widespread adoption of the new forms of production, the interviewees considered it a priority to address the need for training in digital skills, so that companies can take advantage of the benefits of Industry 4.0 and increase competitiveness as a country, as show in Figure 12.

Leaders point out that it is not only a matter of acquiring technology, but also of having the capacity to adapt it to the needs and generate strategies oriented to local conditions and contexts in a holistic and integrated vision.

Some alternative solutions included in the opinions of the interviewees refer to adequate planning and openness to collaborative work, understanding new situations, and adapting the culture. Finally, considering cybersecurity issues and relevant regulations, as well as favoring investment.

Table 3. Limiting factors for the adoption of Industry 4.0 in Mexico

Group	Answers
Education (6)	P1-F "...internal leaders are needed in organizations to lead transformation processes..." P3-M "...lack of digital competencies in the workforce..." P5-M "...training...", "...ease of experimentation, Agile Adaptation..." P7-M "...knowledge..." P12-M "...poor understanding of I4.0 in the different sectors..." P13-M "...lack of experts in the implementation and management of the technology..."
Domestic conditions of the Mexican industry (7)	P2-F "...culture..." P4-M "...cultural and ecosystems..." P5-M "...mindset..." P10-M "...cultural change including resistance to change..." P1-F "...no vision for the future..." P11-F "...lack of planning in the industries..." P12-M "...regional heterogeneity (states with different characteristics)...", "... scarce understanding of I4.0 in the different sectors..."
Infrastructure / investment (5)	P3-M "...lack of connectivity on the production floor..." P6-M "...low investment in research and development..." P7-M "...investment..." P9-M "...the economic...", "...stopping investments. ..." P12-M "...government (lack of interest), lack of resources..."

** P1 to P13 corresponding to response of leader interviewed, M for Male, and F for Female.*

Source: Authors

Figure 12. Enabling conditions for adoption of Industry 4.0

NOTE: Thickness of the lines represents the terms that appeared most frequently in the responses.

Source: Authors

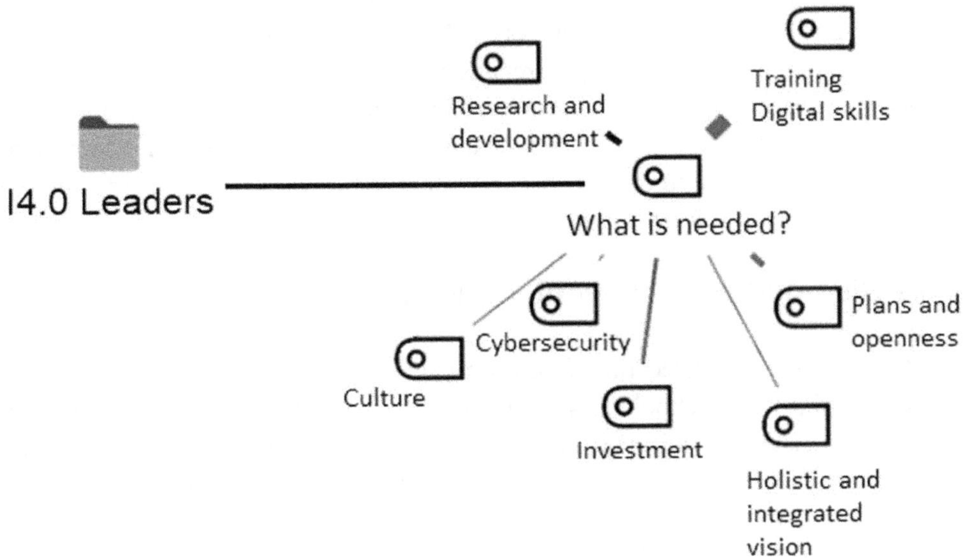

The analysis of the responses of the leading promoters of Industry 4.0 in Mexico made it possible to define three groups of enabling factors:

1. ***Education*** refers to more and better training for the acquisition and updating of digital skills, adequate spaces for testing, training for entrepreneurs, new professional skills, 352 research, and human capital expertise in new technologies.
2. ***Internal conditions of the industry***, the answers state that good strategic plans must be generated, business plans with digital transformation initiatives and responsible innovation. As well as being open to technological proposals and that solution integrators must have a holistic vision so that it is possible to generate protected industrial networks with functional cybersecurity strategies.
3. ***Infrastructure and investment***, answers were related to the need for direct investment from the government in research, science, and technology, generation of incentives for companies, and greater investment in efficient infrastructure to meet the requirements of Industry 4.0 technologies.

These three groups of factors were related to needs that must be resolved for Industry 4.0 adoption plans to success in Mexico.

National Strategies for the Promotion of Industry 4.0

The Mexican strategy most cited by leaders was Nuevo Leon 4.0. Others known were MIND 4.0, Jalisco Digital Initiative, Bajio Industrial Corridor Guanajuato - Hannover Messe, automotive, aerospace, software, biotechnology clusters, Mexican Association of Private Industrial Parks (AMPIP), and Vallejo-i.

As show in Figure 13, each expert mentioned the strategy or initiative they were part of, or the one developed in their region of proximity and influence. Nuevo Leon 4.0 is a strategic initiative that involve local and state government, companies, and academia to execute different projects, geographically is near to US. It was known and signaled by all the interviewees.

Skills Required to Implement Industry 4.0

Leaders were asked to reflect on the competencies required to promote the adoption and exploitation of the technologies involved in this new industrial revolution. Figure 14 shows the result of analysis of the responses that made it possible to group the competencies into human, specialized - conceptual, and technological.

Figure 13. Initiatives of Industry 4.0 in Mexico
NOTE: Thickness of the lines represents the terms that appeared most frequently in the responses.
Source: Authors

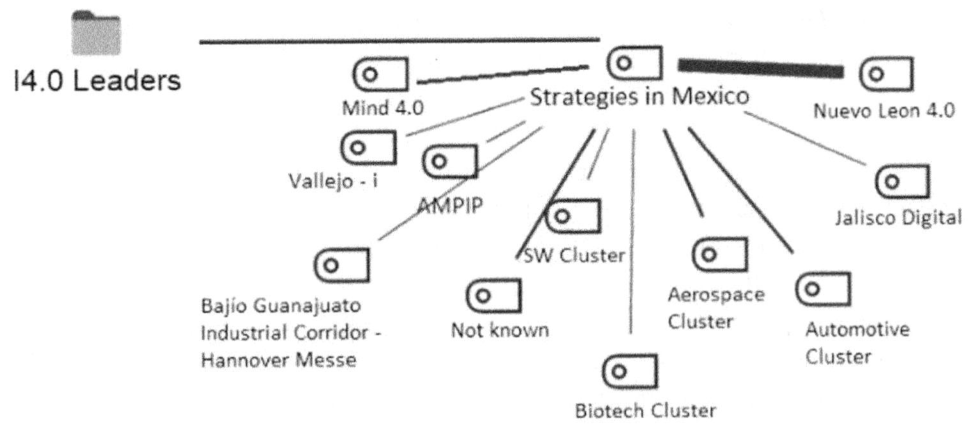

Figure 14. Skills required to adoption of Industry 4.0.
Source: Authors

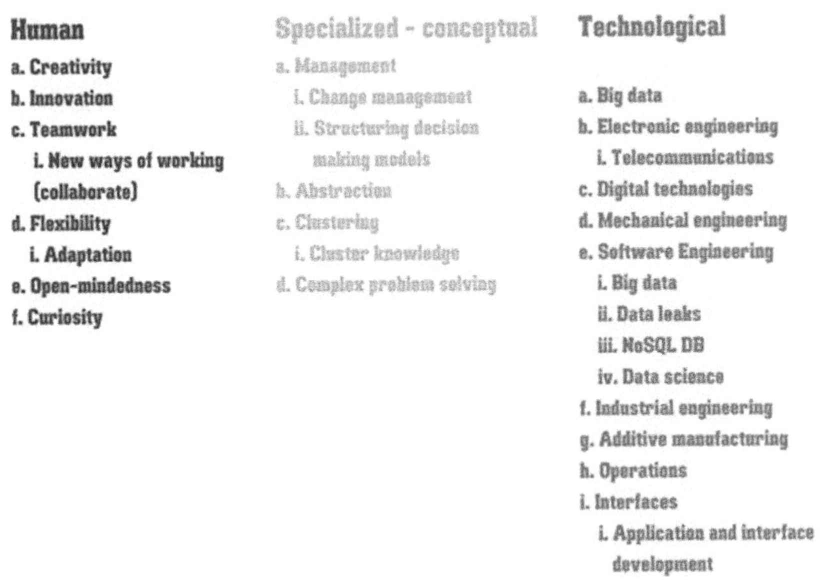

- ***Human competencies*** were related to personal characteristics and skills that allow individuals to develop in intense work environments in collaboration with other people and integrated with the Industry 4.0 technologies.

- ***Specialized or conceptual competencies*** associated with thinking skills linked to the conceptualization and abstraction of business processes for their operation, maintenance, and improvement.
- ***Technological competencies*** were linked to the technical knowledge and experience that allow people to efficiently operate the technologies that promote Industry 4.0.

About availability of competencies for Industry 4.0 adoption in Mexico, interviewees state that it was not easy to find the competencies for the adoption of Industry 4.0. Arguments linked to their answers where they need to be trained, i.e., they are not yet in the market, knowledge is needed, the current education system does not help, development focused on devices is needed, there are not, but training is starting and there are still few people who understand and know about Industry 4.0.

There is a small group of leaders who believe that the competencies can be found that there are professionals with the vision to be able to carry out these implementations and that organizations and companies have a supply of these skills, as well as the universities in the state of Nuevo Leon.

About collaboration, the leader's comments are shown in Figure 15.

Figure 15. Collaborations that leaders participated in
NOTE: Thickness of the lines represents the terms that appeared most frequently in the responses.
Source: Authors

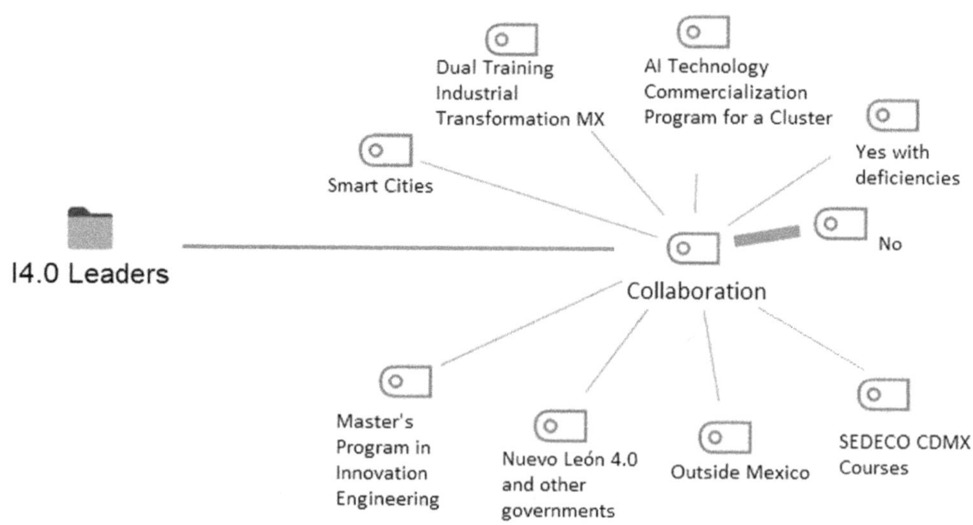

Optimistic responses were found of good experiences with interventions in Nuevo Leon 4.0, Industrial Transformation MX, CIDESI-Conacyt; abroad with Spain, Germany, and the Autonomous University of Catalonia.

Pessimistic opinions were since they have found activities of questionable quality, they have faced the lack of understanding of basic concepts and the lack of infrastructure, and they found lack of knowledge of the topic from local entrepreneurs. Despite the interest of the companies, it is still not clear how to develop skills for Industry 4.0 and they were uncertain regarding how to initiate a diagnosis to implement it in their company.

Future of Industry 4.0

It is worth highlighting the position of the promoter leaders in considering that a planned and global digital transformation should be motivated. Regarding the next steps to be taken, expert opinions were grouped as shown in Figure 16.

Figure 16. Future about adoption of Industry 4.0
NOTE: Thickness of the lines represents the terms that appeared most frequently in the responses.
Source: Authors.

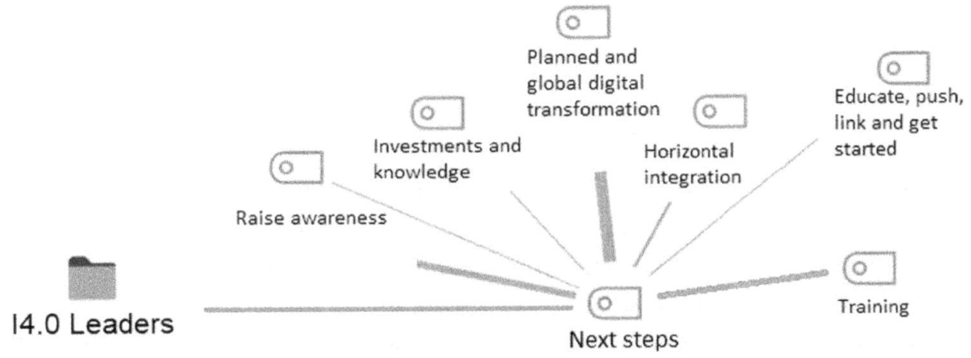

Mention was made of the need to improve training issues to train a new generation of professionals who must be prepared to trigger Industry 4.0. For this to happen, the different sectors and levels of the innovation ecosystem must be made aware so that this leads to more and better investment plans that motivate knowledge and horizontal integration of industries and sectors so that it is a global and coherent strategy.

The answers focus on training, integration of efforts and organized, and strategic plans that will lead the country along a clear and defined path in collaboration with the entire innovation ecosystem.

Leaders promoting Industry 4.0 consider that the sector of greatest opportunity is in industry as is shown in Figure 16. In addition, the following areas were mentioned: Big Data, research and development, tourism, education, human capital, health, work ecosystems, IoT, and AI.

Figure 17. Sectors with opportunities for Industry 4.0
NOTE: Thickness of the lines represents the terms that appeared most frequently in the responses.
Source: Authors

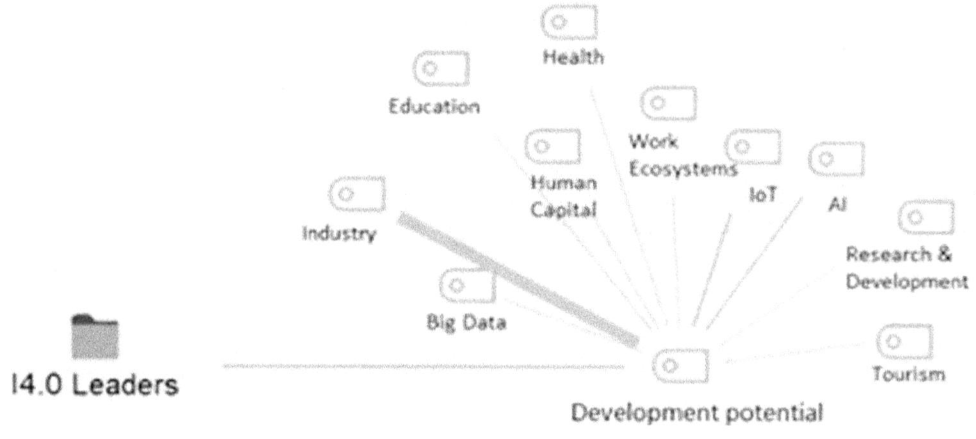

Responses of the interviewees allow us to identify an agreement in considering *manufacturing* as the sector with the greatest growth potential. Based on the needs detected, the area of *knowledge transfer/education* can be considered as the second in terms of development potential.

Interviewees highlighted the following areas as having great potential for development: production and maintenance chains, cold chain, continuous improvement in production chains, creation of ecosystems, broad global vision of trends and real needs, IoT, shipment tracking, AI, Big Data, statistics, telephony, security, aerospace, tourism, and health.

Initiatives in which the interviewees intend to venture were in the sectors of education, volunteering, and the development of technological applications. Coding generated in connection with this topic are shown in Figure 18.

Figure 18. Initiatives in interest of leaders of Industry 4.0
NOTE: Thickness of the lines represents the terms that appeared most frequently in the responses.
Source: Authors

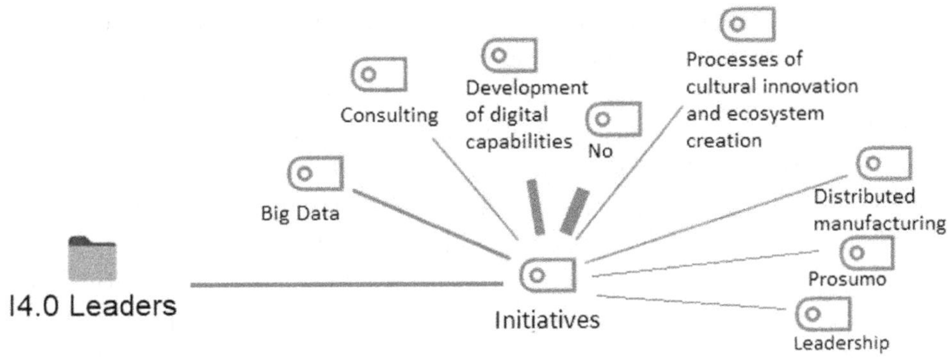

In addition, other sectors that were mentioned: cultural innovation processes and creation of flexible 4.0 ecosystems, researching social and ecological impact, independent entrepreneurship, distributed manufacturing, prosumo, consulting, technological leadership, open innovation, and continuing to promote the adoption of these technologies in Mexico.

CONCLUSION

From the answers obtained to the questions related to obstacles to the adoption of Industry 4.0, limitations to the adoption, availability of competencies and development perspectives, it is identified that from the point of view of the leaders, the lack of qualification of human capital in the manufacturing industry hinders the adoption of I4.0.

Responses of leading Industry 4.0 promoters in Mexico were analyzed and three main enabling factors were identified. Firstly, the need for better *education and training*, including acquiring and updating digital skills, adequate testing spaces, training for entrepreneurs, and investing in human capital experts in new technologies. Secondly, the industry must have good *strategic plans*, be open to technological proposals, and have a holistic approach to innovation. Finally, there must be *government investment* in research and technology, incentives for companies, and efficient infrastructure to support Industry 4.0 technologies. These three groups of factors are crucial for the successful implementation of Industry 4.0 in Mexico.

Leaders promoting Industry 4.0 do not have a clear knowledge of national or regional strategies for its promotion and adoption. There is also a lack of a clear route or national bodies to guide companies towards adopting new ways of working. Additionally, there is no adequate strategy for disseminating information about Industry 4.0 initiatives to interested agents.

About the competencies required to promote the adoption and exploitation of the technologies involved in this new industrial revolution, leaders say: *human, specialized conceptual and technological.* These factors highlight the importance of combining human and technical skills to drive creativity, innovation, teamwork, flexibility, and efficient problem-solving in the digital age.

Leaders promoting Industry 4.0 in Mexico agree that the manufacturing sector has the greatest potential for growth in the country. To achieve this growth, they believe that a planned and global digital transformation should be motivated, and that training is key to preparing a new generation of professionals.

Leaders see education as the second area with the greatest potential for development, followed by other areas such as Big Data, research and development, tourism, human capital, health, work ecosystems, IoT, and AI. The interviewees also plan to venture into initiatives in the sectors of education, volunteering, and technological application development. The goal is to promote the adoption of Industry 4.0 technologies in Mexico to drive growth and innovation across multiple sectors.

LIMITATIONS

As described in the methodological part, there was not official source that recognizes the leadership of agents promoting the adoption of Industry 4.0. Therefore, although an effort was made to use a method that would guarantee the objectivity of the process, there is a possibility that some researchers of the subject may disagree with the selection of expert leaders included in this research.

Based on the work of Guest et al. (2006), 13 cases studied were considered enough; however, a larger sample of subjects included in the process could be desirable to cover a greater number of points of view and reflections on the subject.

FUTURE RESEARCH

Adoption processes are evolutionary and constantly changing, so new research will be needed in the future, since the maturity of the process and its protagonists will provide another perspective, as well as new and valuable information to support good adoption practices. Comparative studies between geographic zones, clusters, and strategic sectors will be very interesting work to be done soon.

Also, research that links the degree of adoption of Industry 4.0 with strategic issues such as competitiveness, social responsibility and sustainability, education, among others.

ABBREVIATIONS

CR[1-7]: CR Criteria of inclusion, # number of criteria;
P[1-13][F, M]: **P** Person, ***#*** number of interviewed; ***F*** Female, ***M*** Male.

REFERENCES

Acosta, M. (2018). *Skills needed in industry 4.0 and related challenges A response to Industry 4.0 challenges. Master, Aalto University School of Business*. Finlandia.

Agostini, L., Galati, F., & Gastaldi, L. (2020). The digitalization of the innovation process: Challenges and opportunities from a management perspective. *European Journal of Innovation Management*, *23*(1), 1–12. doi:10.1108/EJIM-11-2019-0330

Agostini, L., & Nosella, A. (2020). The adoption of Industry 4.0 technologies in SMEs: Results of an international study. *Management Decision*, *58*(4), 625–643. doi:10.1108/MD-09-2018-0973

Ammar, M., Haleem, A., Javaid, M., Bahl, S., & Verma, A. S. (2022). Implementing Industry 4.0 technologies in self-healing materials and digitally managing the quality of manufacturing. *Materials Today: Proceedings*, *52*, 2285–2294. doi:10.1016/j.matpr.2021.09.248

Ancarani, A., Di Mauro, C., & Mascali, F. (2019). Backshoring strategy and the adoption of Industry 4.0: Evidence from Europe. *Journal of World Business*, *54*(4), 360–371. doi:10.1016/j.jwb.2019.04.003

Arboleda, G. Y. C., Perdomo, S. B. H., & Álvarez, E. F. E. (2020). Pymes colombianas y los Retos de la Industria 4.0. *Revista Innova ITFIP*, *7*(1), 10–17. doi:10.54198/innova07.06

Battaglia, M., Sampling, N., & Lavrakas, P. J. (2008). *Encyclopedia of survey research methods* (2nd ed.). Sage Publications.

Baum, M., & Kabst, R. (2014). The effectiveness of recruitment advertisements and recruitment websites: Indirect and interactive effects on applicant attraction. *Human Resource Management*, *53*(3), 353–378. doi:10.1002/hrm.21571

Bissola, R. & Imperatori, B. (2019). HRM in the Industry 4.0 Era: Are Workers Still in The Center? *HRM 4.0 For Human-Centered Organizations.* Advancedseries in management.

Caiado, R. G. G., Scavarda, L. F., Azevedo, B. D., de Mattos Nascimento, D. L., & Quelhas, O. L. G. (2022). Challenges and benefits of sustainable industry 4.0 for operations and supply chain management—A framework headed toward the 2030 agenda. *Sustainability (Basel), 14*(2), 830. doi:10.3390u14020830

Corò, G., & Volpe, M. (2020). Driving factors in the adoption of Industry 4.0 technologies: An investigation of SMEs. In L. De Propris & D. Bailey (Eds.), *Industry 4.0 and regional transformations* (pp. 112–132). Routledge. doi:10.4324/9780429057984-7

daSilva, L. B. P., Soltovski, R., Pontes, J., Treinta, F. T., Leitão, P., Mosconi, E., de Resende, L. M., & Yoshino, R. T. (2022). Human Resources Management 4.0: Literature Review and Trends. *Computers & Industrial Engineering, 108111*, 108111. doi:10.1016/j.cie.2022.108111

Dieste, M., Sauer, P. C., & Orzes, G. (2022). Organizational tensions in industry 4.0 implementation: A paradox theory approach. *International Journal of Production Economics, 108532*, 108532. doi:10.1016/j.ijpe.2022.108532

Gospel, H. (2015). Varieties of qualifications, training, and skills in long-term care: A German, Japanese, and UK Comparison. *Human Resource Management, 54*(5), 833–850. doi:10.1002/hrm.21714

Grabowska, S., & Saniuk, S. (2022). Business models in the industry 4.0 environment—Results of web of science bibliometric analysis. *Journal of Open Innovation, 8*(1), 19. doi:10.3390/joitmc8010019

Guest, G., Bunce, A., & Johnson, L. (2006). How many interviews are enough? An experiment with data saturation and variability. *Field Methods, 18*(1), 59–82. doi:10.1177/1525822X05279903

Habraken, M., & Bondarouk, T. (2017). Smart industry research in the field of HRM: Resetting job design as an example of upcoming challenges. In *Electronic HRM in the smart era.* Emerald Publishing Limited. doi:10.1108/978-1-78714-315-920161009

Hernández-Sampieri, R., & Torres, C. P. M. (2018). *Metodología de la investigación: las rutas cuantitativa, cualitativa y mixta* (1st ed.). McGraw-Hill Interamericana.

Inkulu, A. K., Bahubalendruni, M. R., & Dara, A. (2022). Challenges and opportunities in human robot collaboration context of Industry 4.0-a state of the art review. *Industrial Robot: the international journal of robotics research and application*, *49*(2), 226-239. doi:10.1108/IR-04-2021-0077

Ivlev, I., Kneppo, P., & Bartak, M. (2015). Method for selecting expert groups and determining the important of experts' judgments for the purpose of managerial decision-making tasks in health system. *E+M. Ekonomie a Management*, *18*(2), 57–72. doi:10.15240/tul/001/2015-2-005

Javaid, M., Haleem, A., Singh, R. P., & Suman, R. (2022a). Enabling flexible manufacturing system (FMS) through the applications of industry 4.0 technologies. *Internet of Things and Cyber-Physical Systems*, *2*, 49–62. doi:10.1016/j.iotcps.2022.05.005

Javaid, M., Haleem, A., Singh, R. P., & Suman, R. (2022b). Artificial intelligence applications for industry 4.0: A literature-based study. *Journal of Industrial Integration and Management*, *7*(01), 83–111. doi:10.1142/S2424862221300040

Javaid, M., Haleem, A., Singh, R. P., Suman, R., & Gonzalez, E. S. (2022). Understanding the adoption of Industry 4.0 technologies in improving environmental sustainability. *Sustainable Operations and Computers*, *3*, 203–217. doi:10.1016/j.susoc.2022.01.008

Johnson, R. B., & Christensen, L. B. (2013). *Educational research: Quantitative, qualitative, and mixed approaches* (5th ed.). SAGE Publications.

Kagermann, H., Lukas, W. D., & Wahlster, W. (2011). Industrie 4.0: Mit dem Internet der Dinge auf dem Weg zur 4. industriellen Revolution. *VDI nachrichten*, *13*(1), 2-3.

Kagermann, H., & Wahlster, W. (2022). Ten years of Industrie 4.0. *Sci*, *4*(3), 26. doi:10.3390ci4030026

Kagermann, H., Wahlster, W., & Helbig, J. (2013). *Recommendations for implementing the strategic initiative INDUSTRIE 4.0: Final report of the Industrie 4.0 Working Group. Final Report of the Industrie 4.0*. National Academy of Science and Engineering.

Karasar, N. (2012). *Scientific Research Method* (15th ed.). Nobel Publishing.

Khan, I. H., & Javaid, M. (2022). Role of Internet of Things (IoT) in adoption of Industry 4.0. *Journal of Industrial Integration and Management*, *7*(04), 515–533. doi:10.1142/S2424862221500068

Klitou, D., Conrads, J., Rasmussen, M., Probst, L., & Pedersen, B. (2017). Digital Transformation Monitor Germany: Industrie 4.0. *European Union.*

Kluemper, D. H., Mitra, A., & Wang, S. (2016). Social Media use in HRM. *Research in Personnel and Human Resources Management, 34*, 153–207. doi:10.1108/S0742-730120160000034011

Kohnke, O. (2017). *It's not just about technology: The people side of digitization. Shaping the digital enterprise, 69-91*. Springer.

Liboni, L. B., Cezarino, L. O., Jabbour, C. J. C., Oliveira, B. G., & Stefanelli, N. O. (2019). Smart industry and the pathways to HRM 4.0: Implications for SCM. *Supply Chain Management, 24*(1), 124–146. doi:10.1108/SCM-03-2018-0150

Lorenz, M., Rüßmann, M., Strack, R., Lueth, K. L., & Bolle, M. (2015). *Man, and machine in industry 4.0: How will technology transform the industrial workforce through 2025.* The Boston Consulting Group, 2. http://hdl.voced.edu.au/10707/405644

Marnewick, A. L., & Marnewick, C. (2019). The ability of project managers to implement industry 4.0-related projects. *IEEE Access : Practical Innovations, Open Solutions, 8*, 314–324. doi:10.1109/ACCESS.2019.2961678

MAXQDA. (2023) *Home.* Maxqda. https://www.maxqda.com

Mikhailov, A. N., Rodin, A. B., & Smirnova, M. I. (2018). Humanization of Engineering Education in Conditions of the Process of Industry 4.0 Forming. *4th International Conference on Information*. IEEE. 10.1109/INFORINO.2018.8581805

Miles, M. B., & Huberman, A. M. (1994). *Qualitative data analysis: An Expanded Sourcebook* (2nd ed.). SAGE.

Nguyen, X. T., & Luu, Q. K. (2020). Factors affecting adoption of industry 4.0 by small-and medium-sized enterprises: A case in Ho Chi Minh city, Vietnam. *The Journal of Asian Finance. Economics and Business, 7*(6), 255–264. doi:10.13106/jafeb.2020

Okkonen, J., Vuori, V., & Palvalin, M. (2019). Digitalization changing work: Employees' view on the benefits and hindrances. *International Conference on Information Technology & Systems,* (pp. 165-176). Springer, Cham. 10.1007/978-3-030-11890-7_17

Ollington, N., Gibb, J., & Harcourt, M. (2013). Online social networks: An emergent recruiter tool for attracting and screening. *Personnel Review, 42*(3), 248–265. doi:10.1108/00483481311320390

Pérez-Rojas, J. G. (2020). Retos de las instituciones de educación superior para su articulación en la Industria 4.0. *Revista CEA*, *6*(11), 9–11. doi:10.22430/24223182.1584

Ponce Romero, J. M., Hallett, S. H., & Jude, S. (2017). Leveraging big data tools and technologies: Addressing the challenges of the water quality sector. *Sustainability (Basel)*, *9*(12), 2160. doi:10.3390u9122160

Pozzi, R., Rossi, T., & Secchi, R. (2023). Industry 4.0 technologies: Critical success factors for implementation and improvements in manufacturing companies. *Production Planning and Control*, *34*(2), 139–158. doi:10.1080/09537287.2021.1891481

Radanliev, P., De Roure, D., Nicolescu, R., Huth, M., & Santos, O. (2022). Digital twins: Artificial intelligence and the IoT cyber-physical systems in Industry 4.0. *International Journal of Intelligent Robotics and Applications*, *6*(1), 171–185. doi:10.100741315-021-00180-5

Raj, A., Dwivedi, G., Sharma, A., de Sousa Jabbour, A. B. L., & Rajak, S. (2020). Barriers to the adoption of industry 4.0 technologies in the manufacturing sector: An inter-country comparative perspective. *International Journal of Production Economics*, *224*, 107546. doi:10.1016/j.ijpe.2019.107546

Rose, J., Lukic, V., Milon, T., & Cappuzzo, A. (2016). *Sprinting to value in Industry 4.0*. BCG Perspectives. https://www.bcg.com/publications/2016/lean-manufacturing-technology-digital-sprinting-to-value-industry-40

Saniuk, S., Grabowska, S., & Straka, M. (2022). Identification of Social and Economic Expectations: Contextual Reasons for the Transformation Process of Industry 4.0 into the Industry 5.0 Concept. *Sustainability (Basel)*, *14*(3), 1391. doi:10.3390u14031391

Schneider, P., & Sting, F. J. (2020). Employees' perspectives on digitalization-induced change: Exploring frames of industry 4.0. *Academy of Management Discoveries*, *6*(3), 406–435. doi:10.5465/amd.2019.0012

Senna, P. P., Ferreira, L. M. D., Barros, A. C., Roca, J. B., & Magalhaes, V. (2022). Prioritizing barriers for the adoption of Industry 4.0 technologies. *Computers & Industrial Engineering*, *171*, 108428. doi:10.1016/j.cie.2022.108428

Sigov, A., Ratkin, L., Ivanov, L. A., & Xu, L. D. (2022). Emerging enabling technologies for industry 4.0 and beyond. *Information Systems Frontiers*, 1–11. doi:10.100710796-021-10213-w

Silva, V. L., Kovaleski, J. L., Pagani, R. N., Corsi, A., & Gomes, M. A. S. (2020). Human factor in smart industry: A literature review. *Future Studies Research Journal: Trends and Strategies*, *12*(1), 87–111. doi:10.24023/FutureJournal/2175-5825/2020.v12i1.473

Sony, M., Antony, J., Mc Dermott, O., & Garza-Reyes, J. A. (2021). An empirical examination of benefits, challenges, and critical success factors of industry 4.0 in manufacturing and service sector. *Technology in Society*, *67*, 101754. doi:10.1016/j.techsoc.2021.101754

Tang, J., Zhang, J., Yao, L., Li, J., Zhang, L., & Su, Z. (2008). Arnetminer: extraction and mining of academic social networks. In *Proceedings of the 14th ACM SIGKDD international conference on Knowledge discovery and data mining*. ACM. 10.1145/1401890.1402008

Terán, A. (2013). *El capital humano, la estrategia y el desarrollo y acumulación de capacidades, competencias organizacionales: su impacto en innovación en el sector biofarmacéutico en México.*

Tutty, L. M., Rothery, R. M., & Grinnell, J. R. (1996). *Analyzing Your Own Data, Qualitative Research for Social Workers; Publisher.* Allyn and Bacon.

Van-Dijck, J., & Poell, T. (2013). Understanding social media logic. *Media and Communication*, *1*(1), 2–14. doi:10.17645/mac.v1i1.70

World Economic Forum. (2020). *The Future of Jobs Report 2020.* WEF. https://www.weforum.org/reports/the-future-of-jobs-report-2020

Xu, X., Lu, Y., Vogel-Heuser, B., & Wang, L. (2021). Industry 4.0 and Industry 5.0—Inception, conception and perception. *Journal of Manufacturing Systems*, *61*, 530–535. doi:10.1016/j.jmsy.2021.10.006

Yang, F., & Gu, S. (2021). Industry 4.0, a revolution that requires technology and national strategies. *Complex & Intelligent Systems*, *7*(3), 1311–1325. doi:10.100740747-020-00267-9

Zangiacomi, A., Pessot, E., Fornasiero, R., Bertetti, M., & Sacco, M. (2020). Moving towards digitalization: A multiple case study in manufacturing. *Production Planning and Control*, *31*(2-3), 143–157. doi:10.1080/09537287.2019.1631468

Zhang, J., Tang, J., & Li, J. (2007). *Expert finding in a social network.* In *12th International Conference on Database Systems for Advanced Applications*, Bangkok, Thailand.

Chapter 3
Analysis of IoT Wearable Sensors to Monitor Chronic Diseases

M. S. Nidhya
Jain University, India

ABSTRACT

There are a number of diseases that are attacking human beings every day and everywhere. Chronic disease patients should be monitored continuously, and actions should be taken if it goes beyond the level of tolerability before one has to pay some large amount of money to health care centre. Financially struggling people cannot afford that amount, or even the equipment for monitoring purposes only. Generally chronic diseases are affected only by adults; they can be controlled but not cured. The authors propose chapter to monitor chronic disease patients instead of admitting them in hospital—a wearable bio sensor will monitor the patient efficiently and continuously, changes in their reading will be seen sooner by the caretaker or medical advisor. This chapter analyse wearables applied for chronic disease like cardiovascular disease and respiratory disease. Analysis on these devices is based on the parameters like convenience, cost, and efficiency of the device.

INTRODUCTION

The term "Internet of Things" refers to a network of physical objects that communicate with one another and with users by using sensors, microcontrollers, and transceivers. The Internet of Things is made up of appropriate protocol stacks that enable this interaction, making the objects a fundamental component of the Internet. Several

DOI: 10.4018/978-1-6684-9267-3.ch003

facets of the prospective user's daily life are now impacted by the internet. With these factors in mind, a number of IoT-based apps have been created in which every physical item is linked to the Internet via sensor devices. Healthcare is becoming more and more dependent on IoT as a means of improving access to treatment, enhancing the quality of care, and ultimately lowering the cost of care.

The integrated practise of wellness, healthcare, and patient support is characterised as tailored healthcare depending on an individual's particular biological, behavioural, and cultural traits. By adhering to the fundamental healthcare tenet of "the appropriate care for the right person at the right time," which produces more acceptable outcomes and increases satisfaction while lowering costs, this empowers each and every individual. An effective healthcare system should focus on homecare, early pathology identification, and prevention rather than expensive clinical care.

By preserving each patient's digital identity, IoT enables the personalization of healthcare services. Numerous health issues go undiagnosed in traditional healthcare systems because ready-to-access healthcare services are not readily available. But ubiquitous, non-intrusive, potent IoT-based devices have proved useful in efficiently monitoring and assessing the patient data. In IoT-based healthcare, real-time medical data is collected, processed, and sent to the cloud by a variety of distributed devices, making it feasible to capture, store, and process massive data streams in a number of innovative ways and trigger context-sensitive warnings. Through the Internet, any linked device may continuously and widely access medical devices thanks to this ground-breaking data collecting paradigm.

HEALTH CARE

Healthcare is one of the top priorities for all governments. It is primarily affected by population growth, rural urbanisation, declining birthrates, population ageing, economic growth, and socially unbalanced resource utilisation. Some social problems have emerged in the healthcare sector that IoT may help to prevent or combat in a most effective manner:

- The level of health management and responsiveness to crises;
- There is a critical shortage of both medical professionals and institutional resources, especially in more remote areas. There are also insufficient healthcare systems and a lack of medical facilities.
- An inadequate disease prevention system is unable to satisfy the demands of the national policy to protect citizen health, placing a significant cost on the economy, people, families, and the State;
- Inadequate capabilities for illness prevention and early detection

LITERATURE SURVEY

Applications for wearable sensors include monitoring and diagnostics. Physiological, biochemical, and motion sensing are now among their capabilities (Bonato, 2010; Teng et al., 2008). The scope of the issues that these technologies might potentially aid with is difficult to emphasise. A sizable percentage of patients may respond favourably to physiological treatment of epilepsy, hypertension, dysrhythmias, and asthma as well as other neurological, cardiovascular, and pulmonary conditions. Motion detection at home has the potential to lessen the occurrence of falls, boost a person's sense of autonomy, and propel them to engage more actively in their communities.

Systems for remote monitoring might help with troublesome patient access concerns. Only 9% of doctors work in rural regions in the US, despite the fact that almost 20% of people do. Due to the fact that millions of additional patients would have access to insurance thanks to the health care reform, many organisations anticipate a shortage of primary care physicians over time (Caudill et al., 2011). The differences in care that rural inhabitants must contend with are well-documented in the literature (2005). Rural residents travel 2 to 3 times further to visit a doctor than their urban counterparts, see fewer specialists, and have poorer results for illnesses like diabetes and heart attacks, which are both frequent (Caudill et al., 2011; Chan et al., 2006). Wearable sensors and remote monitoring technologies may make it possible for professionals in metropolitan regions to reach rural areas and reduce these inequities.

Figure 1 displays a conceptual delineation of a remote checking framework. Wearable sensors are used to collect mobility and physiological data, allowing for the monitoring of patients' conditions. Depending on the therapeutic application of interest, sensors are installed. Sensors measuring vital indicators (such as heart rate and breathing rate) would be used, for instance, to keep an eye on patients undergoing therapeutic intervention for chronic obstructive pulmonary disease or congestive heart failure.

Sensors that collect movement data, for instance, could be used in applications. like tracking the success of stroke survivors' home-based rehabilitation therapies or the usage of older individuals' mobility aids. Patient information is sent wirelessly to an access point or mobile device, which then forwards it to a remote centre through the Internet. When critical events (like falls) are detected, the system uses data processing to send a notification to a central location where help may be delivered immediately. In the case of an emergency, loved ones will be notified, but they may also be notified when the patient has a more routine need for assistance, such as remembering to take their medications.

Medical professionals may remotely monitor a patient's health and get updates when a decision must be made about that patient's treatment.

Despite the obvious advantages, there are still several hurdles to jump over before On a large scale, a wearable-based remote monitoring system like the one described above can be used.

These difficulties include both cultural and technical obstacles, such as the limits of presently available battery technology and the stigma attached to using medical equipment for in-home clinical monitoring.

Figure 1. Wearable sensors to monitor patients and inform caretaker through network

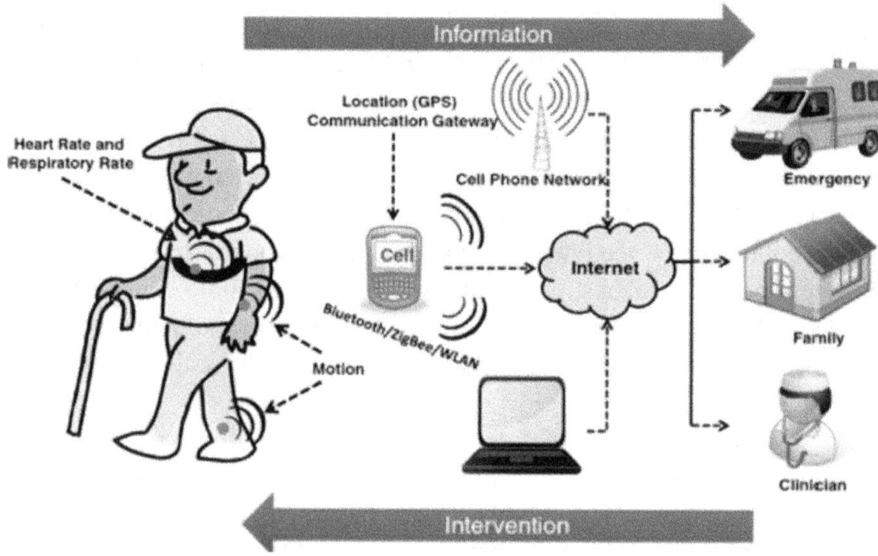

The drawn out checking of physiological pointers in ongoing patients is appropriate to wearable methods due to their inborn usability and minimal expense (Dieffenderfer, Goodell, Mills, McKnight, Yao, Lin, Beppler, Bent, Lee, Misra, Zhu, Oralkan, Strohmaier, Muth, Peden, & Bozkurt, 2016; Yang et al., 2019).There are now a large variety of wearable gadgets available on the market that may be used to keep tabs on all sorts of crucial physiological data. Proof from reads up on home consideration for patients with constant infections (Dieffenderfer, 2015; Huang et al., 2020) suggests that wearable devices may be used to recognise a variety of often seen side effects and indications of deteriorating, like the checking of cardiovascular capabilities and breathing as a pressing necessity. In circumstances of arrhythmia, cardiovascular breakdown, and other cardiovascular sicknesses, online ECG examination may be utilised to monitor the patient's progress. Wearable ECG devices, for instance, may quickly identify the beginning of atrial fibrillation (Randazzo, 2019). An essential physiological marker for identifying chronic respiratory illnesses is breathing rate. A warning that might serve as a reminder to patients to prevent mishaps that are both directly and indirectly brought on by the illness is activated when abnormal conditions (such as a rapid or shallow breathing rate) are detected by respiration monitoring devices (Dieffenderfer, Goodell, Mills, McKnight, Yao, Lin, Beppler, Bent, Lee, Misra, Zhu, Oralkan, Strohmaier, Muth, Peden, & Bozkurt, 2016). The many uses for these wearable technologies indicate potential for significantly lessening the stress that patients and their families experience due to unexpected progression of illnesses.

Figure 2 provides an overview of the strategies used.

Figure 2. Wearables for monitoring cardiovascular

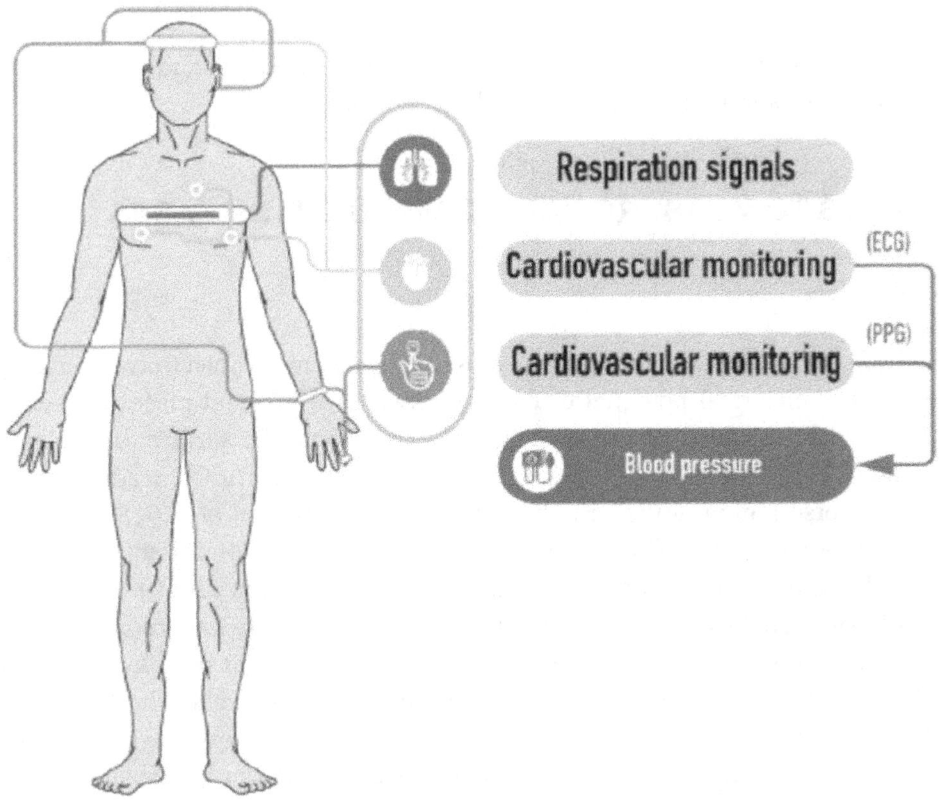

CARDIOVASCULAR DISEASES

Due to the difficulties of radical therapy, cardiovascular disorders, like coronary illness and cardiovascular breakdown, have a high demise rate and may quickly develop into chronic conditions, placing a significant financial and psychological burden on patients (Wei-Wei, 2017). Cardiovascular functions are important physiological markers of the health of the body and are closely related to the signs of chronic cardiovascular illnesses. Wearable cardiovascular monitoring technologies have therefore been long-standing research foci. There have been several developments and releases in the consumer market of wearable devices, such as algorithms.

These tools allow for direct monitoring of the main clinical indices, including heart rate (HR) and heart rate variability (HRV). Through indirect measurement and machine learning-based physiological modelling, it is possible to deduce other important characteristics including blood pressure (BP), oxygen saturation of the blood (SpO2), and blood sugar. Based on these derived signs, early warning and identification of heart dysfunction may also aid in clinical decision-making for diagnosis and therapy.

Features that may be used to predict cardiovascular disease include Blood sugar, blood oxygen saturation, heart rate, and heart rate variability.

DEVICES TO MEASURE CARDIOVASCULAR DISEASES

1. ECG

The gold standard for monitoring the heart is usually regarded as electrocardiography (ECG). The electrical potential difference between two distant places is used to calculate the projection of cardiac electric activity onto the surface of the body. Numerous earlier investigations have shown the reliability of ECG readings taken from the torso region, utilising either vests (singlets) (Liu et al., 2018; Ozkan et al., 2019) or chest bands (Span'o et al., 2016) to firmly and comfortably fasten the electrodes to the skin in both adults and children (Huang et al., 2020). Other user-friendly alternatives have also been pursued by researchers by incorporating ECG monitoring into a number of everyday accessories. In a fascinating investigation, ECG electrodes were tried to be included inside a cap for long haul checking while at the same time cycling (Von Rosenberg, Chanwimalueang, Goverdovsky, Looney, Sharp, & Mandic, 2016). Also, ear-telephone styled gadgets have been shown to be efficient for ECG collection and may give good signal quality even with vigorous daily head movement (Xu et al., 2020; Zhang et al., 2017). Wearable ECG also often uses wristbands (watches) (Beach et al., 2018). With confidence, it is simple to assess many physiological ECG signs from two fingers, including HR. However, some conditions, including infarct localisation, need multi-lead ECG recordings (using an ECG vest, for instance). Due to their demanding measurement requirements, long-term continuous monitoring has a very restricted range of applications. For the majority of wristband-styled systems, the measuring process—during which the user's fingers from the contrary hand ought to be placed on the band (watch) to lay out a cardiovascular circle —weakens the systems' ability for long-term continuous monitoring. A single-sided wristband with specifically crafted electrodes and strategically placed placement might be an ideal option (Rachim & Chung, 2016). Typical designs for wearable ECG monitoring are shown in Figure 3.

Figure 3. Wrist-worn ECG monitor (a) (Rachim & Chung, 2016), smart-vest ECG monitoring system (b) (Ozkan et al., 2019), singlet ECG monitoring system (c) (Von Rosenberg, Chanwimalueang, Goverdovsky, Looney, Sharp, & Mandic, 2016), and ear-ECG device (d) (Beach et al., 2018; Xu et al., 2020)

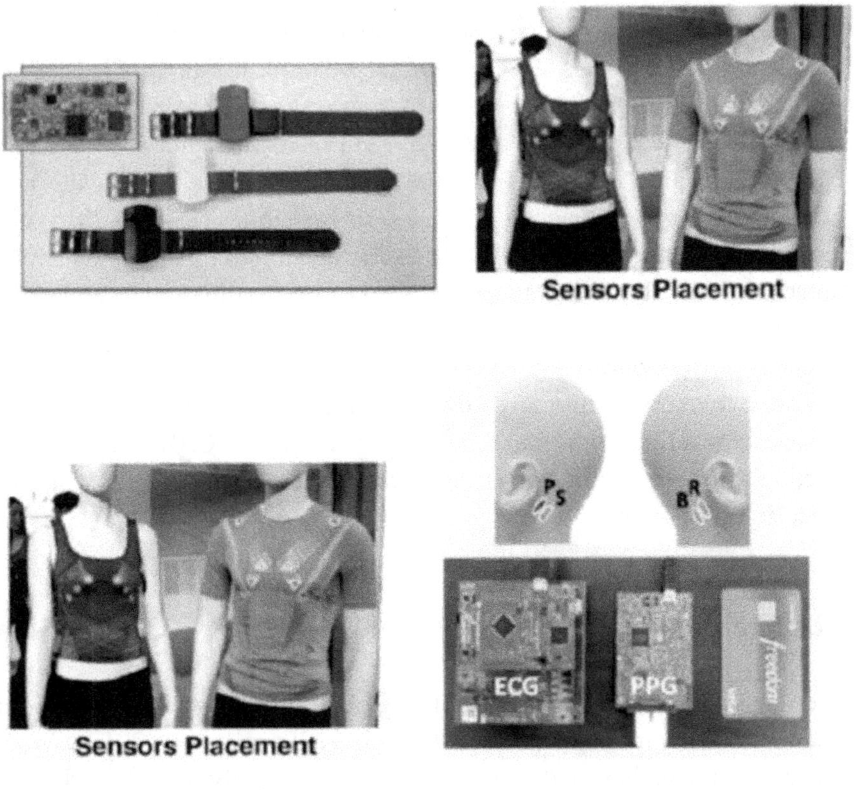

Wearable Electrodes for ECG Monitoring

Wearable ECG systems often utilise textile electrodes (TEs) or dry metal electrodes rather than hydrogel electrodes, as is the case with traditional ECG systems (Chung et al., 2018). TEs have been recognised as a potentially effective replacement for conventional electrodes because to their skin-friendly qualities.

The compelling utilization of multi-lead ECG in wearables has likewise acquired a great deal of consideration in the writing. Remembering standard 12-lead cathodes for a vest, for instance, adds a tangle of cables and connections that makes the system bulkier and more complicated than it has to be. New analog-signal compatible communication buses and textile printed circuits (Liu et al., 2018) are two examples of recent approaches to solve this issue (Xu et al., 2020).

2. Photoplethysmography (PPG)

Photoplethysmography (PPG) is one more sort of fundamental sign procured by infusing photons into human body tissues and deciding how much light is communicated (transmission-type PPG) or reflected (reflection-type PPG) (Allen, 2007). Reflection-type PPG is often utilised for wearable technology because it may be monitored from any portion of the body as long as the subdermal vascular network is sufficiently perfused (Castaneda, 2018). PPG measurements using a headband have been documented (Azhari, 2017), and comparable tools may alternatively take the shape of gloves (Beach et al., 2018) or earphones (Da He, 2015) (Figure 3). Because it is simple to include into commercial fitness wristbands (watches), wrist-worn PPG has also attracted a great deal of interest in the literature (Biswas et al., 2019). Due to motion artefacts, PPG has a rather poor level of resilience and reliability during vigorous activity (Chung et al., 2019). In recent years, work has been done to extract extremely precise average HRs (Chung et al., 2018) and restore waveform information (Xu et al., 2019; Xu et al., 2020) from flawed PPG data, making it usable in clinical settings. Researchers have been looking towards viable solutions for decades.

WEARABLE DEVICES FOR CHRONIC RESPIRATORY DISEASE

The third most common cause of mortality globally is chronic respiratory illness, which accounts for around 6% of all fatalities (Sleurs et al., 2019). Wearable respiratory monitoring has made it feasible to automatically provide early warning of lung damage associated with disruption in ventilation at home.

Figure 4. Glove-style wearable PPG device (Sleurs et al., 2019)

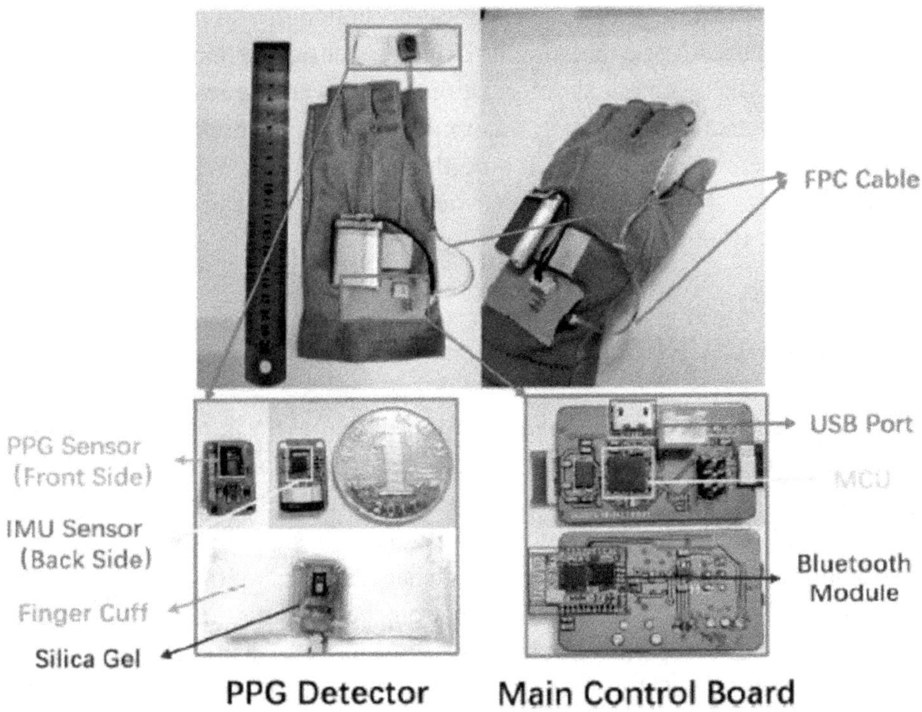

Respiration, one of the most fundamental physiological processes, provides considerable biological data that may be used to determine the physical well-being of an individual. As a marker of hypercarbia and pneumonic embolism, it has also been used to detect sepsis and pneumonia (Charlton et al., 2017).

DIRECT RESPIRATION METHOD

Based on various concepts like pressure-sensitivity, humidity-sensitivity, change in thoracic impedance, electromyography (EMG), and acceleration induced by chest motions, numerous wearable devices have been developed for direct respiration monitoring (Charlton et al., 2017; Lin et al., 2020). According to the research on personal oxygen monitors, flexible pressure-sensitive materials have received a lot of interest because of how well they fit the body and how well they can detect body movement brought on by breathing. Based on diverse processes, such as piezoresistivity (Lin et al., 2020), piezoelectricity (Wang et al., 2017), and triboelectricity (Zhang et al., 2019), they have been transformed into a variety of wearable respiration sensors.

Self-powered pressure sensors are well suited for long-term monitoring of chronic illnesses since they are thought to outperform typical wearable breathing sensors' bulky construction (Wang et al., 2017). But it is challenging to determine precise timings of inward breath or exhalation utilizing the tension delicate sensor in light of the fact that of the mechanical deformation process' erratic reaction time (Pang et al., 2018). By detecting the shift in water molecules during breathing and exhale, humidity-sensitive sensors may overcome this shortcoming. With materials like WS2, MoS2, and graphene, humidity sensors are thus often utilised in respiratory signal assessment (Pang et al., 2018).

INDIRECT MEASUREMENT

Thoracic impedance estimation is a technique that measures the impedance variations between electrodes on the skin in order to infer the change in lung capacity brought on by breathing (Piuzzi et al., 2018). Thoracic impedance measurement, a technique that has been utilized in the respiratory observing of basic patients or children, is also commonly used in wearable devices (Acharya & Basu, 2020) because of its non-invasive and pleasant qualities. The electrode belt and thorax directly relate to one another. In order to record breathing data from chest expansion during sleep and mild activity, Jayarathna et al. employed stretchy resistive bands made of polymer that were affixed to a T-shirt (Fig. 4(a)) (Jayarathna et al., 2020). To assess respiration and other performance variables, the business item Breeze BioPatch is a basic, off-the- shelf choice (as illustrated in Figure 4(b)). Other investigations (Mlynczak, 2017) have shown that the cardiac signal may also be collected simultaneously provided the impedance sensor is sensitive enough.

Figure 5. Wearable respiratory monitoring

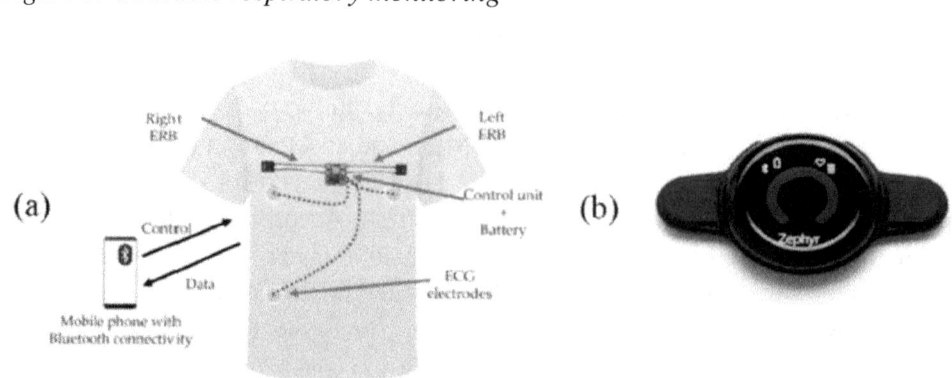

Chest wall electromyography (CWEMG), a technique for detecting respiratory effort, has been included into sleep monitoring and is very accurate at classifying various obstructive, mixed, or central apneas diseases (Berry, 2016). Given its inexpensive cost and wide patient accessibility, the accelerometer is another intriguing alternative tool to monitor respiratory rate. The reliability of RR signal extraction from accelerometer data acquired during ordinary exercises, like sitting, strolling, running, and dozing, has been shown in earlier research (Jin, 2009; Liu et al., 2011).

Epilepsy, stroke, and other diseases of the brain are not just chronic illnesses but also major causes of death across the globe. Electroencephalography (EEG), electromyography (EMG), and inertial measurement unit (IMU)-based wearable devices have been shown in recent advanced techniques to be able to detect and evaluate chronic brain illnesses. For instance, EEG may be used to automatically identify seizures and is the gold standard for diagnosing epilepsy. It is necessary to design an EEG collecting device that can be easily worn in everyday life after reaching the best seizure detection performance in clinical settings. A brilliant protective cap for gathering ceaseless EEG recordings has been developed by Rosenberg et al. (2016), and a covert and inconspicuous ear-centered EEG collection device has been presented by Bleichner et al. (2017). Meanwhile, multiple investigations have shown that seizure detection may be accomplished using behind-the-ear EEG, which catches epileptic releases much the same way to scalp EEG (Gu, 2018; You, 2020).

Figure 6. Smart helmet for continuous EEG recording

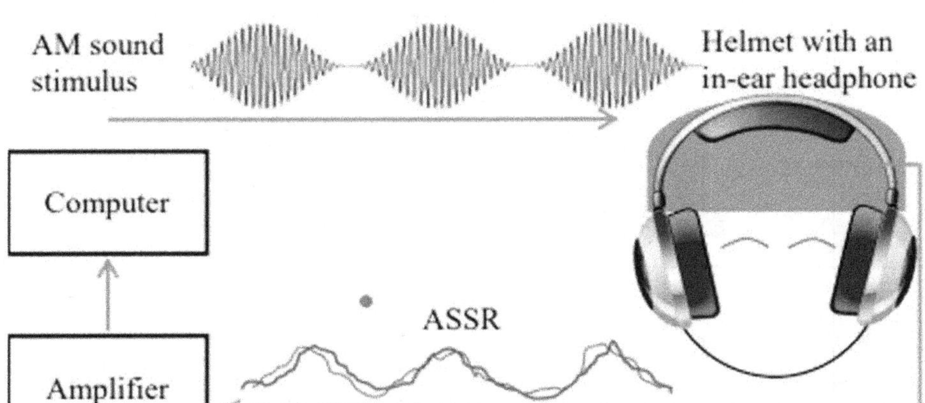

Table 1. Wearable technology

Chronic Diseases	Features	Wearable's	Parameter	Prediction & Warning	Convenient	Cost	Efficiency
Cardiovascular	ECG	Chest Band	HR	Tachycardia, Arrhythmia, AMI	Fair	Medium	Good
		Vest					
		Armband	HRV		Good	Low	Poor
		Helmet					
		Wristband					
		Forehead band					
	PPG	Ear phones	SPO_2	AP	Good	Low	Poor
		Glove	Blood Glucose	AMI			
Respiratory	Direct	Thoracic impedence	Breathing rate	Pneumonia, Sepsis, Hypercarbia, Pulmonary Embolism	Fair	Medium	Good
		Thermo-sensitivity, Humidity change, Triboelectricity, Piezoelectricity CM-EMG Acceleration caused by chest movements	Breathing depth, Respiratory modes		Fair	Medium	fair
	Indirect	BW, AM FM from EEG and PPG	Breathing rate and Breathing Sound	Deterioration Wheeze & Crackle Seizure	Good	Medium	poor

Additionally, since they are user-friendly and affordable, EMG and IMU might be utilized for ischemic stroke appraisal. Li et al. (2016) propose a lower-appendage movement order technique for the assessment of hemiparetic patients, in view of the capacity of IMU and EMG information to predict the Brunnstrom stage for the affected lower limb. In order to measure the activity of the muscles in the lower leg, Isezaki et al. (2019) developed a sock-shaped wearable sensor that can monitor distal EMG signals. Sock-style electromyography (EMG) measurement devices may collect the electrical activity of the target muscle, which is spread all the way around the shin. In order to provide a consistent assessment of the methods outlined above and to aid the two doctors and gadget producers in choosing ideal answers for clinical use, we have assembled the wearable methodologies into Table 1 and appointed scores for comfort, expenses, and vigor.

In the above Table-1 we analysed wearable devices used for cardiovascular diseases and respiratory diseases. Cost, Convenient and Efficiency are the parameters applied to analyse wearable devices . Cost means amount spend to buy the device which will be categorised in low, medium and high. Convenient means whether patient feel free to wear it or they feel some inconvenient in wearing those devices. It will be categorised in fair, good, poor. Efficiency of the device, value predicted by the device is also categorised as fair, good, poor.

For convenience, we labelled methods that do not directly touch the skin and can be readily incorporated into everyday accessories as "a€oeGood," whereas methods that do need skin-contact or are challenging to wear are labelled "a€oeFair" or "a€oePoor." The market pricing of relevant commercial gadgets are used to estimate costs. Since they are still in the experimental stages and are not yet commercially accessible, several innovative materials-based gadgets and procedures could likewise be regarded as expensive. Both intrinsic signal quality and noise resistance are used to evaluate robustness. Table 1 provides examples of the methods for various wearable technology. In conclusion, the results of the literature review show that wearable technologies may enable long term noticing and finish of steady cardiovascular, aspiratory, and other clinical problems with a strong foundation.

CONCLUSION

In this chapter we analysed wearable devices used for cardiovascular diseases and respiratory diseases. Cost, Convenient and Efficiency are the parameters applied to analyse wearable devices . Cost means amount spend to buy the device which will be categorised in low, medium and high. Convenient means whether patient feel free to wear it or they feel some inconvenient in wearing those devices. It will be categorised in fair, good, poor. Efficiency of the device, value predicted by the device is also categorised as fair, good, poor. It is predicted that if the cost of the device is low then the convenient is medium and efficiency is also in poor. Spending high amount for the wearable's will give patient satisfaction and convenient to them, efficiency is also high.

REFERENCES

Acharya, J., & Basu, A. (2020). Deep neural network for respiratory sound classification in wearable devices enabled by patient specific model tuning. *IEEE Transactions on Biomedical Circuits and Systems*, *14*(3), 535–544. doi:10.1109/TBCAS.2020.2981172 PMID:32191898

Allen, J. (2007). Photoplethysmography and its application in clinical physiological measurement. *Physiological Measurement*, *28*(3), R1–R39. doi:10.1088/0967-3334/28/3/R01 PMID:17322588

Azhari, A. (2017). *A patch-type wireless forehead pulse oximeter for spo 2 measurement, in: 2017 IEEE Biomedical Circuits and Systems Conference (BioCAS)*. IEEE. doi:10.1109/BIOCAS.2017.8325557

Beach, C., Krachunov, S., Pope, J., Fafoutis, X., Piechocki, R. J., Craddock, I., & Casson, A. J. (2018). An ultra low power personalizable wrist worn ecg monitor integrated with iot infrastructure. *IEEE Access : Practical Innovations, Open Solutions*, *6*, 44010–44021. doi:10.1109/ACCESS.2018.2864675

Berry, R. B. (2016). *Use of chest wall electromyography to detect respiratory effort during polysomnography*. National Institute of Health.

Biswas, D., Simoes-Capela, N., Van Hoof, C., & Van Helleputte, N. (2019). Heart rate estimation from wrist-worn photoplethysmography: A review. *IEEE Sensors Journal*, *19*(16), 6560–6570. doi:10.1109/JSEN.2019.2914166

Bleichner, M. G., & Debener, S. (2017). Concealed, unobtrusive ear-centered eeg acquisition: Ceegrids for transparent eeg. *Frontiers in Human Neuroscience*, *11*, 163. doi:10.3389/fnhum.2017.00163 PMID:28439233

Bonato, P. (2010). Wearable sensors and systems. From enabling technology to clinical applications. *IEEE Engineering in Medicine and Biology Magazine*, *29*(3), 25–36. doi:10.1109/MEMB.2010.936554 PMID:20659855

Cardiovascular diseases (heart attack, stroke). (n.d.). WHO. https://www.who.int/westernpaci fic/health-topics/cardiovascular-diseases.

Castaneda, D. (2018). A review on wearable photoplethysmography sensors and their potential future applications in health care. *International Journal of Biosensors & Bioelectronics*, *4*(4), 195. PMID:30906922

Caudill, T. S., Lofgren, R., Jennings, C. D., & Karpf, M. (2011). Commentary: Health care reform and primary care: training physicians for tomorrow's challenges. *Academic Medicine*, *86*(2), 158–160. doi:10.1097/ACM.0b013e3182045f13 PMID:21270552

Chan, L., Hart, L. G., & Goodman, D. C. (2006). Geographic access to health care for rural Medicare beneficiaries. *The Journal of Rural Health*, *22*(2), 140–146. doi:10.1111/j.1748-0361.2006.00022.x PMID:16606425

Charlton, P. H., Birrenkott, D. A., Bonnici, T., Pimentel, M. A. F., Johnson, A. E. W., Alastruey, J., Tarassenko, L., Watkinson, P. J., Beale, R., & Clifton, D. A. (2017). Breathing rate estimation from the electrocardiogram and photoplethysmogram: A review. *IEEE Reviews in Biomedical Engineering*, *11*, 2–20. doi:10.1109/RBME.2017.2763681 PMID:29990026

Chen, H., Chen, W., Bao, S., Lu, C., Wang, L., Ma, J., Wang, P., Lu, H., Shu, F., & Bambang Oetomo, S. (2020). Design of an integrated wearable multi-sensor platform based on flexible materials for neonatal monitoring. *IEEE Access : Practical Innovations, Open Solutions*, *8*, 23732–23747. doi:10.1109/ACCESS.2020.2970469

Chung, H., Lee, H., & Lee, J. (2018). Finite state machine framework for instantaneous heart rate validation using wearable photoplethysmography during intensive exercise. *IEEE Journal of Biomedical and Health Informatics*, *23*(4), 1595–1606. doi:10.1109/JBHI.2018.2871177 PMID:30235152

Chung, H. U., Kim, B. H., Lee, J. Y., Lee, J., Xie, Z., Ibler, E. M., Lee, K. H., Banks, A., Jeong, J. Y., Kim, J., Ogle, C., Grande, D., Yu, Y., Jang, H., Assem, P., Ryu, D., Kwak, J. W., Namkoong, M., Park, J. B., & Rogers, J. A. (2019). Binodal, wireless epidermal electronic systems with in-sensor analytics for neonatal intensive care. *Science*, *363*(6430), 6430. doi:10.1126cience.aau0780 PMID:30819934

Da He, D. (2015). An ear-worn vital signs monitor, IEEE (Inst. Electr. Electron. Eng.). *IEEE Transactions on Biomedical Engineering*, *62*(11), 2547–2552. doi:10.1109/TBME.2015.2459061

Dieffenderfer, J., Goodell, H., Mills, S., McKnight, M., Yao, S., Lin, F., Beppler, E., Bent, B., Lee, B., Misra, V., Zhu, Y., Oralkan, O., Strohmaier, J., Muth, J., Peden, D., & Bozkurt, A. (2016). Low-power wearable systems for continuous monitoring of environment and health for chronic respiratory disease. *IEEE Journal of Biomedical and Health Informatics*, *20*(5), 1251–1264. doi:10.1109/JBHI.2016.2573286 PMID:27249840

Dieffenderfer, J., Goodell, H., Mills, S., McKnight, M., Yao, S., Lin, F., Beppler, E., Bent, B., Lee, B., Misra, V., Zhu, Y., Oralkan, O., Strohmaier, J., Muth, J., Peden, D., & Bozkurt, A. (2016). Low-power wearable systems for continuous monitoring of environment and health for chronic respiratory disease. *IEEE Journal of Biomedical and Health Informatics*, *20*(5), 1251–1264. doi:10.1109/JBHI.2016.2573286 PMID:27249840

Dieffenderfer, J. P. (2015). Wearable wireless sensors for chronic respiratory disease monitoring. *2015 IEEE 12th International Conference on Wearable and implantable Body Sensor Networks (BSN)*. IEEE.

Gu, Y. (2018). Comparison between scalp eeg and behind-the-ear eeg for development of a wearable seizure detection system for patients with focal epilepsy. *Sensors (Basel)*, *18*(1), 29. PMID:29295522

Guo, H., Lan, C., Zhou, Z., Sun, P., Wei, D., & Li, C. (2017). Transparent, flexible, and stretchable ws 2 based humidity sensors for electronic skin. *Nanoscale*, *9*(19), 6246–6253. doi:10.1039/C7NR01016H PMID:28466937

Huang, C., Wang, Y., Li, X., Ren, L., Zhao, J., Hu, Y., Zhang, L., Fan, G., Xu, J., Gu, X., Cheng, Z., Yu, T., Xia, J., Wei, Y., Wu, W., Xie, X., Yin, W., Li, H., Liu, M., & Cao, B. (2020). Clinical features of patients infected with 2019 novel coronavirus in wuhan, China. *Lancet*, *395*(10223), 497–506. doi:10.1016/S0140-6736(20)30183-5 PMID:31986264

Isezaki, T., Kadone, H., Niijima, A., Aoki, R., Watanabe, T., Kimura, T., & Suzuki, K. (2019). Sock-type wearable sensor for estimating lower leg muscle activity using distal emg signals. *Sensors (Basel)*, *19*(8), 1954. doi:10.339019081954 PMID:31027302

Jayarathna, T., Gargiulo, G. D., & Breen, P. P. (2020). Continuous vital monitoring during sleep and light activity using carbon-black elastomer sensors. *Sensors (Basel)*, *20*(6), 1583. doi:10.339020061583 PMID:32178307

Jin, A. (2009). Performance evaluation of a tri-axial accelerometry-based respiration monitoring for ambient assisted living. *2009 Annual International Conference of the IEEE Engineering in Medicine and Biology Society*. IEEE. 10.1109/IEMBS.2009.5333116

Li, H.-T. (2016). Lower-limb motion classification for hemiparetic patients through imu and emg signal processing. *2016 International Conference on Biomedical Engineering (BME-HUST)*. IEEE. 10.1109/BME-HUST.2016.7782096

Lin, X., Zhang, T., Cao, J., Wen, H., Fei, T., Liu, S., Wang, R., Ren, H., & Zhao, H. (2020). Flexible piezoresistive sensors based on conducting polymer-coated fabric applied to human physiological signals monitoring. *Journal of Bionics Engineering*, *17*(1), 55–63. doi:10.100742235-020-0004-9

Liu, C., Zhang, X., Zhao, L., Liu, F., Chen, X., Yao, Y., & Li, J. (2018). Signal quality assessment and lightweight qrs detection for wearable ecg smartvest system. *IEEE Internet of Things Journal*, *6*(2), 1363–1374. doi:10.1109/JIOT.2018.2844090

Liu, G., Guo, Y.-W., Zhu, Q.-S., Huang, B.-Y., & Wang, L. (2011). Estimation of respiration rate from three-dimensional acceleration data based on body sensor network. *Telemedicine Journal and e-Health*, *17*(9), 705–711. doi:10.1089/tmj.2011.0022 PMID:22035321

Mlynczak, M. (2017). *Decomposition of the cardiac and respiratory components from impedance pneumography signals*. BIOSIGNALS.

Ozkan, H., Ozhan, O., Karadana, Y., Gulcu, M., Macit, S., & Husain, F. (2019). A portable wearable tele-ecg monitoring system. *IEEE Transactions on Instrumentation and Measurement*, *69*(1), 173–182. doi:10.1109/TIM.2019.2895484

Pang, Y., Jian, J., Tu, T., Yang, Z., Ling, J., Li, Y., Wang, X., Qiao, Y., Tian, H., Yang, Y., & Ren, T.-L. (2018). Wearable humidity sensor based on porous graphene network for respiration monitoring. *Biosensors & Bioelectronics*, *116*, 123–129. doi:10.1016/j.bios.2018.05.038 PMID:29879538

Piuzzi, E., Pisa, S., Pittella, E., Podesta, L., & Sangiovanni, S. (2018). Low-cost and portable impedance plethysmography system for the simultaneous detection of respiratory and heart activities. *IEEE Sensors Journal*, *19*(7), 2735–2746. doi:10.1109/JSEN.2018.2887303

Rachim, V. P., & Chung, W.-Y. (2016). Wearable noncontact armband for mobile ecg monitoring system. *IEEE Transactions on Biomedical Circuits and Systems*, *10*(6), 1112–1118. doi:10.1109/TBCAS.2016.2519523 PMID:27214910

Randazzo, V. (2019). Ecg watch: a real time wireless wearable ecg. *IEEE International Symposium on Medical Measurements and Applications (MeMeA)*. IEEE 10.1109/MeMeA.2019.8802210

Sleurs, K., Seys, S. F., Bousquet, J., Fokkens, W. J., Gorris, S., Pugin, B., & Hellings, P. W. (2019). Mobile health tools for the management of chronic respiratory diseases. *Allergy*, *74*(7), 1292–1306. doi:10.1111/all.13720 PMID:30644567

Span'o, E., Di Pascoli, S., & Iannaccone, G. (2016). Low-power wearable ecg monitoring system for multiple-patient remote monitoring. *IEEE Sensors Journal*, *16*(13), 5452–5462. doi:10.1109/JSEN.2016.2564995

Teng, X.-F., Zhang, Y.-T., Poon, C. C. Y., & Bonato, P. (2008). Wearable medical systems for p-Health. *IEEE Reviews in Biomedical Engineering*, *1*, 62–74. doi:10.1109/RBME.2008.2008248 PMID:22274900

Von Rosenberg, W., Chanwimalueang, T., Goverdovsky, V., Looney, D., Sharp, D., & Mandic, D. P. (2016). Smart helmet: Wearable multichannel ecg and eeg. *IEEE Journal of Translational Engineering in Health and Medicine*, *4*, 4. doi:10.1109/JTEHM.2016.2609927 PMID:27957405

Von Rosenberg, W., Chanwimalueang, T., Goverdovsky, V., Looney, D., Sharp, D., & Mandic, D. P. (2016). Smart helmet: Wearable multichannel ecg and eeg. *IEEE Journal of Translational Engineering in Health and Medicine*, *4*, 4. doi:10.1109/JTEHM.2016.2609927 PMID:27957405

Wang, B., Liu, C., Xiao, Y., Zhong, J., Li, W., Cheng, Y., Hu, B., Huang, L., & Zhou, J. (2017). Ultrasensitive cellular fluorocarbon piezoelectret pressure sensor for self-powered human physiological monitoring. *Nano Energy*, *32*, 42–49. doi:10.1016/j.nanoen.2016.12.025

Wei-Wei, C. (2017). China cardiovascular diseases report 2015: A summary. *Journal of Geriatric Cardiology : JGC*, *14*(1), 1. PMID:28270835

Xu, K., Jiang, X., & Chen, W. (2019). Photoplethysmography motion artifacts removal based on signal noise interaction modeling utilizing envelope filtering and time-delay neural network. *IEEE Sensors Journal*, *20*(7), 3732–3744. doi:10.1109/JSEN.2019.2960370

Xu, K., Jiang, X., Lin, S., Dai, C., & Chen, W. (2020). Stochastic modeling based nonlinear bayesian filtering for photoplethysmography de-noising in wearable devices. *IEEE Transactions on Industrial Informatics*, *16*(11), 7219–7230. doi:10.1109/TII.2020.2988097

Yang, G., Pang, G., Pang, Z., Gu, Y., Mantysalo, M., & Yang, H. (2019). Non-invasive flexible and stretchable wearable sensors with nanobased enhancement for chronic disease care. *IEEE Reviews in Biomedical Engineering*, *12*, 34–71. doi:10.1109/RBME.2018.2887301 PMID:30571646

You, S. (2020). *Unsupervised automatic seizure detection for focal-onset seizures recorded with behind-the-ear eeg using an anomaly-detecting generative adversarial network, Comput*. Methods Progr. Biomed. doi:10.1016/j.cmpb.2020.105472

Zhang, H., Zhang, J., Hu, Z., Quan, L., Shi, L., Chen, J., Xuan, W., Zhang, Z., Dong, S., & Luo, J. (2019). Waist-wearable wireless respiration sensor based on triboelectric effect. *Nano Energy*, *59*, 75–83. doi:10.1016/j.nanoen.2019.01.063

Zhang, Q., Zeng, X., Hu, W., & Zhou, D. (2017). A machine learning-empowered system for long-term motion tolerant wearable monitoring of blood pressure and heart rate with ear-ecg/ppg. *IEEE Access : Practical Innovations, Open Solutions*, *5*, 10547–10561. doi:10.1109/ACCESS.2017.2707472

Zhao, J., Li, N., Yu, H., Wei, Z., Liao, M., Chen, P., Wang, S., Shi, D., Sun, Q., & Zhang, G. (2017). Highly sensitive mos2 humidity sensors array for noncontact sensation. *Advanced Materials*, *29*(34), 1702076. doi:10.1002/adma.201702076 PMID:28692765

Chapter 4
Analysis and Comparison of Hash Functions in Terms of Security and Performance

Öner Özer
https://orcid.org/0009-0004-0162-2677
Ahmet Yesevi University, Turkey

Gurkan Tuna
https://orcid.org/0000-0002-6466-4696
Trakya University, Turkey

ABSTRACT

Hash functions map data of arbitrary size to fixed-size values, though there are some that support variable length output. The values returned by a hash function are called hash values. Hash functions are used in many fields such as generation of passwords and encryption which are the building blocks of cryptology, digital signatures, data integrity control, crypto storage, and random number generation. However, some hash functions like MD5 and SHA1 are now broken. Using a weak or broken hash function can leave data vulnerable, and thus should not be used. This chapter presents definitions, structures, performances, as well as security vulnerabilities, of hash functions.

DOI: 10.4018/978-1-6684-9267-3.ch004

INTRODUCTION

Cryptology is the science of secret messages/communications (Qian et al., 2022). It is a mathematics-based science that studies and develops methods for the safe transmission of data, safe storage of data and verification of data integrity in the field of communications. It is divided into two branches as cryptography and cryptanalysis (Mihailescu & Nita, 2023). Cryptography is the science of information security (Ural & Örenç, 2019). Cryptanalysis, on the other hand, is a branch of cryptology that deals with information security and the process of cracking secure/encrypted information (Banoth & Regar, 2023). Cryptanalysts often try to decrypt and test the security of cryptographic functions.

Today, due to the rapidly developing technology and rapid developments in communication systems, ensuring information security has become an increasingly important issue. Hash functions are among the most important cryptographic algorithms and form the basis of various cryptological functions used in many fields (Sadeghi-Nasab & Rafe, 2023). As shown in Figure 1, they take an input of arbitrary length and produce a hashed representation of that input (Paar & Pelzl, 2009). They take different sizes of input and generate fixed size outputs each time, and the output of the hash function is called "hash value".

Hash functions are used in many fields such as generation of passwords and encryption which are the building blocks of cryptology, digital signatures, data integrity control, crypto storage, and random number generation. Standardized hash functions have been readily available and used in software programming frameworks and hardware devices for years. However, some of those hash functions like MD5 and SHA-1 were broken (Leurent & Peyrin, 2020). Using a weak or broken hash function can leave data vulnerable, and should not be used.

It may be necessary to use different hash functions depending on the purpose and situation of use. Because hash functions differ greatly in speed and security between those used for data validation and those used to derive and store keys. Although most of the existing hash functions have lost their reliability, they are still actively used. Because the vast majority of programmers generally use standard cryptology libraries and frameworks for the security part of the software they produce. In this chapter, firstly commonly used hash functions are described. Then, security and speed of the hash functions are evaluated.

Figure 1. How hash functions operate?

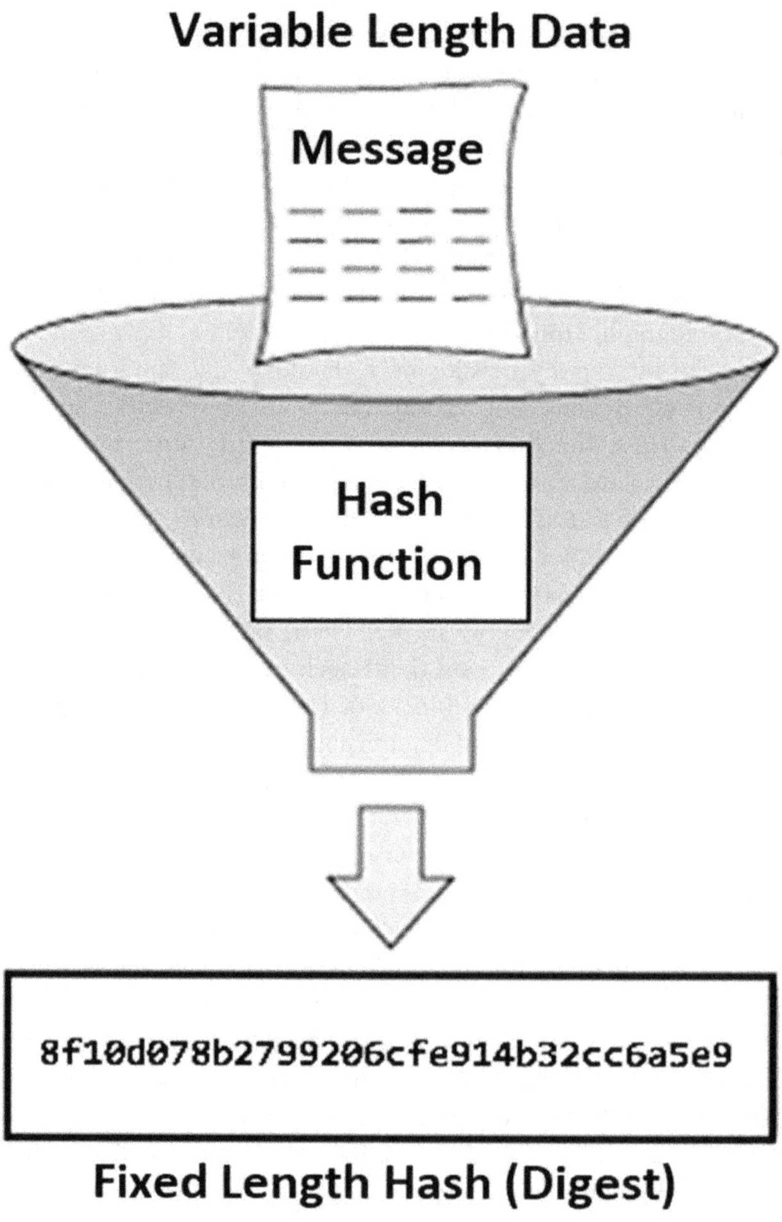

CRYPTOGRAPHIC HASH FUNCTIONS AND THEIR COMMON FEATURES

Regardless of the length of the input size of the data, all hash functions should have the same output length, and the same data should always yield the same hash output value. Since the hash value will be a fixed length value; it will be easier and more convenient to reserve space for summary in database, file format or network protocol area. It also helps prevent hackers from knowing how big the original input is because all output is fixed length and does not change, no matter how long or short the original input is.

Hash functions are one-way functions. You cannot decode the original input data using the hash value. This means that you can easily convert a data to a hash value, but it will be impossible to derive the data from the hash value (Paar & Pelzl, 2009). In other words, given a hash value h, it should be difficult to find any m messages such that h = hash (m). Summary functions should always be one-way. This feature protects against an attacker targeting only the hash and original message data. Similarly, it should be difficult to find a different entry with the same hash (Paar & Pelzl, 2009) . Given an m1 data entry, it should be difficult to find a different m2 entry such that hash (m1) = hash (m2). It provides protection against an attacker who wants to introduce a different value as a legitimate value instead of the original data with a hash value.

In hash functions, when two different data inputs produce the same hash values, it is called a cryptographic hash collision. Collision resistance means that it should be difficult to find two different data with the same hash value. This feature is sometimes referred to as strong crash durability (Paar & Pelzl, 2009). In a secure hash function, it should be difficult to find two different messages m1 and m2 where hash (m1) = hash (m2). If there is a way to generate collisions, cybercriminals can mislead the system into believing that they have the original input data, even if they do not have the original input data (Gauravaram & Knudsen, 2010).

Suppose the attackers learned that the hash of your password and another entry match the same hash. They can now use the hash, which is not the same thing as the login (password) itself, as your password hash to log into your account, even if you don't have the password in the first one. Therefore, all hash functions need to be resistant to collisions. One way to reduce collisions in password hashing and reduce the risk of rainbow table attacks is to use salting (Hatzivasilis, 2017). The MD5 has been technically proven to produce collisions and it was broken in March 2005 (Wang et al., 2005).

The avalanche effect in hash functions is the name given to the slightest change in the data entered into the function, large changes in the summary output. This feature ensures that the original text cannot be deciphered (Paar & Pelzl, 2009). The name comes from the concept that the slightest change or shift in snowdrifts on a mountainside triggers an avalanche. For example, in the text and bit string examples given below, it can be seen that the SHA-1 hashes have changed enormously with the slightest change. As seen in the examples below, a slight change in the data (even 1-bit) changes the summary output value drastically.

```
" You, me and Ali" => SHA-1 =
155949473ec7d782fb840e0aca3b915ae7890b2c
" You, me and Ali" => SHA-1 =
0dbf495f27473d77909a3cb2295f46ed3bdd4030
010011011011 0 => SHA-1 Digest =
b20860facf81dcff36b74318664a4694686a8e51
010011011011 1 => SHA-1 Digest =
4a095b078eabce3b04c67735c84d6c28c4ecbb03
```

Based on their design, hash functions are expected to produce results quickly or slowly. In most cases, the hash function should calculate hash values quickly; this is considered an ideal property of a cryptographic hash. MD5, SHA-1 group are examples of fast-running hash functions (Martin, 2020). When a user enters their account's password, it expects to log in in micro seconds. This can only happen if the hash function runs at extremely high speed in hash generation. However, this feature is a bit more subjective, not all hash functions are expected to produce hashes quickly. In some cases, such as digital signature certificate generation or the Password Based Key Derivation Function (PBKDF), it requires hash functions to be slow. In this case, you want the computation to be slower to slow down and make it harder for attackers to brute force attacks on user passwords. Argon2, bcrypt, scrypt hash functions, which are designed specifically for the key derivation function, are examples of slow-running hash functions (Martin, 2020). In addition, hash functions may show speed differences according to the hardware they are working with. While algorithms such as MD5 and SHA-1 are designed for 32-bit microprocessors, hash functions such as SHA-2 and SHA-3 that can be considered relatively new work better on 64-bit microprocessors (Martin, 2020).

Today, there are many applications that use hash functions that have been proven to be broken. Although hash functions are basically designed for hash value generation, there are serious speed and security differences between them depending on where they are used. Due to their native support, ready-to-use cryptological frameworks and libraries have now become software that also supports outdated and insecure algorithms. Most software developers working in the industry prefer to use old hash functions that they think are reliable, although they do not follow current developments in the field of cryptology, which is generally seen as a difficult branch of mathematics. As a result of the studies carried out by institutions such as the National Institute of Standards and Technology (NIST) and The International Association for Cryptologic Research (IACR), along with the rapidly developing microprocessor, graphics processing unit (GPU) and field-programmable gate array (FPGA) processors, new functions that are safer and more efficient are emerging every year.

Keccak algorithm, developed by Bertoni et al., which was successful in the competition organized by NIST in 2012 due to the need for reliable hash functions, has been standardized as the SHA-3 family of functions (Bertoni et al., 2013). But Wei et al. reported that successful attacks can be made by reducing the probability space against the Keccak algorithm (Wei et al., 2022). Compared to their competitors, bcrypt, scrypt and argon2 work slowly due to more processing cycles and memory (Ertaul et al., 2016). In spite of hardware improvements, these algorithms are currently resistant to brute force attacks. However, it is thought that the functions used today will no longer be reliable thanks to the qubit system, which will replace the binary system we use today with the development of quantum computers (Dong et al., 2020). Already, the National Security Agency (NSA) has been working for a long time to research and develop the future relevance of post-quantum cryptographic standards.

STRUCTURES OF CRYPTOGRAPHIC HASH FUNCTIONS AND THEIR USES

Hash functions can be grouped into three main structures as FrAmework (HAIFA), Merkle-Damgard (MD) and Sponge (Koçak, 2009). Others are formed by differentiating these three main structures. Hash functions are used for various purposes, including data integrity verification (Silva, 2003), password storage and verification, digital/electronic signature (Kerry & Gallagher, 2013; Schneier, 2007).

Keyed Digest Message Authentication Code (HMAC) is a specific type of Message Authentication Code (MAC) that includes a cryptographic hash function and a secret key. It can be used to simultaneously verify both the data integrity and authenticity of a message (Pittalia, 2019). Before communication, the parties agree on a key over a confidential channel. Key distribution algorithms for this (Diffie-Hellman key exchange) is also available. Using HMAC eliminates the need for an asymmetric private and public key infrastructure. To put it simply; we can say that the hash value of the message is generated by the hash function according to a secret password (key).

```
Message = "Hello"
Secret_Key = "Ankara"
```

```
SHA-1(Message) = df53c2b7c583ab7c9b0a5ad5912b2db83aa80571
HMAC-SHA-1(Secret_Key, Message) =
10307bd7c08001944a8af147d1feb0fe8c5059de
```

While the standard hash function is transmitted with the message, the communication can be captured by the person listening and the message content can be changed, the digest value can be recalculated according to the changed message and transmitted to the receiver. If the hash function HMAC is used here, even if the interlocutor recalculates the hash value of the changed message, he will have produced an incorrect hash value because he does not know the secret key. Any cryptographic hash function such as MD5, SHA-1, SHA-2, or SHA-3 can be used in an HMAC calculation; the resulting MAC algorithm is called HMAC-X, where X is the hash function used (for example, HMAC-SHA_256 or HMAC-SHA-3_512).

The hash values of malicious software written by the attackers to damage the system or to take over the system are calculated very quickly by the current antivirus software and compared in their own database. Thus, malware can be detected quickly and using less hardware power before performing detailed code analysis of large pieces of code (Aumasson, 2017).

The security of blockchains is based entirely on hash functions. It is a structure that is planned to be used in the future to prevent data changes in institutions such as banks, notaries and land registry offices. It writes the hash value of the preceding blockchain to the first line of the next blockchain, and the hash value is calculated again with the data in this blockchain, and then it is written to the first line of the next blockchain to form interconnected data chains. Thus, it becomes impossible to change any retroactive block in the blockchain (Aumasson, 2017).

Merkle-Damgard (MD)

MD structure is the oldest structure used to construct a hash function. In this structure, message blocks are split and each block is processed by sending it to the compression function. The compression function generates an output for each message block and these outputs are used in the next iteration. In this way, a hash value is obtained. In summary functions, a certain repetition logic is used, usually by changing the blocks (Coron et al., 2005). It consists of two stages:

- **Padding:** In the MD structure, message blocks are processed by sending them to the compression function. However, since the length of the message is usually not an exact multiple of the block size expected by the compression function, a padding must be added to the end of the message. Padding is a series of “1” and “0” bits appended to the end of the message. Padding ensures that the message is an exact multiple of the block size expected by the compression function. Padding is also used to make the message more conflict resistant.

- **Iteration:** It is performed by dividing the message into blocks and processing each block by sending it to the compression function. In this structure, the compression function also uses the outputs generated in the previous iteration of the message block and thus generates an output in the next iteration. This Iteration process ends when the entire message has been processed, resulting in a hash value.

Hash Iterative Framework (HAIFA)

Merkle-Damgard structure was proven to be resistant to collision; however, it was shown to be weak against methods such as herding attack, so its structure was modified slightly (Dunkelman & Biham, 2006). In the HAIFA structure, there are two new variables that enter the f function. These are the salt value (pure) s and the number of message bits hashed so far b.

Sponge

The sponge structure is newer and consists of two basic parts. These parts are Absorbing and Squeezing. The sponge structure produces an output for an input string and places some of that output back into the input string. This structure allows it to be capable of handling an input string in a variety of ways and to accommodate various input lengths. r and c bits affects the speed and reliability of the sponge construction. The c bits are expressed as "capacity" and determine the security of the structure. To the r bits, the message is added with different approaches and mixed in it thanks to the f function. Increasing r bits speeds up the build, while decreasing it slows it down. Usually, the function is run empty several times without adding anything to the r bits, thus ensuring reliability between the suction and squeezing parts. If the f function is a permutation, the sponge structure is called P - Sponge, in other cases it is called T-Sponge (Bertoni et al., 2007). If the f function is random, the structure is a random seer (random oracle) is considered to be. Examples of these structures are Panama, RadioGatun, Grindahl, and Keccak.

COMMONLY USED HASH FUNCTIONS AND THREATS TO HASH FUNCTIONS

Some hash functions are block cipher-based. Examples of this type are MD4, MD5, SHA-0, and SHA-1. This structure can be created in 64 different ways, depending on whether the message and intermediate values will be added and if they are added, in which places they will be added. However, only 12 of these constructs have been identified as safe (Preneel et al., 1994). On the other hand, some hash functions are streaming password-based. Panama and RC4-Hash are examples of such hash functions aimed at speed and efficiency. On the other hand, they do not have as strong proofs of reliability as those based on block ciphers.

In order to replace MD4, MD5 was designed in 1991 and specified as RFC 1321 in 1992 (Rivest, 1992). Nowadays, collisions against MD5 can be calculated in seconds, so it has been deprecated and designated as unsafe as it is no longer safe. Secure Hash Algorithm (SHA-1) was developed by NIST in 1993 in the USA as FIPS-PUB-180 Secure. It was later withdrawn by the NSA and replaced by a version developed as SHA-1, published in FIPS-PUB-180-1 in 1995. Theoretically degraded since 2005, officially deprecated by NIST in 2011. In 2017, SHA-1 was broken (Stevens et al, 2017). But SHA-1 is still safe for HMAC.

SHA-2 is a set of cryptographic hash functions and was first published in 2001. They were built using the MD structure from a one-way compression function, condensation, and the Davies - Meyer structure from a classified special block cipher (Pittalia, 2019). The SHA-2 family has four other SHA variants, SHA-224, SHA-256, SHA-384, and SHA-512, based on the number of bits in the hash values. No successful attack has yet been reported on the SHA-2 hash function. As expected, SHA-512 is more secure than SHA-256. But it can run faster than SHA-256 on systems with 64-bit microprocessors. Although SHA-2 is a powerful hash function, its basic design is still similar to SHA-1. Due to previous successful attacks on MD5 and SHA-1, NIST perceived the need for an alternative, different cryptographic hash for the future, called SHA-3. SHA-3 was not originally intended to replace SHA-2. Because no significant attack has been shown yet on SHA-2. In 2006, NIST began holding the Hash Algorithm competition to create the new hash standard, SHA-3. Keccak algorithm was the winner of the competition (Dworkin, 2015). Keccak uses the sponge structure to generate cryptographic primitives. SHA-3 provides the same output sizes as SHA-2: 224, 256, 384 and 512 bits. SHA-3 is one of the most up-to-date and considered safest functions today.

Race Integration Primitives Evaluation Message Digest (RIPEMD) is a hash function developed by Preneel and first published in 1996 (Dobbertin et al., 1996). It is commonly known as a family of European hash functions. This family of hash functions; RIPEMD includes RIPEMD-128 and RIPEMD-160. 256 and 320 bit versions of RIPEMD are also available. The original RIPEMD relies on the design principles of MD4; therefore, its security is questionable. Consequently, RIPEMD-128 and RIPEMD-160 were developed against vulnerabilities in the original RIPEMD. However, its 256 and 320 bit versions are also available.

BLAKE hash function was submitted to the NIST hash function competition, making it to the final round of five candidates. But it lost the competition (Aumasson et al., 2014). It newer version, BLAKE2 provides better security than SHA-2 and is resistant to length extension. It provides output sizes equal to SHA-3. BLAKE3 is an enhanced version of BLAKE2. BLAKE3 is a Merkle tree, with higher degrees of parallelism (both SIMD and multi-threading) than BLAKE2. In contrast to the multivariate families BLAKE and BLAKE2, parallelism is an algorithm with many desirable properties such as XOF, KDF, PRF and MAC (O'Connor et al., 2020).

Table 1. Parametric properties of commonly used hashing functions (Venedikian, 2022)

Hash Function	Output (bits)	Internal State Size	Block Size	Message Length	Word Size	Number of Repetitions
BLAKE2b	512	512	1024	128	64	12
BLAKE2s	256	256	512	64	32	10
BLAKE3	Variable	256	512	64	32	7
AIR	256/224/192/160/128	256	1024	64	32	3/4/5
MD4	128	128	512	64	32	3
MD5	128	128	512	64	32	64
RIPEMD	128	128	512	64	32	48
RIPEMD-128, -256	128/256	128/256	512	64	32	64
RIPEMD-160	160	160	512	64	32	80
RIPEMD-320	320	320	512	64	32	80
SHA-0	160	160	512	64	32	80
SHA-1	160	160	512	64	32	80
SHA-224, -256	224/256	256	512	64	32	64
SHA-384, -512, -512/224, -512/256	384/512/224/256	512	1024	128	64	80
SHA-3	224/256/384/512	1600	1600 - 2* bits	Variable	64	24
SHA-3_224	224	1600	1152	Variable	64	24
SHA-3_256	256	1600	1088	Variable	64	24
SHA-3_384	384	1600	832	Variable	64	24
SHA-3_512	512	1600	576	Variable	64	24
Tiger-192/160/128	192/160/128	192	512	64	64	24
Whirlpool	512	512	512	256	8	10

The BLAKE3 compression function, with a close look at that of the BLAKE2's, the biggest difference is that the lap rating has been reduced from 10 to 7. BLAKE3, the latest version of the BLAKE family of functions, is the fastest hash algorithm for current hardware today. It is considered the fastest hash algorithm for 64-bit multi-core processors today (O'Connor et al., 2020). Table 1 lists the parametric properties of commonly used hashing functions, some of them described above.

Hashcat software, which is used in brute force attack and continuously developed, is used for breaking hash functions. Hashcat is an open source application and has microprocessor and GPU support. It can run on Linux, Windows and macOS. It comes with Linux distributions such as Kali Linux, which are distributed for many security purposes. Hashcat supports almost all hash functions, as well as a wide range of application scenarios, such as HMAC and salt (only) messages, or finding nested hash values in a loop. The speed of hash functions when an Nvidia RTX 4090 GPU has been used for a sample data is listed in Table 2 (Hashcat v6.2.6, 2022). Due to the ever increasing processing power of GPUs, it is not sufficient to store only the hash value, especially for values such as passwords. Although methods such as HMAC and salting are used together, hash calculation and slower password-based key generation functions should be used. Currently, Argon2, bcrypt, and scrypt are the most up-to-date password-based key generation functions.

Table 2. Nvidia RTX 4090 GPU hash function calculation results

Hashing Function	Calculated in Seconds
MD5	164.1 GH/s
md5($sal.$ pass)	84654.8 MH/s
md5(md5(md5($ pass)))	31672.4 MH/s
SHA-1	50638.7 MH/s
Sha-1($ pass .$salt)	51432.6 MH/s
Sha-1($sal.$ pass)	39126.2 MH/s
BLAKE2b-512	13072.3 MH/s
SHA-2-224	21353.8 MH/s
SHA-2-256	21975.5 MH/s
sha256($ pass .$salt)	21960.3 MH/s
SHA-2-512	7483.4 MH/s

Units;
1 MH/s = 1 million hash per second
1 GH/s = 1 billion hash per second

A CASE STUDY

Microsoft Visual Code application has been used for coding. In the application, Hashcat has been preferred to show how long a password given MD5 value is cracked. For this example, it is specified that the password is alphanumeric and has a minimum of 6 and a maximum of 8 characters, and it will be searched as MD5 digest.

```
Example Password: "z7YRk9"
MD5 Value: 4ea7acaa537883c944638c443b00a2bb
```

As can be seen in Figure 2, hashcat (v6.2.6) has started to run. Hashcat has been expected to find passwords consisting of random 8-digit alphanumeric and symbols in 4 to 5 hours from their hash values. Figure 3 shows the outputs of hashcat. Figure 4 lists the randomly selected passwords and their hash values. These passwords have been used in the case study. Gigabyte in all applications A computer with Nvidia RTX3070 GPU, Intel 11th generation i7 microprocessor and 16 Gb DDR RAM has been used in the case study. Windows 10 version 22H2 has been used as the operating system. When hashcat has been used, time needed to find passwords hashed with different hash functions is listed in Table 3. In this study, brute force attack has been used by Hashcat. However, passwords can be found in a much shorter time using other attach types such as birthday attack.

A one-time password (OTP), also known as a one-time PIN, one-time authorization code (OTAC), or dynamic password, is a one-time password created to authenticate a system login or transaction. It changes continuously depending on a counter or time. With a fixed username and password on platforms such as banks, Paypal, WhatsApp, Facebook, Instagram, Twitter and Amazon use Two-Factor Authentication (2FA). When Time Based OTPs (TOTP) is used, temporary passwords are created based on time, with the secret public key shared between the server and the user. The generated passwords are valid for a certain period of time. OTPs are also used in mobile devices such as smart phones, hardware keys, and in environments such as computers and the web. Different from the 2FA systems found on Android and iOS platforms, Twilio Authy includes all-in-one with multi-platform support (See Figure 5). While installing these applications, the server side shares a secret key with the user, often asking them to scan this key as a QR code on their mobile phone for convenience. After that, the mobile phone uses this secret key and time to generate a dynamic password valid only for a certain period of time.

Figure 2. Hashcat outputs-1

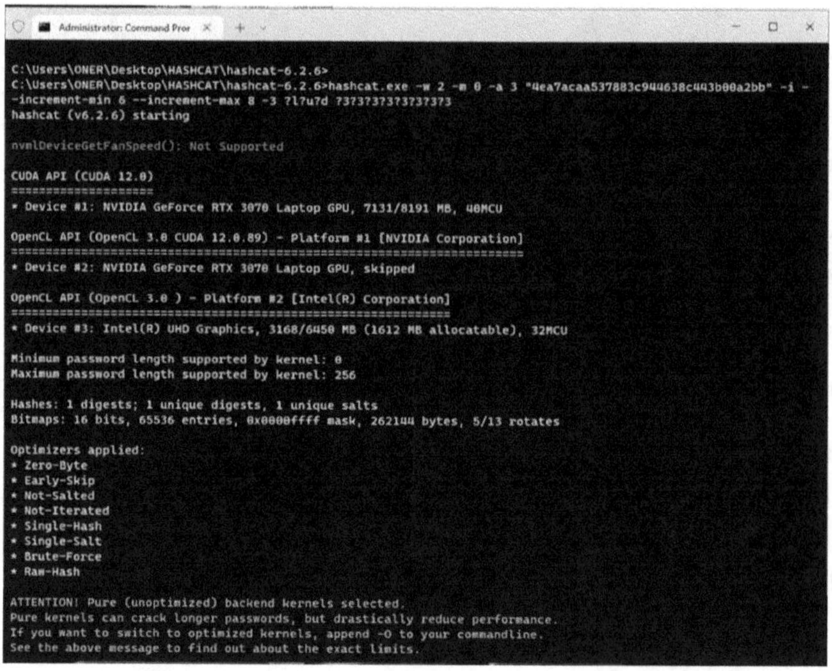

Figure 3. Hashcat outputs-2

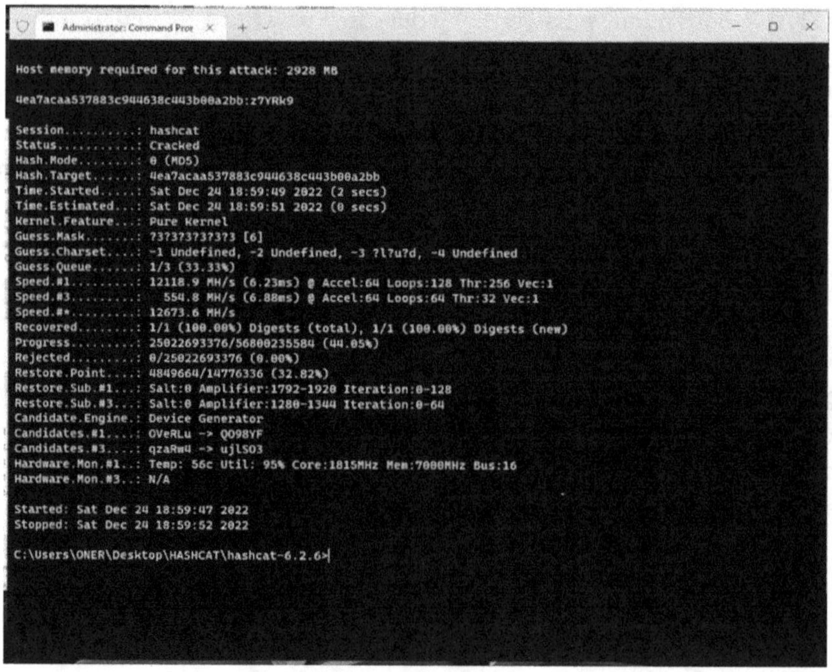

Figure 4. Randomly selected passwords and hash values

PASSWORD	Number of characters	MD5 Digest	SHA1 Digest
z7YRk9	6	4ea7acaa537883c9 44638c443b00a2bb	9e9f210425b5967b1398 46fb751f11ebda6f96c0
7uV4Vz	6	39705cc5297cd926 598c66540988191	4eb026a2ad85bddcac2d 1d66b8b013bc88b1f46b
p5q5E#	6	e988e16226e9ea2a 415c9568ce978217	b2ab27ae345ab9b17b05 e1041f0b4f0ad32223bf
yG&3x5Pz	8	d8efa7f77f621b5c f50c111a3e6ade52	ec88416297582721718e a0618458533e73be4159
p@ub4h!4	8	31e08affc572f0f0 736c076ffee724e2	c066f4df9732412dc919 254dbaba76ed18ea1fa4

Table 3. Time to find passwords hashed with different hash functions

Password	Number of Characters	Character Set	Hash Function	HashCat Breaking Time
z7YRk9	6	Alphanumeric	MD5	5 sec
7uV4Vz	6	Alphanumeric	SHA-1	32 sec
p5q5E#	6	Alphanumeric + Symbols	MD5	960 sec
yG&3x5Pz	8	Alphanumeric + Symbols	MD5	16211 sec
p @ub4h!4	8	Alphanumeric + Symbols	MD5	19089 sec

In this part of the case study, details of a TOTP application designed in this case study are presented. Visual Studio output of the code block given below is presented in Figure 6.

```
################### TOTP APPLICATION
###########################
import hmac
```

Figure 5. Twilio supporting multiple platforms for 2FA authy screenshot

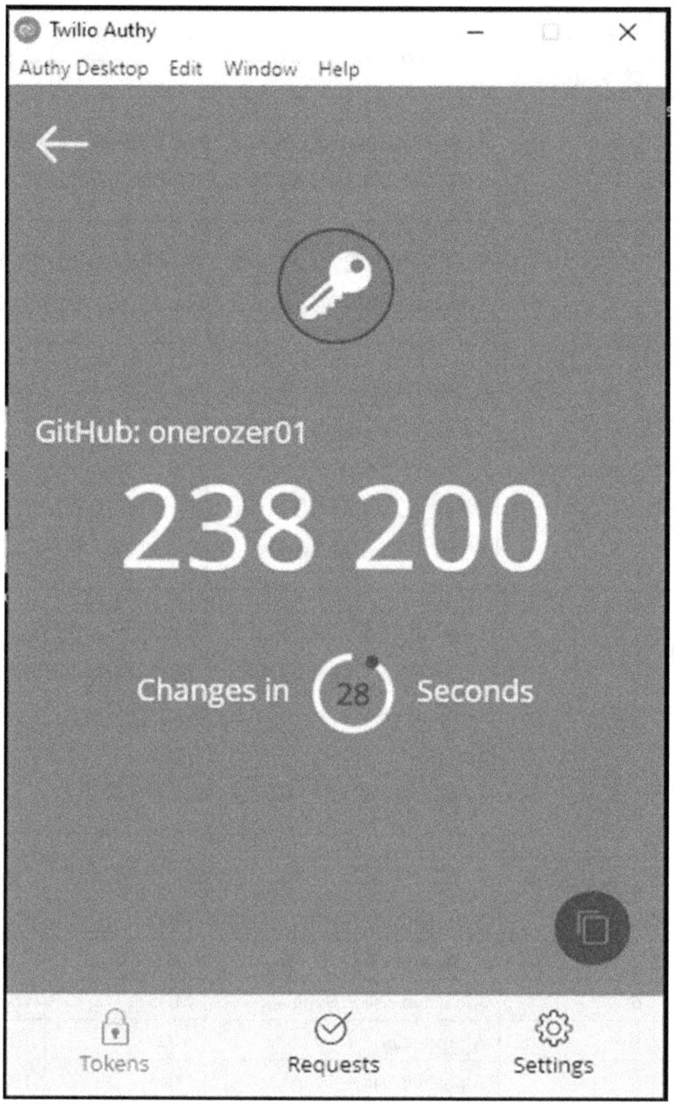


```
import time
# A secret key is shared between the server and the user.
SECRET = 'ZYTY-YE5F-OAGW-5'
def main():
    print ("Temporary 2FA Password: ", hotp (SECRET, 30, 6,
'sha3_512'))
# time_step = Indicates how long the generated password will be
```

```
valid. In general use, it is in the #direction for 30 seconds.
# digits = How many digits will the generated password be?
# digest = which digest algorithm to use. The following
# algorithms can be used optionally.
#                   blake2b
#                   blake2s
#                   md5
#                   sha1
#                   sha224
#                   sha256
#                   sha384
#                   sha512
#                   sha3_224
#                   sha3_256
#                   sha3_384
#                   sha3_512
#
# time.time () function to convert instantaneous time to Unix
Epoch time
# For example;        18/12/2022 02:28:06 GMT+03:00 --->
1671319686
Def hotp (key, time_step =30, digits =6, digest ='sha256'):
converting
instantaneous time to Unix Epoch time,    # convert to password
expiry time.
    counter = int (time.time () / time_step)
# time value hmac using it as key, byte for it
    # is converting.
    counter _byte = counter.to_bytes (8, byteorder = ' big ')
    key = hmac.new (key.encode (), counter_byte, digest)
# We convert the big hexeminal result from the
hmac algorithm to the digit number we want.
    mac = int (key.hexdigest (), 16) % (10 ** digits)
# 0 per if the result is less than the number of digits we want
    # we add.
    return str (mac). zfill (digits)
if __name__ == '__main__':
main()
##############################################################
##
```

Figure 6. Visual Studio output of the code block given above

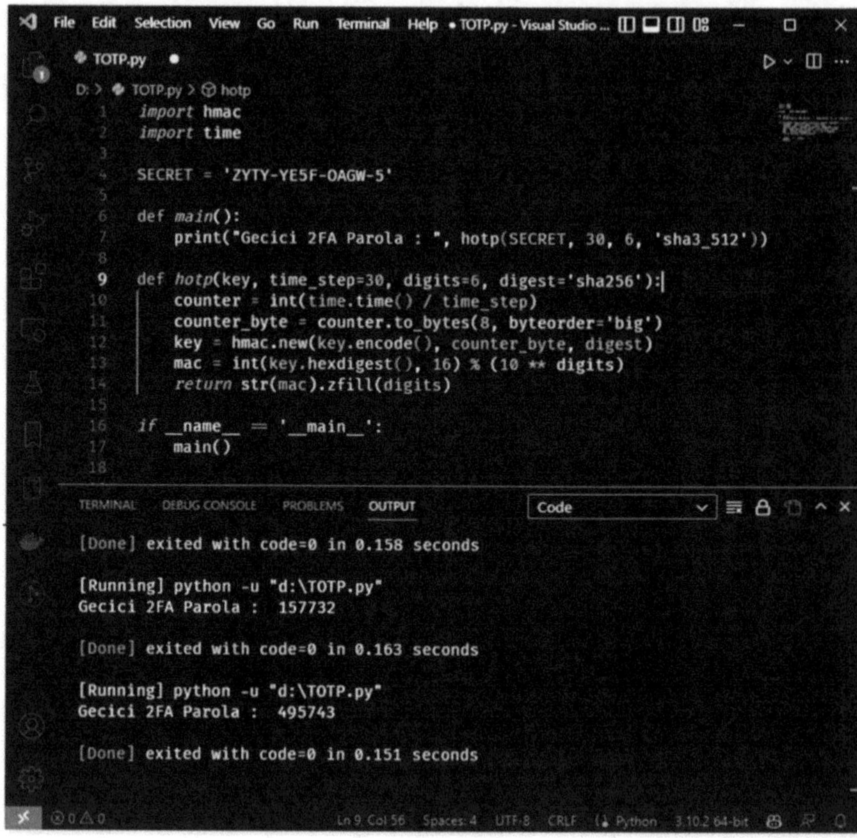

Although traditional hash functions are generally expected to work fast, newer ones such as argon2, bcrypt, scrypt, which work slower and use hardware structures such as microprocessor and main memory, have been developed especially for password conversion and storage. The purpose here is to slow down the brute-force attack on powerful hardware computers as much as possible, and to reduce the hashing rate that can be tried per second. In addition, x86/x64 microprocessors such as Intel, AMD or arm64 microprocessors such as Apple m1 have a ready-made instruction set in processors to accelerate functions such as SHA and AES. In this part of the case study, a notebook with Intel i5 microprocessor and 16GB main memory has been used. Like the previous ones, Python has been preferred. Speed comparisons of the hash functions used in this study are presented in Figure 7. Test results of the hash functions are listed in Table 4.

```
####################################################################
##
import timeit
import hashlib
import os
import_module = """
import hashlib
import os
import argon2
import bcrypt
data = os.urandom (1024*1024)
salt = b'value of salt '
"""
rounds = 1000          # how many times to run.
blake 2b = "hashlib.blake2b(data)"
blake 2s = "hashlib.blake2s(data)"
md 5 = "hashlib.md5(data)"
sha 1 = "hashlib.sha1(data)"
sha 224 = "hashlib.sha224(data)"
sha 256 = "hashlib.sha256(data)"
sha 384 = "hashlib.sha384(data)"
sha 512 = "hashlib.sha512(data)"
sha 3_224 = "hashlib.sha3_224(data)"
sha 3_256 = "hashlib.sha3_256(data)"
sha 3_384 = "hashlib.sha3_384(data)"
sha 3_512 = "hashlib.sha3_512(data)"
shake _128 = "hashlib.shake_128(data)"
shake _256 = "hashlib.shake_256(data)"
scrypt = " hashlib.scrypt (data, salt=salt, n=2**14, r=8, p=1,
dklen =64)"
argon 2 = "argon2.hash_password(password =data, salt=salt)"
bcrypt = " bcrypt.hashpw (data, bcrypt.gensalt ())"
t _blake2b = timeit.timeit (stmt =blake2b, setup= import_
module, number = rounds)
t _blake2s = timeit.timeit (stmt =blake2s, setup= import_
module, number = rounds)
t _md5 = timeit.timeit (stmt =md5, setup= import_module, number
= rounds)
t _sha1 = timeit.timeit (stmt =sha1, setup= import_module,
number = rounds)
```

```
t _sha224 = timeit.timeit (stmt =sha224, setup= import_module, 
number = rounds)
t _sha256 = timeit.timeit (stmt =sha256, setup= import_module, 
number = rounds)
t _sha384 = timeit.timeit (stmt =sha384, setup= import_module, 
number = rounds)
t _sha512 = timeit.timeit (stmt =sha512, setup= import_module, 
number = rounds)
t _sha3_224 = timeit.timeit (stmt =sha3_224, setup= import_
module, number = rounds)
t _sha3_256 = timeit.timeit (stmt =sha3_256, setup= import_
module, number = rounds)
t _sha3_384 = timeit.timeit (stmt =sha3_384, setup= import_
module, number = rounds)
t _sha3_512 = timeit.timeit (stmt =sha3_512, setup= import_
module, number = rounds)
t _shake_128 = timeit.timeit (stmt =shake_128, setup= import_
module, number = rounds)
t _shake_256 = timeit.timeit (stmt =shake_256, setup= import_
module, number = rounds)
t _scrypt = timeit.timeit (stmt = scrypt, setup= import_module, 
number = rounds)
t _argon2 = timeit.timeit (stmt =argon2, setup= import_module, 
number = rounds)
t _bcrypt = timeit.timeit (stmt = bcrypt, setup= import_module, 
number = rounds)
print ("blake2b:", round (t_blake2b, 4))
print ("blake2s:", round (t_blake2s, 4))
print ("md5:", round (t_md5, 4))
print ("sha1:", round (t_sha1, 4))
print ("sha224:", round (t_sha224, 4))
print ("sha256:", round (t_sha256, 4))
print ("sha384:", round (t_sha384, 4))
print ("sha512:", round (t_sha512, 4))
print ("sha3_224:", round (t_sha3_224, 4))
print ("sha3_256:", round (t_sha3_256, 4))
print ("sha3_384:", round (t_sha3_384, 4))
print ("sha3_512:", round (t_sha3_512, 4))
print ("shake_128:", round (t_shake_128, 4))
print ("shake_256:", round (t_shake_256, 4))
```

```
print (" scrypt:", round (t_scrypt, 4))
print ("argon2:", round (t_argon2, 4))
print (" bcrypt:", round (t_bcrypt, 4))
#################################################################
##
```

Figure 7. Speed comparisons of the hash functions

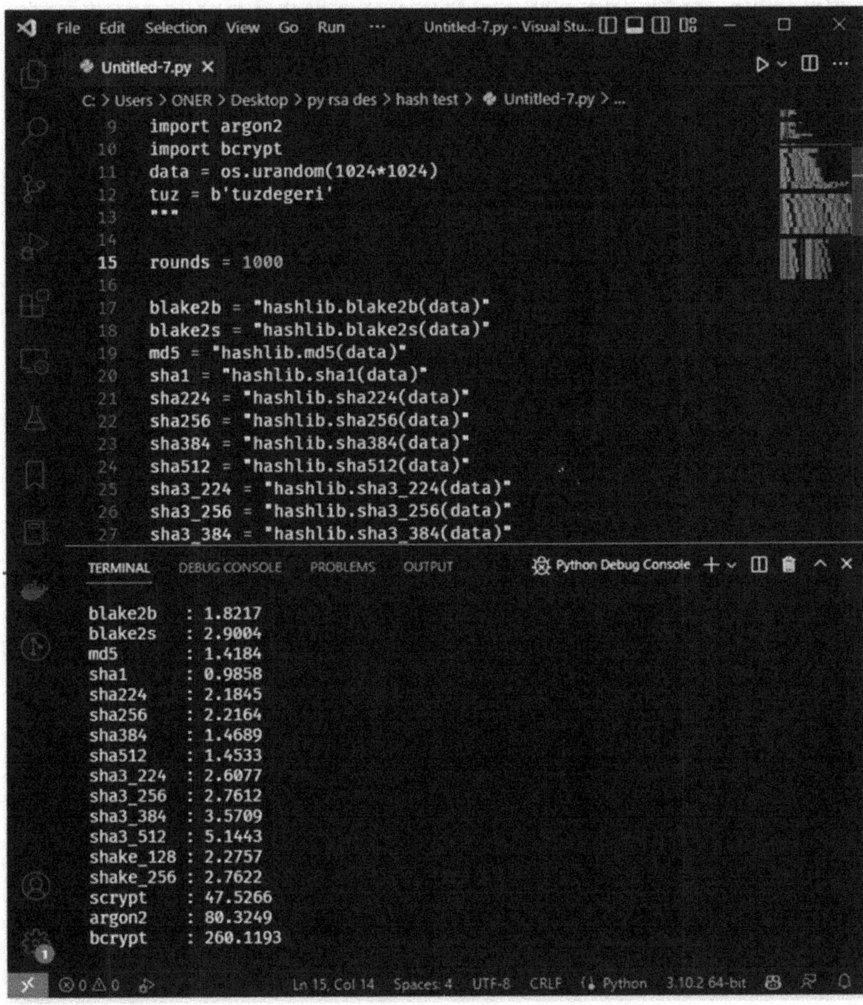

As can be seen in Table 4, new generation functions such as argon2, bcrypt, and scrypt have needed much more time to calculate. They are designed to be used extensively by hardware such as microprocessor, GPU and main memory. While they are extremely useful for relatively small word-based data such as passwords, they should not be used for data file integrity checking. With argon2, bcrypt or scrypt, it can take minutes or even hours to hash 1 Gbyte of data, while the microprocessor will use 100% of power. The reason why SHA-1 function has shown the best performance is due to the presence of the SHA-1 instruction set directly on the processor used. In systems with special Application-Specific Integrated Circuit (ASIC) or Field Programmable Gate Arrays (FPGA) processors cryptographic algorithms can be run faster. Nowadays, ASIC-based special processors are used in BitCoin mining.

Table 4. Test results of the hash functions

Hash Function	Time to Hash 1MB Data 1000 Times (sec.)
blake2b	1.8217
blake2s	2.9004
md5	1.4184
sha1	0.9858
sha224	2.1845
sha256	2.2164
sha384	1.4689
sha512	1.4533
sha3_224	2.6077
sha3_256	2.7612
sha3_384	3,5709
sha3_512	5.1443
shake_128	2.2757
shake_256	2.7622
scrypt	47.5266
argon2	80.3249
bcrypt	260.1193

Suggestions

Despite the developments, it is a matter of debate that many hash functions developed on binary systems for the quantum computer working with the qubit system will lose their validity in the future. Quantum computers developed for now are still at the prototype stage and due to their limited short-term operation, the hypotheses put forward remain only theoretical, applications and tests cannot be done. These studies are especially important for the future of blockchain structures based on the security of standard hash functions such as the SHA family and crypto digital assets such as Bitcoin. NSA has been working for a long time to investigate and improve cryptography standards.

Instead of some key derivation fuctions such as PBKDF1 and PBKDF2, bcrypt, scrypt and argon2 functions may be preferred. Because, these functions are resistant to brute force and rainbow table attacks as they are required to be used purely by nature. In cases where it is mandatory to use PBKDF2, the minimum hash cycle number should be 100,000 or more.

CONCLUSION

The reliability rate of hash functions decreases with the development of technology and new attack methods. The main problem with hash functions is not that even if they cannot be broken, some of them require a lot of processing power, memory and time. However, thanks to rapidly developing hardware technologies, more powerful and reliable hash functions have been constantly being designed.

Hash functions differ in speed, purpose, and system requirements. In mission-critical applications, it is more efficient in terms of security to use more than one nested different hash functions in applications.

In cases such as password storage, it is recommended to use consecutive or multiple hash functions instead of a single one. Although theoretical attacks have been designed in SHA-2 and SHA-3 functions today, no collision has been found. In cases when it is necessary to use an algorithm from the SHA-2 family, a minimum of 256 bits and a salt feature should be used with HMAC. SHA-3_256 and SHA-3_512 bit functions are safe for common applications such as tokens, tickets and e-signatures. BLAKE3 shows high performance especially in applications such as data integrity control and database indexing. Although it is considered safe for quantum computing, 256-bit and higher versions should be used to resist brute force attacks.

ACKNOWLEDGMENT

Some parts of this chapter were produced from the first author's master's project. This research received no specific grant from any funding agency in the public, commercial, or not-for-profit sectors.

REFERENCES

Aumasson, J.-P. (2017). *Serious cryptography: a practical introduction to modern encryption*. No Starch Press.

Aumasson, J.-P., Meier, W., Phan, R. C.-W., & Henzen, L. (2014). The Hash Function BLAKE (1st ed. 2014.). Springer Berlin Heidelberg: Imprint: Springer. doi:10.1007/978-3-662-44757-4

Banoth, R., & Regar, R. (2023). An Introduction to Classical and Modern Cryptography. In *Classical and Modern Cryptography for Beginners*. Springer. doi:10.1007/978-3-031-32959-3_1

Bertoni, G., Daemen, J., Peeters, M., & Van Assche, G. (2007). Sponge functions. *ECRYPT Hash Workshop*.

Bertoni, G., Daemen, J., Peeters, M., & Van Assche, G. (2013). Keccak. *32nd Annual international conference on the theory and applications of cryptographic techniques*. Springer.

Coron, J.-S., Dodis, Y., Malinaud, C., & Puniya, P. (2005). Merkle-Damgård revisited: How to construct a hash function. In Shoup, V. (eds) Advances in Cryptology – CRYPTO 2005. Springer, Berlin, Heidelberg. doi:10.1007/11535218_26

Dobbertin, H., Bosselaers, A., & Preneel, B. (1996). RIPEMD-160: A strengthened version of RIPEMD. In D. Gollmann (Ed.), Lecture Notes in Computer Science: Vol. 1039. *Fast Software Encryption. FSE 1996*. Springer. doi:10.1007/3-540-60865-6_44

Dong, X., Sun, S., Shi, D., Gao, F., Wang, X., & Hu, L. (2020). Quantum Collision Attacks on AES-Like Hashing with Low Quantum Random Access Memories. In *Advances in Cryptology – ASIACRYPT 2020: 26th International Conference on the Theory and Application of Cryptology and Information Security*. Springer-Verlag. 10.1007/978-3-030-64834-3_25

Dunkelman, O., & Biham, E. (2006). A framework for iterative hash functions: HAIFA. *2nd NIST cryptographich hash workshop*. IEEE.

Dworkin, M. (2015). SHA-3 Standard: Permutation-Based Hash and Extendable-Output Functions. National Institute of Standards and Technology. doi:10.6028/NIST.FIPS.202

Gauravaram, P., & Knudsen, L. R. (2010). Cryptographic hash functions. In *Handbook of Information and Communication Security* (pp. 59–79). Springer. doi:10.1007/978-3-642-04117-4_4

Hatzivasilis, G. (2017). Password-Hashing Status. *Cryptography*, *1*(2), 10. doi:10.3390/cryptography1020010

Kerry, C. F., & Gallagher, P. D. (2013). *Digital signature standard (DSS)*. FIPS 186-4. doi:10.6028/NIST.FIPS.186-4

Leurent, G., & Peyrin, T. (2020). *SHA-1 is a Shambles*. International Association for Cryptologic Research.

Martin, K. (2020). *Cryptography: The key to digital security, how it works, and why it matters*. WW Norton & Company.

Mihailescu, M. I., & Nita, S. L. (2023). Getting Started in Cryptography and Cryptanalysis. In *Pro Cryptography and Cryptanalysis with C++23*. Apress. doi:10.1007/978-1-4842-9450-5_1

O'Connor, J., Aumasson, J.-P., Neves, S., & Wilcox-O'Hearn, Z. (2020). BLAKE3: one function, fast everywhere. GitHub. https://github.com/BLAKE3-team/BLAKE3

Paar, C., & Pelzl, J. (2009). *Understanding cryptography: a textbook for students and practitioners*. Springer Science & Business Media.

Pittalia, P. P. (2019). A comparative study of hash algorithms in cryptography. *International Journal of Computer Science and Mobile Computing*, *8*(6), 147–152.

Preneel, B., Govaerts, R., & Vandewalle, J. (1994). Hash functions based on block ciphers: a synthetic approach. In D. R. Stinson (Ed.), Lecture Notes in Computer Science: Vol. 773. *Advances in Cryptology—CRYPTO' 93*. Springer. . doi:10.1007/3-540-48329-2_31

Qian, Y., Ye, F., & Chen, H.-H. (2022). Cryptographic Techniques. In Y. Qian, F. Ye, & H.-H. Chen (Eds.), *Security in Wireless Communication Networks*., doi:10.1002/9781119244400.ch4

Rivest, R. (1992). RFC1321: The MD5 message-digest algorithm. In RFC Editor.

Sadeghi-Nasab, A., & Rafe, V. (2023). A comprehensive review of the security flaws of hashing algorithms. *J Comput Virol Hack Tech*, *19*(2), 287–302. doi:10.100711416-022-00447-w

Schneier, B. (2007). *Applied cryptography: protocols, algorithms, and source code in C*. John Wiley & Sons.

Silva, J. E. (2003). *An overview of cryptographic hash functions and their uses*. (White paper). SANS. https://www.sans.org/white-papers/879/

Stevens, M., Bursztein, E., Karpman, P., Albertini, A., & Markov, Y. (2017). The First Collision for Full SHA-1. In J. Katz & H. Shacham (Eds.), Lecture Notes in Computer Science: Vol. 10401. *Advances in Cryptology – CRYPTO 2017. CRYPTO 2017*. Springer., doi:10.1007/978-3-319-63688-7_19

Ural, N., & Örenç, Ö. (2019). *Uygulamalı Şifreleme ve Şifre Çözme Yöntemleri*. Pusula.

Venedikian, G. (2022). *Message Digest vs Secure Hash Algorithm vs RIPEMD*. CUNY New York City College of Technology. https://www.coursehero.com/file/56888501/hash-comparisondocx/

Wang, X., Yin, Y. L., & Yu, H. (2005). Finding Collisions in the Full SHA-1. In Shoup, V. (eds) Advances in Cryptology – CRYPTO 2005. Springer, Berlin, Heidelberg. doi:10.1007/11535218_2

Wei, C., Wu, C., Fu, X., Dong, X., He, K., Hong, J., & Wang, X. (2022). Preimage attacks on 4-round keccak by solving multivariate quadratic systems. *International Conference on Information Security and Cryptology*. Springer. 10.1007/978-3-031-08896-4_10

ADDITIONAL READING

Buchanan, W. J. (2017). *Cryptography*. River Publishers.

Perez, A. (2014). *Network security*. Wiley-ISTE.

Pieprzyk, J., & Sadeghiyan, B. (1993). *Design of hashing algorithms*. Springer-Verlag. doi:10.1007/3-540-57500-6

KEY TERMS AND DEFINITIONS

Cryptography: It is the study and practice of techniques for secure communication in the presence of adversarial behaviour.

Data Integrity: It is the overall accuracy, completeness, and consistency of data.

Hash Function: It is a mathematical function that can convert any digital data into an output string with a fixed number of characters.

Chapter 5
Analysis and Monitoring of Design Challenges in the Cyber Physical Systems in Manufacturing Sectors Using Cloud Computing

L. Natrayan
https://orcid.org/0000-0003-1001-3422
Department of Mechanical Engineering, Saveetha School of Engineering, SIMATS, Chennai, India

ABSTRACT

The integration of additive manufacturing (AM) technology, commonly known as 3D printing, into industrial processes has led to the adoption of Industry 4.0 principles. This includes the integration of dynamic changes in the user experience, data-driven smart systems, and simpler production processes. In this research, a consumer-driven cyber-physical systems (CPS) framework is proposed, utilizing customer demand data to automate the entire process. The use of cloud computing, specifically Microsoft Azure, is utilized to optimize the process parameters and improve the efficiency of the 3D printing process using fused deposition modeling (FDM) technology. Additionally, machine learning techniques, specifically the multi-layer perceptron (MLP) neural network model, are implemented to further optimize the process parameters. A raspberry pi is used to connect to the Azure IoT hub and machine learning studio, where the algorithm is automatically reviewed, and the best value is selected.

DOI: 10.4018/978-1-6684-9267-3.ch005

INTRODUCTION

The integration of cyber physical systems (CPS) in manufacturing sectors has significantly improved efficiency and productivity. However, the design and implementation of CPS in these industries also pose various challenges that need to be addressed. These challenges include security, reliability, scalability, and integration with existing systems (Kathiravelu et al. 2019).

In this research article, we aim to comprehensively analyse the design challenges associated with CPS in the manufacturing sector. We will examine the current state of CPS in manufacturing industries and identify the key challenges that need to be illustrated (Shahriar et al. 2018; Liu et al. 2020). We will also investigate the potential solutions that have been proposed to address these challenges and evaluate their effectiveness.

One potential solution we propose in this research is using cloud computing to address CPS design challenges in the manufacturing industry. Cloud computing offers several benefits such as scalability, flexibility, and cost-effectiveness, which make it a suitable option for addressing CPS design challenges. We will explore how cloud computing can address these challenges and evaluate this approach's potential benefits and limitations (Ma et al. 2018; Broy et al. 2019).

In this research, we will conduct a detailed analysis of CPS in manufacturing sectors and thoroughly explain the associated design challenges. We will also examine the potential solutions that have been proposed to address these challenges, and evaluate their effectiveness (Tsigkanos et al. 2016; Tyszberowicz and Faitelson 2020). We will also propose using cloud computing as a solution for addressing CPS design challenges in the manufacturing industry and provide an in-depth analysis of this approach's potential benefits and limitations (Cong et al. 2016; Mosterman and Zander 2016).

The results of this research will have implications for industry practitioners and researchers in CPS and manufacturing. Industry practitioners will be able to understand the design challenges associated with CPS in manufacturing sectors, and the potential solutions that can be used to address them. Researchers in the field of CPS and manufacturing will gain insight into the current state of CPS in these industries, and the potential directions for future research (Mueller et al. 2017; Liu and Wang 2020; Herdt and Drechsler 2022).

In summary, this research article will provide a comprehensive analysis of the design challenges associated with CPS in the manufacturing sector, and propose using cloud computing to address these challenges. The results of this research will have important implications for industry practitioners and researchers in CPS and manufacturing (White et al. 2010; Jha et al. 2021; Verma 2022).

DESIGN CHALLENGES

Cyber-Physical Systems (CPS) is an interdisciplinary field that aims to connect the cyber world of information transfer with the physical world. To accomplish technical innovations and reap the foremost social and financial benefits, CPS requires the cooperation of computer scientists and network specialists with professionals from various research and technology sectors. These outstanding trials include blackout-free energy production and circulation, energy-conscious constructions and metropolises, high-yield cultivation, eternal life support for the elderly and incapacitated, safe automotive systems, and continuous monitoring and control of patient conditions (Mishra et al. 2022).

CPS is a hybrid design that combines computation, communiqué devices and instruments. It can be found in various applications, such as hifi cities, intelligent grids, motorized controls, engineering, and therapeutic devices. For example, incorporating mathematical control and sensor systems into a motorized approach is essential for constructing safe vehicle systems that defend customers from potentially harmful circumstances. Similarly, a cultured medical system uses interacting and announcement capabilities to monitor and regulate the patient's condition. Energy production and distribution are crucial for providing uninterrupted electricity and the development of energy-efficient buildings and municipalities. Current security research focuses on determining verification, critical administration, and honeypot folders. CPS-based solutions must be resilient to unanticipated occurrences and adaptable to subsystem flaws. They must be developed with components that are predictable and reliable. (Alaghbari et al. 2022).

In order to achieve the goals of CPS, it is important to focus on research and development in key areas such as proper verification, computer software engineering methods, enterprise patterns, artificial intelligence approaches, competition and simulation methodologies, and constituent design knowledges. Additionally, it is important to consider the real-time response and performance of CPS systems to ensure they can meet their intended applications' needs.

One of the key challenges facing CPS is the need for resilience and adaptability in the face of unexpected occurrences and subsystem flaws. This requires the development of robust and flexible systems that can withstand and recover from unexpected failures. Additionally, it is important to focus on security research to protect critical national assets such as power, oil, water, transportation, communication, and networks.

Overall, CPS has the potential to bring about major societal and economic benefits through breakthroughs in areas such as energy production and distribution, healthcare, transportation, and more. However, to achieve these goals, it is important to bring together experts from a wide range of disciplines and focus on research and development in key areas such as formal confirmation, software production methods, design patterns, machine learning approaches, emulation and simulation methodologies, and constituent design knowledges (Darwish and Hassanien 2018).

CPS WITH CLOUD ENABLED

The integration of cloud computing, on-device systems, and physical systems has the potential to benefit from technological advancements in various fields. High-tech companies are researching cloud-based Cyber-Physical Systems (CCPS) due to resource flexibility and scalability advantages. Business functions can be exposed as services through CCPS, which can be utilized for various purposes, including traffic management, vehicle position identification, robot surgery, and healthcare. It can also assist soldiers in determining their location in a battle zone and identifying adversaries. In the future, consumers may use the cloud as their facility, paying for services as they use them. One of the most pressing issues today is the real-time dispensation of Big Data for choice making. Big data is bigger and has a larger storage capacity but is less flexible and unstructured. To process this type of data, a new expertise is necessary. Big Data can be collected using a variety of sensors, including medical, traffic, and social data. One example of Big Data is a healthcare system database.

CCPS requires a new project, expansion, and process approach because of its advanced capabilities. The cloud and physical parts must work closely to satisfy client requests. The level of reliance on processing power, the landscape of the features, network connection, and operative circumstances all play a role in determining whether they are hosted on-device or in the cloud. Data security is critical at both the system and network levels, and security rules and techniques must be implemented at the local stage for each physical approach and substructure. A failure in security at the system level can impact network infrastructure, emphasizing the importance of studying Cyber-Physical approach.

CPS WITH CLOUD COMPUTING

Cyber-Physical Systems (CPS) is an interdisciplinary field that aims to connect the cyber world of information transfer with the physical world. This is accomplished by integrating computation, communication devices, sensors, and actuators into systems that can monitor, control, and interact with the physical world. CPS has the potential to bring about major societal and economic benefits through breakthroughs in areas such as energy production and distribution, healthcare, transportation, and more.

Figure 1. Cloud architecture

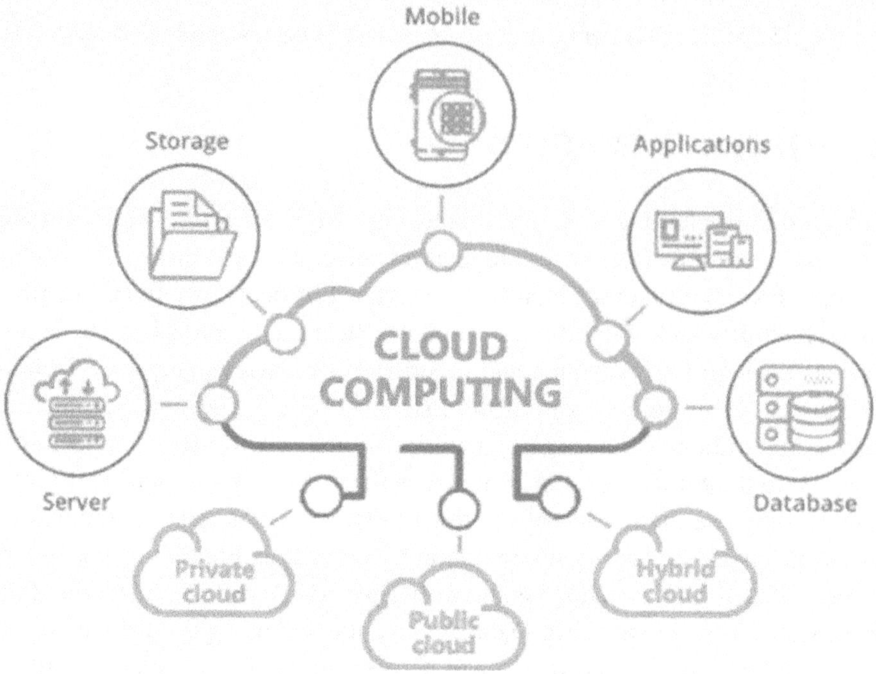

Cloud computing is a technology that allows for delivering computing resources, including storage, processing power, and software, over the internet. Integrating CPS and cloud computing can bring significant advancements in various fields. Cloud-based CPS (CCPS) is being researched by high-tech companies such as Microsoft, Schneider Electric, and Honeywell due to the advantages of resource flexibility and scalability. One of the key advantages of combining CPS with cloud computing is the ability to process large amounts of data in real-time. With cloud-based storage and processing power, handling big data and making decisions quickly becomes possible. This is particularly important in healthcare, transportation, and defence, where real-time decision making can save lives and improve efficiency. CCPS also allows for the exposure of business functions as services, which can be utilized for various purposes, including traffic management, vehicle position identification, robot surgery, and healthcare. It can also assist soldiers in determining their location in a battle zone and identifying adversaries. In the future, consumers may use the cloud as their facility, paying for services as they use them.

Another advantage of CCPS is the ability to scale resources as needed. This is particularly important for applications requiring high-performance computing or handling variable workloads. Utilizing cloud resources makes it possible to scale up or down as needed, without the need for large investments in hardware or software.

Proposed Attack Model

The CPS (Cyber-Physical Systems) network is susceptible to physical and virtual attacks. As shown in Figure 1, there is a connection between physical devices and the virtual world of cyberspace. Cybersecurity breaches can occur due to damage to physical components, and vice versa. A common method used in detecting attacks is through sensor data analysis. Research in this area addresses mathematical modeling, emergency management, and intrusion detection challenges. The foundation of the CPS network is the integration of physical components, such as sensors, into a network and eventually into cyberspace. Figure 2 illustrates a cooperative method to building faith with a node along the trail of transmission.

Figure 2. Topology of wireless network

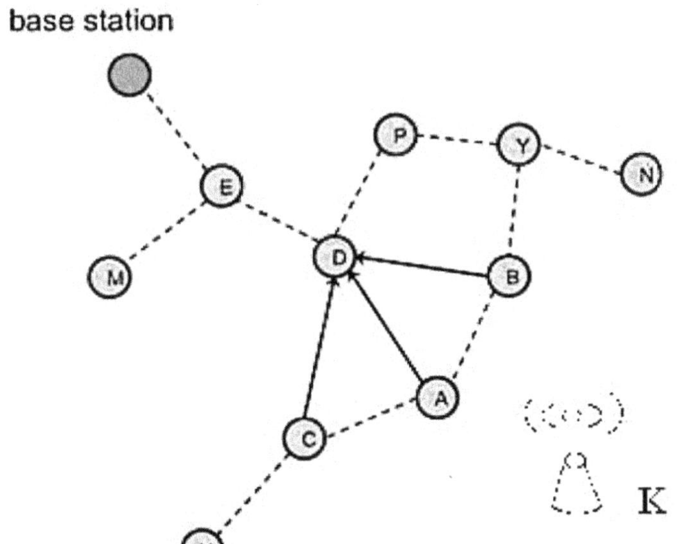

One specific technique, the Sporas formula, has been used in sensor networks to evaluate a node's confidence in packet transmission. This strategy could also be applied to CPS networks as it has been successful in other infrastructure systems.

Figure 3a illustrates the assessments for Node A. Though the evaluations of additional nodes remain constant, the rating of Node A declines. This suggests that Node A may be compromised and further research is needed to investigate the possible causes, such as energy consumption or communication path. Figure 3b shows nodes communicating through Node D have a value less than 0.85. This indicates that Node D is potentially evil and may be disconnected from other nodes. The base station will then determine an alternative path for package transmission. The trust-based approach is an efficient way to identify and remove rogue nodes from the CPS network.

Figure 3. Output node measured at A

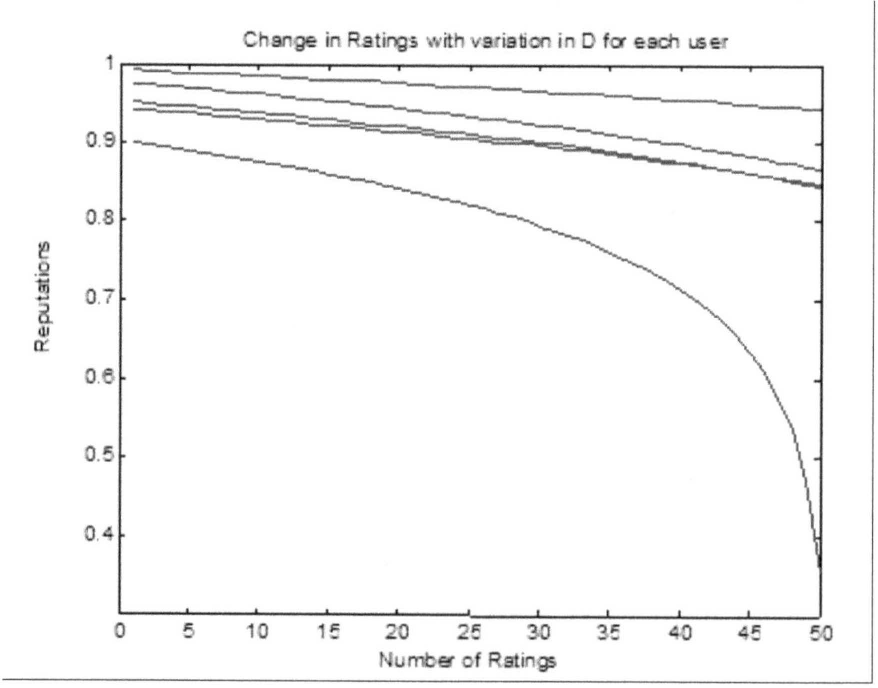

Virus and Threat

Rootkit vulnerabilities present a significant threat to Linux-based operating systems. These attacks allow an attacker to gain unauthorized access to a board computer deprived of the holder's knowledge. Rootkits are notoriously problematic to eliminate owing to the method they are installed and used, and they are becoming increasingly prevalent as a means to conceal malware payloads. Traditional antivirus software may have limitations when used in a cloud computing environment (Natrayan L & Kumar MS. 2020).

The process of removing a rootkit is dependent on both the hardware and operating system in use. Once an operating system has been installed on a computer, rootkits can be removed using specialized virus scanning methods. Figure 4 shows service providers use established security criteria and techniques to eliminate infections. It is important to have a security service within the operating system and to regularly update security settings and controls for all versions of the OS.

Figure 4. Rating recorded at different node A,Y,C,B and P

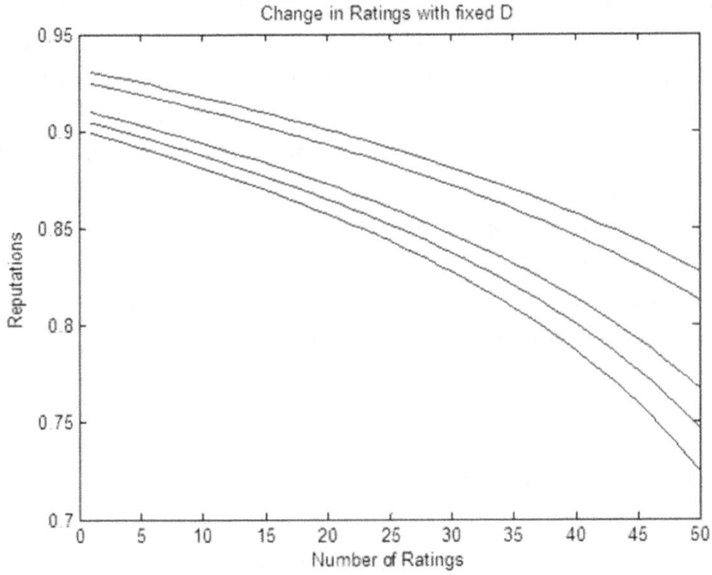

Figure 5. Service provider for the security

Proposed Security Enhancement System-Based Cloud

A dynamic and integrated approach is necessary to maintain information safety in cloud-based Cyber-Physical Systems (CPS). Balancing security levels with performance and productivity is important to keep costs low and customer satisfaction high. To achieve this, security systems must have fault tolerance, control, and enforcement capabilities (Veeman et al., 2021). Additionally, the company's workers, assets, services, and network activities should all be fully visible. The cloud environment monitors a variety of settings, including hybrid, public, private, government, and industrial environments (as shown in Figure 5). As different customers interrelate with the cloud atmosphere, it is necessary to integrate security and admittance control at the resident cloud level (Natrayan L et al., 2021). The simulated environment makes a unique safety space for each client, which consumes resources as needed and releases them once the requirement is met. This approach allows for the cloud system to adjust security and access control as needed.

Cloud truthfulness aims to provide criteria for info security inside a business and its system. This includes ensuring that data is dependable, consistent, and accurate throughout its life cycle. The security standards are determined by the value of the information and the number of persons who have access to it (Natrayan L et al., 2018). For example, in a hospital setting, the user's role - such as nurse, administrator, or doctor - determines their level of access to patient information. A doctor may access a patient's medical data to diagnose their condition, while a nurse may only access certain parts of the database (Pragadish et al., 2022). Similarly, access privileges differ within an organisation depending on the employee's role. The virtual security system monitors users who interact with incomes to admission or process data and ensures that they comply with the access privileges (Ramaswamy et al., 2022).

RESULT AND DISCUSSION

The combination of cyber and physical approach, known as Cyber-Physical Systems (CPS), presents a range of research challenges in various fields. Cloud CPS (CCPS) specifically, significantly impacts information security and infrastructure protection (Balaji et al., 2022). One of the major challenges in this field is dealing with security threats such as rootkit viruses, which can be difficult to address and may not have readily available solutions. In healthcare, medical CPS require advanced embedded systems with network access for body monitoring and control (Yogeshwaran et al., 2015). Safety and dependability are crucial considerations as the failure of these devices can harm the patient. Modern GPU (graphics processing unit) technology may also be able to replace supercomputing capabilities for real-time organ modeling. In transportation, sensor systems and computer power have enabled the development of numerous corporal instruments to aid drivers on the road. Communication between automobiles and between vehicles and infrastructure has reduced traffic bottlenecks and accidents. However, new designs for intelligent vehicles are needed for research to avoid collisions, accidents, and the transmission of automated emergency calls (Kumar MS et al., 2018).

Research in CPS also includes constructing biological experiments using mathematical analysis and pattern analysis methodologies and providing elderly individuals with care at home rather than sending them to expensive nursing institutions (Santhosh MS et al., 2018). Telepresence technology in CPS infrastructure can improve the living circumstances of the elderly and allows them to live independently while protecting their privacy (Velmurugan et al., 2022).

Integrating cyber and physical systems also requires a novel real-time resource management paradigm that surpasses existing techniques. Because of the connectivity of mobile devices, dynamic reconfiguration is necessary, and system infrastructure must be able to handle these requirements. CPS dependability, safety, and security must also withstand environmental risks or natural calamities (Nagarajan et al., 2022).

In a heterogeneous hybrid environment, service quality is a crucial criteria. CPS users require a high degree of confidence and trust can be gained through dependability, safety, security, privacy, and usefulness for the user. However, due to the infancy of cloud CPS and security research, the problems and challenges are greater.

CONCLUSION

In conclusion, the combination of cyber and physical systems, known as Cyber-Physical Systems (CPS), presents a range of research challenges in various fields. In this research article, we focused on analyzing and monitoring the design challenges in the CPS in the manufacturing sectors using cloud computing. One of the major challenges in this field is dealing with security threats such as rootkit viruses, which can be difficult to address and may not have readily available solutions. In healthcare, medical CPS require advanced embedded systems with network access for body monitoring and control. Safety and dependability are crucial considerations as the failure of these devices can harm the patient. Modern GPU (graphics processing unit) technology may also be able to replace supercomputing capabilities for real-time organ modeling. In transportation, sensor systems and computer power have enabled the growth of various physical tools to aid drivers on the road. Communication between automobiles and between vehicles and infrastructure has reduced traffic bottlenecks and accidents. However, new designs for intelligent vehicles are needed for research to avoid collisions, accidents, and the transmission of automated emergency calls.

Research in CPS also includes constructing biological experiments using mathematical analysis and pattern analysis methodologies and providing elderly individuals with care at household rather than sending them to expensive nursing institutions. Telepresence technology in CPS infrastructure can improve the living circumstances of the elderly and allows them to live independently while protecting their privacy.

Integrating cyber and physical systems also requires a novel real-time resource management paradigm that surpasses existing techniques. Because of the connectivity of mobile devices, dynamic reconfiguration is necessary, and system infrastructure must be able to handle these requirements. CPS dependability, safety, and security must also withstand environmental risks or natural calamities.

Overall, the research shows that while there are many challenges to implementing CPS in the manufacturing sector using cloud computing, there are also many potential benefits. By addressing the security concerns, developing sophisticated embedded systems, and utilizing modern technology such as GPU and sensor networks, CPS can improve efficiency, safety, and dependability in various industries. Additionally, utilizing citing in CPS can provide cost-effective solutions and enable real-time monitoring and control. However, as cloud CPS and security research are still in their infancy, it is important to continue researching and developing solutions to address the challenges and realize the potential of this technology.

REFERENCES

Alaghbari, K. A., Saad, M. H. M., Hussain, A., & Alam, M. R. (2022). Complex event processing for physical and cyber security in datacentres - recent progress, challenges and recommendations. *Journal of Cloud Computing (Heidelberg, Germany)*, *11*(1), 65. doi:10.118613677-022-00338-x

Balaji, N., Natrayan, L., Kaliappan, S., Patil, P. P., & Sivakumar, N. S. (2022). Annealed peanut shell biochar as potential reinforcement for aloe vera fiber-epoxy biocomposite: Mechanical, thermal conductivity, and dielectric properties. *Biomass Conversion and Biorefinery*, 1–9. doi:10.100713399-022-02650-7

Broy, M., Daembkes, H., & Sztipanovits, J. (2019). Editorial to the theme section on model-based design of cyber-physical systems. *Software & Systems Modeling*, *18*(3), 1575–1576. doi:10.100710270-018-0670-9

Cong S, Jianfeng M, Qingsong Y (2016). On the architecture cyber-physical. *Communications and Information Networks 11(*44), 1–21. doi:10.11959/j.issn.2096-1081.2016.041

Darwish, A., & Hassanien, A. E. (2018). Cyber physical systems design, methodology, and integration: The current status and future outlook. *Journal of Ambient Intelligence and Humanized Computing*, *9*(5), 1541–1556. doi:10.100712652-017-0575-4

Herdt, V., & Drechsler, R. (2022). Advanced virtual prototyping for cyber-physical systems using RISC-V: Implementation, verification and challenges. *Science China. Information Sciences*, *65*(1), 1–17. doi:10.100711432-020-3308-4

Jha, A. V., Appasani, B., Ghazali, A. N., Pattanayak, P., Gurjar, D. S., Kabalci, E., & Mohanta, D. K. (2021). Smart grid cyber-physical systems: Communication technologies, standards and challenges. *Wireless Networks*, *27*(4), 2595–2613. doi:10.100711276-021-02579-1

Kathiravelu, P., Van Roy, P., & Veiga, L. (2019). SD-CPS: Software-defined cyber-physical systems. Taming the challenges of CPS with workflows at the edge. *Cluster Computing*, *22*(3), 661–677. doi:10.100710586-018-2874-8

Kumar, M. S. (2018). Experimental investigations on mechanical and microstructural properties of Al2O3/SiC reinforced hybrid metal matrix composite. []. IOP Publishing.]. *IOP Conference Series. Materials Science and Engineering*, *402*(1), 012123. doi:10.1088/1757-899X/402/1/012123

Liu, B., Zhang, Y., Cao, X., Liu, Y., Gu, B., & Wang, T. (2020). A survey of model-driven techniques and tools for cyber-physical systems. *Frontiers of Information Technology and Electronic Engineering*, *21*(11), 1567–1590. doi:10.1631/FITEE.2000311

Liu, Z., & Wang, J. (2020). Human-cyber-physical systems: Concepts, challenges, and research opportunities. *Frontiers of Information Technology and Electronic Engineering*, *21*(11), 1535–1553. doi:10.1631/FITEE.2000537

Ma, M., Lin, W., Pan, D., Lin, Y., Wang, P., Zhou, Y., & Liang, X. (2018). Data and Decision Intelligence for Human-in-the-Loop Cyber-Physical Systems: Reference Model, Recent Progresses and Challenges. *Journal of Signal Processing Systems for Signal, Image, and Video Technology*, *90*(8–9), 1167–1178. doi:10.100711265-017-1304-0

Mishra, A., Jha, A. V., Appasani, B., Ray, A. K., Gupta, D. K., & Ghazali, A. N. (2022). *Emerging technologies and design aspects of next generation cyber physical system with a smart city application perspective*. International Journal of Systems Assurance Engineering and Management. doi:10.100713198-021-01523-y

Mosterman, P. J., & Zander, J. (2016). Cyber-physical systems challenges: A needs analysis for collaborating embedded software systems. *Software & Systems Modeling*, *15*(1), 5–16. doi:10.100710270-015-0469-x

Mueller, E., Chen, X. L., & Riedel, R. (2017). Challenges and Requirements for the Application of Industry 4.0: A Special Insight with the Usage of Cyber-Physical System. [English Edition]. *Chinese Journal of Mechanical Engineering*, *30*(5), 1050–1057. doi:10.100710033-017-0164-7

Nagarajan. (2022). Combined economic emission dispatch of microgrid incorporating renewable energy sources using improved mayfly optimization algorithm. *Computational Intelligence and Neuroscience*, 2022.

Natrayan, L., & Kumar, M. S. (2020). Optimization of wear behaviour on AA6061/Al2O3/SiC metal matrix composite using squeeze casting technique–Statistical analysis. *Materials Today: Proceedings*, *27*, 306–310. doi:10.1016/j.matpr.2019.11.038

Natrayan, L., Merneedi, A., Bharathiraja, G., Kaliappan, S., Veeman, D., & Murugan, P. (2021). Processing and characterization of carbon nanofibre composites for automotive applications. *Journal of Nanomaterials*, *2021*, 1–7. doi:10.1155/2021/7323885

Natrayan, L., Sivaprakash, V., & Santhosh, M. S. (2018). Mechanical, microstructure and wear behavior of the material AA6061 reinforced SiC with different leaf ashes using advanced stir casting method. *International Journal of Engineering and Advanced Technology*, *8*, 366–371.

Pragadish, N., Kaliappan, S., Subramanian, M., Natrayan, L., Satish Prakash, K., Subbiah, R., & Kumar, T. C. A. (2022). Optimization of cardanol oil dielectric-activated EDM process parameters in machining of silicon steel. *Biomass Conversion and Biorefinery*, 1–10. doi:10.100713399-021-02268-1

Ramaswamy, R., Gurupranes, S. V., Kaliappan, S., Natrayan, L., & Patil, P. P. (2022). Characterization of prickly pear short fiber and red onion peel biocarbon nanosheets toughened epoxy composites. *Polymer Composites*, *43*(8), 4899–4908. doi:10.1002/pc.26735

Santhosh, M. S. (2018). Investigation of mechanical and electrical properties of Kevlar/E-glass and Basalt/E-glass reinforced hybrid composites. *International Journal of Mechanical and Production Engineering Research and Development*, *8*(3), 591–598. doi:10.24247/ijmperdjun201863

Shahriar, M. R., Al Sunny, S. M. N., Liu, X., Leu, M. C., Hu, L., & Nguyen, N. T. (2018) MTComm based virtualization and integration of physical machine operations with digital-twins in cyber-physical manufacturing cloud. *Proceedings - 5th IEEE International Conference on Cyber Security and Cloud Computing 2018* (pp. 46–51). IEEE. 10.1109/CSCloud/EdgeCom.2018.00018

Tsigkanos, C., Kehrer, T., & Ghezzi, C. (2016). Architecting dynamic cyber-physical spaces. *Computing*, *98*(10), 1011–1040. doi:10.100700607-016-0509-6

Tyszberowicz, S., & Faitelson, D. (2020). Emergence in cyber-physical systems: Potential and risk. *Frontiers of Information Technology and Electronic Engineering*, *21*(11), 1554–1566. doi:10.1631/FITEE.2000279

Veeman, Shree, M. V., Sureshkumar, P., Jagadeesha, T., Natrayan, L., Ravichandran, M., & Paramasivam, P. (2021). Sustainable development of carbon nanocomposites: Synthesis and classification for environmental remediation. *Journal of Nanomaterials*, *2021*, 1–21. doi:10.1155/2021/5840645

Velmurugan, G., Siva Shankar, V., Kaliappan, S., Socrates, S., Sekar, S., Patil, P. P., Raja, S., Natrayan, L., & Bobe, K. (2022). Effect of aluminium tetrahydrate nanofiller addition on the mechanical and thermal behaviour of luffa fibre-based polyester composites under cryogenic environment. *Journal of Nanomaterials*, *2022*, 1–10. doi:10.1155/2022/5970534

Verma, R. (2022). Smart City Healthcare Cyber Physical System: Characteristics, Technologies and Challenges. *Wireless Personal Communications*, *122*(2), 1413–1433. doi:10.100711277-021-08955-6 PMID:34462622

White, J., Clarke, S., Groba, C., Dougherty, B., Thompson, C., & Schmidt, D. C. (2010). R&D challenges and solutions for mobile cyber-physical applications and supporting Internet services. *Journal of Internet Services and Applications*, *1*(1), 45–56. doi:10.100713174-010-0004-9

Yogeshwaran. (2015). Mechanical properties of leaf ashes reinforced aluminum alloy metal matrix composites. *International Journal of Applied Engineering Research: IJAER*, *10*(13), 11048–11052.

Chapter 6
Cyber Physical Systems and Network Security:
The Present Scenarios and Its Applications

C. V. Suresh Babu
https://orcid.org/0000-0002-8474-2882
Hindustan Institute of Technology and Science, India

S. Srisakthi
https://orcid.org/0000-0002-4312-0358
Vellore Institute of Technology, Chennai, India

ABSTRACT

Digital technology has transformed the whole world. Starting from the day-to-day activities of almost all of transactions, communication is happening in the digital format. The data that is communicated is also getting stored in a number of devices, starting from sensors to mobile and other computing devices. These devices can be digital or analog in nature. All the devices communicate within them and exchange the needed data among them. A cyber physical system (CPS) combines the digital and analog devices that are present in the physical and cyber world. The CPS integrates sensors and networks of the physical process to the computational components. These are done so as to control the physical process. There have been a number of applications where CPS have been deployed. Though there are vast applications, the CPS faces a lot of issues like the network security. This chapter aims at providing the current scenario the network security plays and its applications.

DOI: 10.4018/978-1-6684-9267-3.ch006

1.0. INTRODUCTION

The CPS can also be called as an intelligent system. This is a computer system where the mechanisms and their components are controlled and also monitored by computer-based algorithms. The notion or the concept of CPS was proposed initially by the National Aeronautics and Space Administration (NASA) in 1992. Later it was discussed in detail by Baheti and Gill (2011). The CPS is the technology that integrates computation, networking, and the physical processes Lee & Seshia, (2017) Cardenas et al., (2008). The fundamental principle of CPS is to bridge the engineering and physical world applications to that of the computer hardware and cyber worlds. The word "Cyber" in the CPS indicate the computations integration, networking and physical processes integration also. CPS belongs to the engineering discipline Poovendran, (2010) and is focussed on the technology.

Recent times have witnessed CPS in a wider range and it is the core technology when the industrial revolution is concerned Ge et al., (2017). Many researchers worked to establish the importance of CPS. It was brought as the top of other information technologies President's Council of Advisors on Science and Technology, (2007). It was used in the German Industry 4.0 Kagermann, (2015) as the industrial Internet in the U.S Annunziata & Evans, (2013). It also made its way to the ARTEMIS (Advanced Research and Technology for Embedded Intelligence and Systems) ARTEMIS Ind. Ass, (2007).

As the use of CPS have grown over the years, the security threats that these systems face are also growing. In general, the security of the CPS can be classified into two major areas:

(i) **Information Security**: The CPS systems store data or information in its components or systems. These information needs to be secured and protected. This kind of security involves securing such information during its process. The data retrieved may go through a lot of process like data aggregation, data processing, and these data can be shared in the network environment. These networks can be an open network, or loosely coupled networks. The information security aims to protect the data by using cryptographic techniques. The next type of security tries to control the security by protecting the dynamic workings of the control systems and defending them against cyber-attacks.

(ii) **Control security:** This technique of security aims to control the issues in the network environment. It focuses in mitigating the control system and protecting it from any possible attacks.

Cyber security and network security vary by a lot of features and characteristics. Cyber security focusses in defending the system from cyber and malicious attacks. The basic idea is to enhance the system's security. By doing so, the unauthorized access to the system can be eliminated. It thus protects cyberspace from attacks and damages. While network security focusses on the measures taken by an organization with an aim to secure its network. The main aim is at securing the confidentiality and accessibility of the data and more importantly on the network. So, there is a possibility of having cyber threats in CPS.

Despite their enormous advantages of the CPS, the CPS are vulnerable to various cyber and other related threats. They face a lot of attacks and challenges on a daily basis. Many research works have been carried out by researchers. Various goals of CPS were listed out Chen, (2010), the different ways by which the security of CPS can be maintained were presented Humayed et al., (2017). The security challenges faced by CPS were discussed Yoo & Shon, (2016).

The CPSs safety and security model were discussed in detail by the authors Wolf & Serpanos, (2017). The authors Pivoto, de Almeida, da Rosa Righi, Rodrigues, Lugli, & Alberti, (2021) have studied the most prominent CPS models and placing a special emphasis on their various characteristics. The industries have started to use CPS Pivoto, de Almeida, da Rosa Righi, Rodrigues, Lugli, & Alberti, (2021) that have been powered to be better, reliable and safe for production. A survey was proposed Radanliev et al., (2021) that discussed design, technologies, and also included principles and policies for the CPS. The CPS has been used in multiple applications, the network security issues in CPS systems during its applications in a real time scenario have been enormous. The applications of CPS are growing into many fields. Some of the main benefits of CPS for the vast growth are listed as follows:

a) Integration - CPS provides network integration characteristics. For instance, there is a possibility to integrate media access control techniques. They also provide a balance so as to establish a coordination on the networks, a way to control the network transactions, and provide fault tolerances Haque et al., (2014).
b) Interaction between humans and system – Modelling of the interaction and simulating it has made it possible to make a quick decision
c) In dealing with uncertainty – It is possible that the CPS will evolve and function normally with a new and unreliable environment.
d) Increased system performance – With respect to feedback and creating an automatic redesign and establishing a close interaction with that of sensors and cyber infrastructure a better performance was attained while using CPS.

e) Scalability and flexibility – The CPS were able to provide a scalable and flexible resources to the users
f) Quick response time – With CPS the response time of the system was enhanced and it also facilitated an early detection of failure.

Though there are numerous applications, the main issue with CPS lies on the its network security. There are a number of attacks and data breeches that have been carried out using the security loophole present in the CPS system. This chapter focusses on the attacks and on the cost incurred by industries due to these data breeches.

2.0. BACKGROUND

The CPS are a new generation system that can interact with humans. A lot number of projects and researches are going on in this field. The Cyber Physical Systems Security (CPSSEC) project focusses on the security concerns on the CPS and IoT devices. These addresses the important factor affecting the functioning and control of the information security at 43%. CPS and IoT also increase cybersecurity risks and the attack surfaces are more. There were many instances of CPS attacks. Recent times reported a cyberattack that attempted to make ransomware attacks. These attacks were made using an aquarium with an internet connection. The hackers made an attempt to steal data from a casino in North America that used sensors that were connected to a computer which regulated the temperature, cleaning and feeding program of a fish tank.

In fact, it is anticipated that billions of new IoT devices will be connected to the internet. If in any scenario security gets overlooked, then there is a risk of getting unintentional faults or malicious attacks. Cybersecurity gets more challenges when a billion number of devices with security vulnerabilities are added. It is not possible to address the security issues unless the solution can be widely deployed on all the systems. Any of the security issues when raised, it must be first analysed, then understood and then it should be addressed in its early stages of design and deployment. Authors Electric Power Research Institute, (2015) presented a vast description that depicted a few cybersecurity failure scenarios.

The Figure 1 gives the scenario of how the CPS system works. The dataflow and the interactions that happen are depicted. As there is a dataflow within the component, there arises a need for security parameters.

Figure 1. Dataflow and interactions in the CPS (Electric Power Research Institute, 2015)

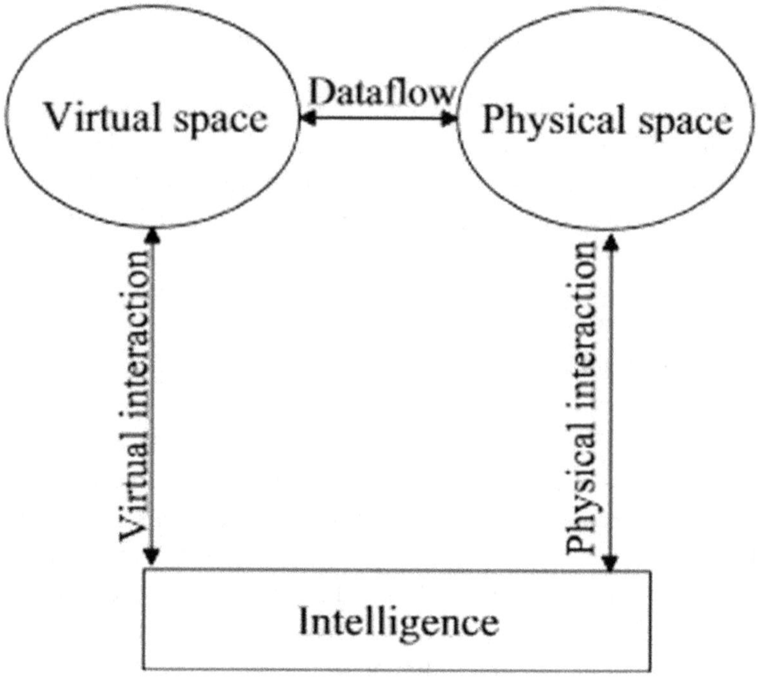

Health Care Applications

Medical cyber-physical systems (MCPS) are healthcare applications. These devices are very critical and is a integration of a number of devices. All these devices and components are connected to each other in a network. These health care related components are widely used in hospitals. These help in achieving a high-quality healthcare service. Though these components and systems are used in a real time scenario, these MCPS components face a design issue. They also face a number of challenges as the data and usage increases. Some of the challenging issues are inoperability, security of the data and privacy of the data IBM, (2022).

With the CPS and the MCPS the feedback and input are received from the physical environment. Due to this there are communication channels and these require security. But this is one of the significant specific characteristics of the CPS Neuman, (2009). The sensor collects the data and transmit them. Hence it is essential to have security during this process in both the CPS and MCPS systems.

All the stages of data placement, monitoring them and services related to system control should be considered for implementing security Baheti & Gill, (2011).

The MCPS facilitates the transfer of the medical records and other related files easily Ackerman et al., (2010), this has been a very helpful service for the doctors and the hospital administrators. Due to the growing advancements of the medical devices, it is possible to replace the individual devices that had been designed for treating patients. These devices can be replaced with those advanced devices that can work in an integrated fashion. These devices can work using the distributed computing approach. A number of trends emerged with the use of this MCPS Alexander & Clarkson, (2000); Goswami et al., (2016); Qian et al., (2009).

Recent advancements in the medical sensors the WSN etc. have paved a way for the CPS system to include a number of healthcare applications, like in the home or in the hospital patient care Milenković et al., (2006). The CPS medical devices and their components are vulnerable and they hold critical information. Hence, there is a need to ensure security, safety, and reliability of these devices. These devices are also interconnected, and have complex structures Jiang et al., (2012); Lee et al., (2012); Otto et al., (2006). In the recent times, social networks, along with the big data analysis and applications are attracting researchers to work upon .

Industrial CPS (ICPS) Applications

The ICPS has been a value technology that is very novel in its creation. This has been used in the modern industrial manufacturing department. Recent times saw the improvements in the future of I-CPS with regard to research and practices. The general architecture of ICPS was given Agrawal & Kumar, (2022) which is depicted as given in Figure 2.

The shift from standard automation solutions to the CPS were presented in Ribeiro & Björkman, (2017). And a new CPS architecture model for the industry 4.0 are discussed in Pivoto, de Almeida, da Rosa Righi, Rodrigues, Lugli, & Alberti, (2021). A survey and an analysis of the state-of-the-art technologies in I-CPS were presented Leitao et al., (2016). Cloud-integrated CPS (CCPS) will pave the way that will allow applications that were previously unachievable o be built, deployed, managed and controlled effectively. It tries to represent the physical system components like sensors, actuators, robots and other devices in the cloud. Accessing of these physical components is also made possible and the data collected from these components are sent to cloud. Integration of cloud techniques like virtualization along with the CPS techniques is also made possible. Similarly cloud integrated manufacturing components and cloud integrated vehicles could be deployed in a very effective manner.

Figure 2. General architecture of ICPS (Agrawal & Kumar, 2022)

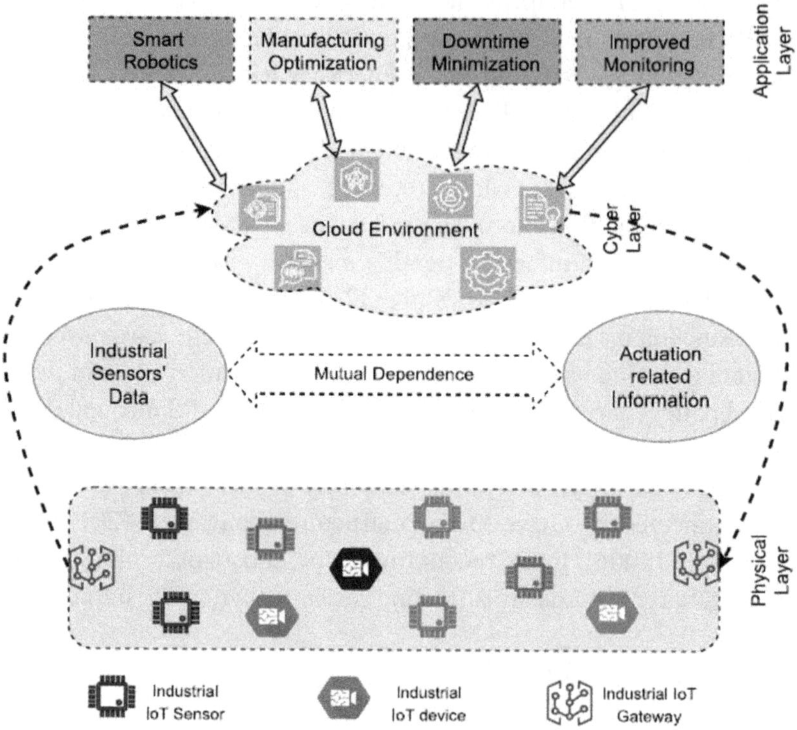

The fundamental reason for using CPS in manufacturing industry is to have an improved production, the time to production could be reduced. These serve as the challenges for deploying CPS in industries. One of the main issues is in bringing forth the current employees to adapt themselves to the newly identified technology (Hummel et al., 2015; Schuh et al., 2015; Yao et al., 2019).

CPS in Automation Applications

The unique design challenges for using CPS in automation are hybrid, heterogeneous, distributed, large scale, dynamic, adaptive. The major design challenges are how these characteristics can be integrated in real time systems. For instance, in order to verify the working of an advanced driver assistance systems (ADAS) in automobiles, these are hybrid systems that operate in a dynamic environment. The developed design tools must be capable of handling this combination and the resulting concerns.

The Internet of Things (IoT) and Internet of Services (IoS) started working together and producing Internet of Things and Services (IoTS) . These help in automating devices and they started sending their data to cloud. The Figure 3 gives the scenario of Industry 4.0 Dafflon et al., (2021) where all the devices, components collaborate with one another and automate the services Suresh Babu, (2023).

Figure 3. Integration of many technologies leading to Industry 4.0 (Dafflon et al., 2021)

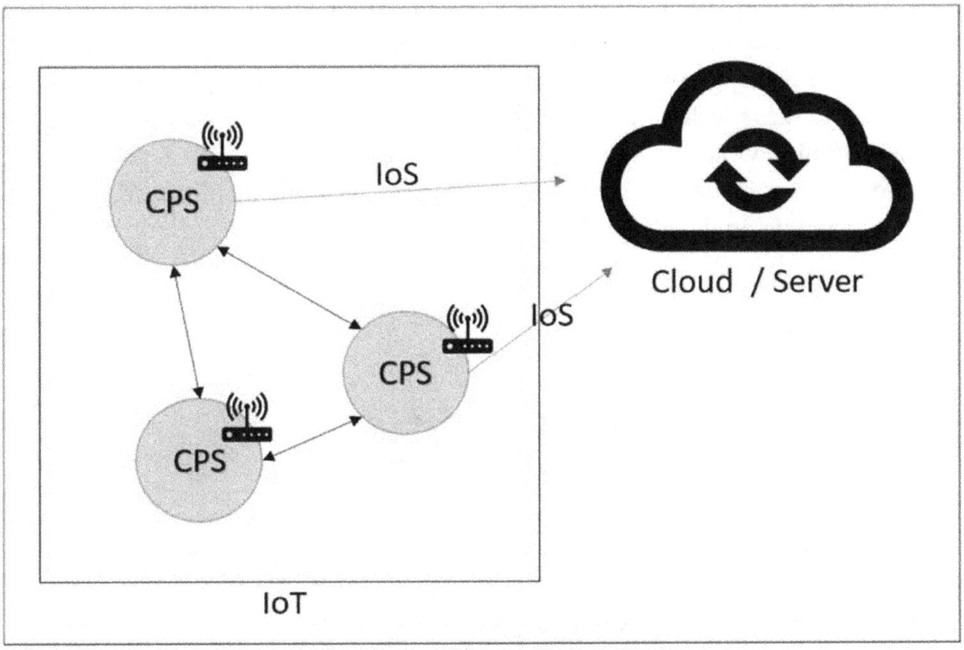

CPS in Robotics

Robots and the CPS together sense the environment, collect the data, process them and based on it react to the information collected. All these components should operate safely though there a number of uncertainties present and have resource constraints. A wide range of challenges starting from visual perception, motor learning, the design and its implementation, and verification of safe and performant CPS. There are a lot of network security issues as data are collected from multiple sources and processed. Robotic Operating System (ROS) plays a vital role in the coming years. Its usage extends to research applications and also in commercial robotics. Hence, the researchers devote their attention to the security issues of ROS and controllers based on ROS Dieber et al. (2017). The fundamentals of any information system starting from a CPS system needed to be protected. The fundamental measures of security are discussed in the following section.

3.0. CIA TRAIDS OF SECURITY

The three letters in "CIA triad" are Confidentiality, Integrity, and Availability. The CIA traid acts as a basic model and serves to be the basis for the security system. These serve as the parameter through which the vulnerabilities are identified and solutions are developed. These three traids of confidentiality, integrity, and availability are crucial to the operation of a business and is depicted in figure 4. There are differences in these techniques and they help guide security teams by pinpointing the ways through which these issues can be addressed. If an organization has met all these three components, it indicates that the security profile of the organization is stronger and it can defend its data against threats and attacks.

Figure 4. CIA traid of security

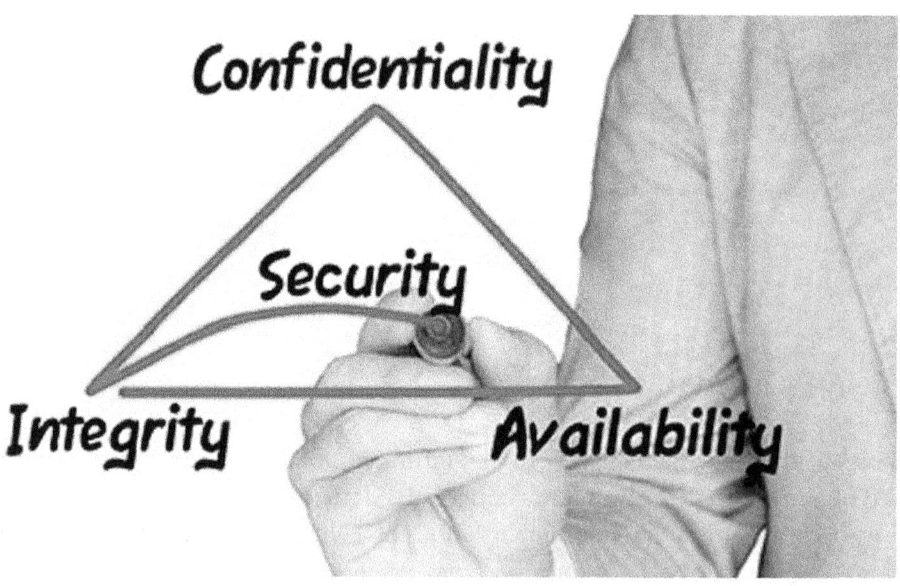

Confidentiality means that data, objects, and resources are protected from unauthorized viewing and other access. The information systems serve as a house to some information which may have some degree of sensitivity. The stored information might be proprietary business information that the other businesses could use to their benefit. This information could be personal in connection to an organization's employees, customers or clients. This confidential information is under frequent attack as criminals hunt for vulnerabilities to exploit. Some of the threat vectors include direct attacks. These kinds of attacks include stealing passwords and capturing network traffic. Some of the confidentiality breaches are done intentional and some are unintentional. A few common accidentals are emailing sensitive information to the wrong recipient, publishing private data to public web servers, and leaving confidential information displayed on an unattended computer monitor.

Example - Healthcare is an industry where the obligation to protect client information is given very high priority. The patients want the healthcare industry to protect their privacy. There are a number of strict regulations governing how healthcare organizations manage security.

Integrity protects the information from an unauthorized alteration. Integrity gives an assurance in maintaining an accuracy and completeness of data. The need to protect data includes both in the place of its storage and during its transmission. There is a need to control access at the system level, and also to further ensure that users are only able to alter information that they are legitimately authorized. The protection of data integrity extends beyond intentional breaches. An effective and good integrity-maintained system must also protect the data against unintentional alteration. These might include such as user errors or data loss that is a result of a system malfunction.

Example: The finance industry has a need to ensure that transactions across its systems are secure and accurate and are defended from tampering. "One of the most notorious financial data integrity breaches occurred in February 2016 when cyber thieves generated $1-billion in fraudulent withdrawals. The hackers executed an elaborate scheme that included obtaining the necessary credentials to initiate the withdrawals, along with infecting the banking system with malware that deleted the database records of the transfers. And then suppressed the confirmation messages which would have alerted banking authorities to the fraud".

Availability states that an information system to be useful it must be available at any time to any authorized users. The availability parameter provides a timely and uninterrupted access to the system. Some of the most fundamental threats to availability can be non-malicious in nature. These include hardware failures, unscheduled software downtime and network bandwidth issues. Malicious attacks may cause a sabotage that is intended to cause harm to an organization by denying users access to the information system.

Disruption of website availability for even a short time can lead to loss of revenue, customer dissatisfaction and reputation damage.

Example - The Denial of Service (DoS) attack is frequently used by hackers to disrupt web service. The DoS attack is caused, when hackers flood a server with superfluous requests. This request overwhelms the server and degrading service for legitimate users. Over the years, service providers have developed sophisticated countermeasures for detecting and protecting against DoS attacks. But hackers also continue to gain in sophistication and such attacks remain an ongoing concern.

When there is a need to use the CPS systems the employees can be trained with the concepts of the CIA. Any hypothetical scenarios or real-life incidents can be used to help employees think in terms of the maintenance of CIA in a system.

4.0. CIA ATTACKS

The figure 5 gives the attacks on the CIA traids of security. The cyber-attacks on CPS can be divided into three classes of availability, integrity and confidentiality ones.

Figure 5. Attacks on the CIA traids of security

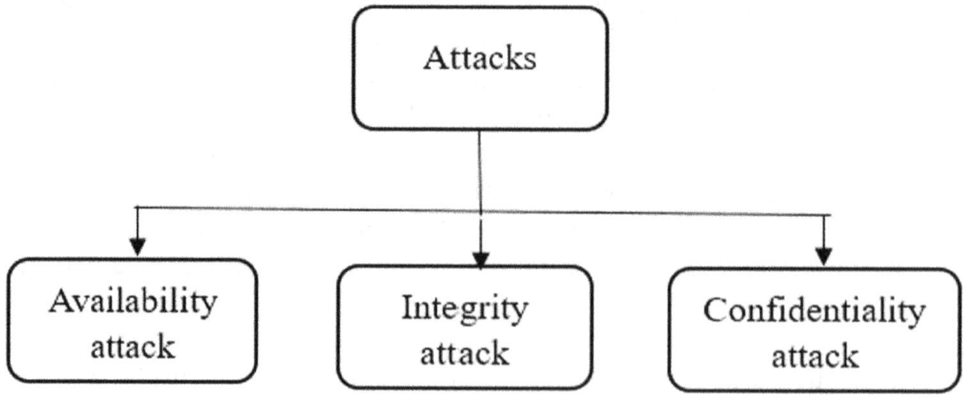

Availability Attacks

The goal of an attacker to go for an availability attack is to create a chaos and make the data, information and resources present in the system to be unavailable. The availability attack is caused by filling the buffers in a user domain, blocking or jamming the communication channel or by altering a routing protocol Suresh Babu, (2023). DoS has been one of the most common availability attacks. In recent years, more works in CPSs have concentrated on the DoS attack. Researchers have analysed it on a time-driven systems, and not on an event-driven systems. The communication via a wireless channel is blocked by the DoS attacks. The energy budget is also limited Peng et al., (2018).

Integrity Attacks

These attacks aim to manipulate and destroy the data integrity of a CPS system. These can be initiated by altering or deleting sensor measurements that were gathered. These are also done by adding incorrect or wrong data into them. Integrity attacks are often very difficult to predict or identify when compared to the availability attacks. The reason is due to the scenario that the falsified data is spread through the sensor network in an epidemic way, thereby leading to negative effects on systems Zhang et al., (2020).

Confidentiality Attacks

These attacks are on CPS, which aims to falsify or steal secret information. Not a lot of research is going on in this type of attack as confidentiality attacks are more complicated and they also involve availability attacks and integrity attacks. For instance, a secret key can be inferred by a inducing a fault injection attack (Bertoni et al., 2002; Blömer & Seifert, 2003).

5.0. RECENT ATTACKS ON THE CYBER INTERNATIONAL WIDE

Recent times have seen a rise in the Cyber-attacks. There are people who use the modern technology to their advantage and try to misuse the conveniences and benefits of it and pose a threat to businesses and data privacy in a global level. The data breaches that happen have a far-reaching impact. It not only affects the company, but also affects the customers, the suppliers of the company and everyone associated to the company. Research experts suggests that the cost of cybercrime will be reaching around $10.5 trillion by 2025.

But there is always a possibility to learn from the past history of cyber attacks.

1. The Melissa Virus

Earliest and one of the biggest cyber threats occurred in 1999 by the Melissa Virus which was carried out by a programmer David Lee Smith. The programmer mailed all the users a file that could be opened via the Microsoft Word that held a virus. As soon as the document was opened the virus was activated. It caused severe damages to a number of companies including Microsoft. An estimation was made that this virus costed $80 million for the affected systems to be repaired.

2. NASA Cyber Attack

The NASA's computers In 1999, were hacked by a 15-year-old James Jonathan and it was shut down for 21 days. Around 1.7 million software were downloaded during the attack, and that costed around $41,000 to repair them.

3. The 2007 Estonia Cyber Attack

Estonia witnessed a first cyber attack in April 2007 on an entire county. Around 58 Estonian websites went offline, including government, bank and media websites.

4. A Cyber Attack on Sony's PlayStation Network

Sony's PlayStation Network in April 2011 compromised the personal information of 77 million users because of a cyber attack.

5. Adobe Cyber Attack

Adobe cyber attack breached the data of 2.9 million users. Along with it the personal data of 38 million users were compromised. It also stated that the passwords and credit card information of the first 2.9 million users were stolen. But remaining 35.1 million users suffered the loss of their passwords and user IDs too.

6. The 2014 Cyber Attack on Yahoo

Yahoo was subject to one of the biggest cyber attacks of the year when 500 million accounts were compromised in 2014. During the attack, basic information and passwords were stolen, but the bank information was not compromised.

7. Ukraine's Power Grid Attack

In 2015 there was a Ukraine's power grid attack and it is believed to be the first cyberattack on a power grid. Due to this attack, many homes in the Ivano-Frankivsk region were without power for a few hours.

8. 2017 WannaCry Ransomware Cyber Attack

2017 saw one of the largest ransomware attacks of all time. It also affected around 200,000 computers in 150 countries. The ransomware made a huge impact on several industries that costed around 6 billion pounds to fix this at a global level.

9. A Cyber Attack on Marriott Hotels

A cyber attack was made on the group of Marriott and Starwood hotels. These attacks were not detected for years, and it came to light only by 2018. But, by this time 339 million guests were attacked and their data was compromised. And the UK's data privacy watchdog had to fine the Marriott Hotels for 18.4 million pounds.

10. The Biggest Password Leak Yet

June 2021, saw a total of about 8.4 billion passwords leaked during the RockYou 2021 attack. Initially the RockYou site was attacked in 2009 that affected 32 million accounts.

The tactics that cyber criminals deploy to trick and cheat the business and employees into clicking on links and documents within emails has been growing. The above are some of the prime highlights through which the business can be protected from cybercrime. China also had attacks that were increased as it made stance towards various communities like the Taiwan and the Uyghur community. In China attacks on government agencies declined to 4.5% from 13.10% last year. But in India, US, and Indonesia, all attacks grew from 6.3% to 13.7%.

The Kon Briefing (Kondruss, 2023) listed the attacks that happened by April 2023. The attack statistics that were published is as follows:

Total: 41 entries
Affected countries: 15

Germany: 14, Israel: 9, Australia: 2, Belgium: 2, France: 2, Netherlands: 2, Portugal: 2, Canada: 1, Cyprus: 1, India: 1, Malaysia: 1, Sint Maarten: 1, Taiwan: 1, UK: 1, USA: 1

Affected industries: 23

Public sector: 18, Government: 8, Universities: 7, Police: 5, IT: 3, Municipality: 3, Education: 2, Telecommunication: 2, Agriculture: 1, Airport: 1, Automotive: 1, Aviation: 1, Building materials: 1, County administration: 1, Finance: 1, Logistics: 1, Maritime: 1, Pharmaceutical: 1, Research: 1, Service provider: 1, Sports: 1, Technology: 1, Utility: 1

On the whole the total may differ as multiple assignments are possible or, conversely, not every entry can be assigned to a country or industry. There were a number of attacks on Russia and this raised by 600% in 2022. This was due to the retaliation to Russia's invasion of Ukraine. This made Russia to be the fifth most targeted country.

Figure 6. Types of cyber-attacks during the recent times (Kondruss, 2023)

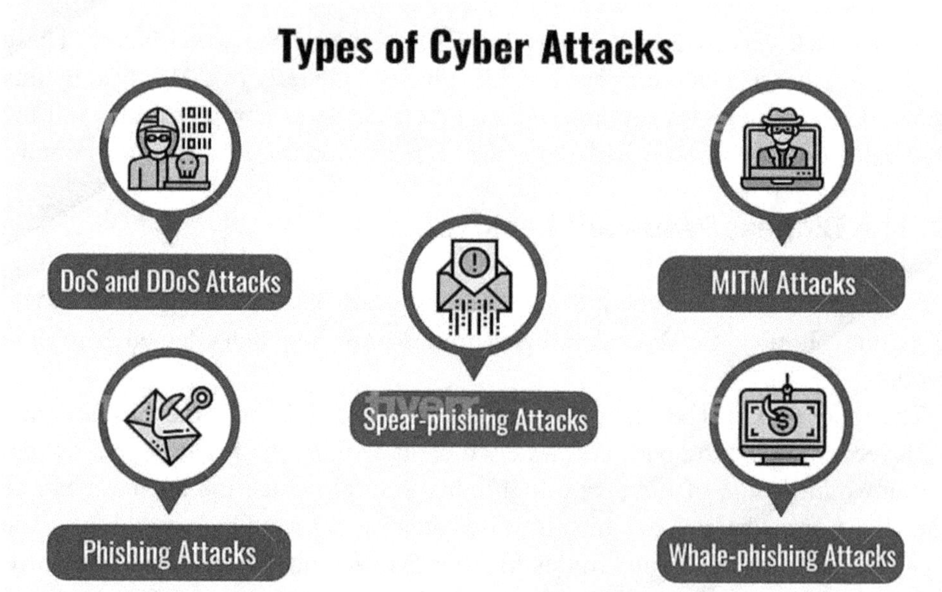

6.0. RECENT ATTACKS ON THE CYBER NATION WIDE

The recent times have seen a lot of cyber-attacks. Figure 6 gives a gist of the types of cyber-attacks that have been happening. India has also seen a number of cyber-attacks. The most recent attack in India has been the biggest cyber-attacks that occurred by 2022. In this attack all the servers of All India Institute of Medical Sciences (AIIMS) were being targeted at for a ransomware. A report published that the healthcare industry was the most targeted industry in India.

However, India, US, Indonesia, and China accounted for 45% of total cyberattacks on government agencies worldwide in the second half of 2022, according to a report by cybersecurity firm CloudSek, that was released. The number of attacks on government agencies were growing and the global cyberattacks also increased by 38% in 2022, when compared to 2021.

India was the most targeted country in 2022 and the attacks on various government agencies were doubled. The CloudSek attributed this attack to an increase in activities of Malaysia-based hacktivist group called Dragon Force. This hacktivist group ran campaigns like #OpIndia and #OpsPatuk that were against India as a retaliation to the controversial comments on Prophet Mohammed by an Indian politician.

The new kind of attack referred to as Hacktivism is a form of cyberattack. The motivation is here was not for financial gains but to promote an agenda that was political or to protest against certain policies of the government. Hacktivism accounted for 9% of the cyberattacks on the government sector by 2022. Many attacks in India were often targeted by various hacker groups that had links to China.

7.0. COST INCURRED DUE TO A DATA BREECH

More than the loss to the customers and the organizations, costs are involved when a data breech happens. IBM (2022) published a report by the end of 2022 that described in detail about the data breech and the cost incurred due to it. Figure 7 depicts the average total cost of a data breech. Figure 8 gives the average per record cost of a data breech. In both the figures of 7 and 8, the cost is measured in USD dollars.

The IBM also published a report that gave the Average cost of a data breech by the country or by the region. United States was the country for an average total cost due to a data breach for the 12th year. Other top five countries or regions with the highest average cost of a data breach are as below:

1. The United States — USD 9.44 million
2. The Middle East — USD 7.46 million
3. Canada — USD 5.64 million
4. The United Kingdom — USD 5.05 million
5. Germany — USD 4.85 million

Though the countries like US, Middle East, Canada, UK and Germany saw a rise in the data breeches. Countries like Japan, France, South Korea, Scandinavia and Turkey saw a reduction in the average total cost of a data breach.

Figure 7. Average total cost of a data breech (IBM, 2022)

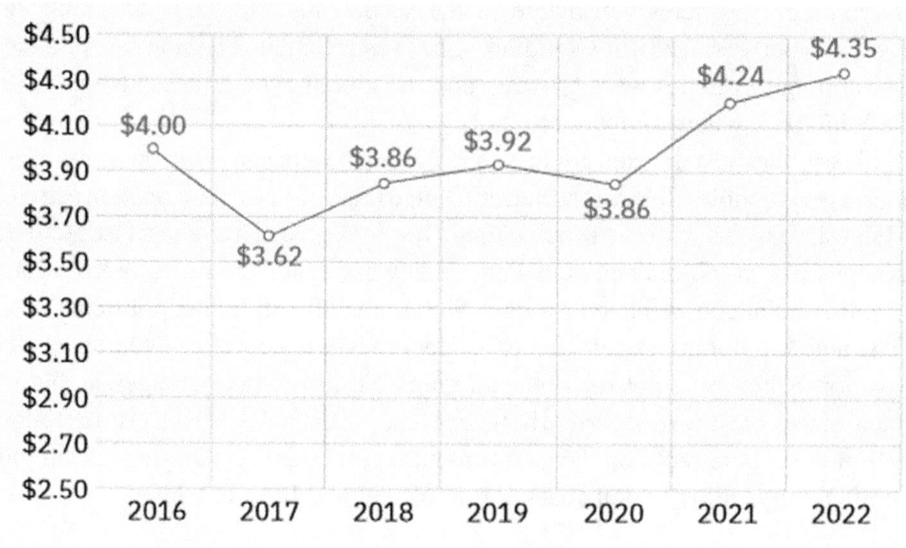

Figure 8. Average per record cost of a data breech (IBM, 2022)

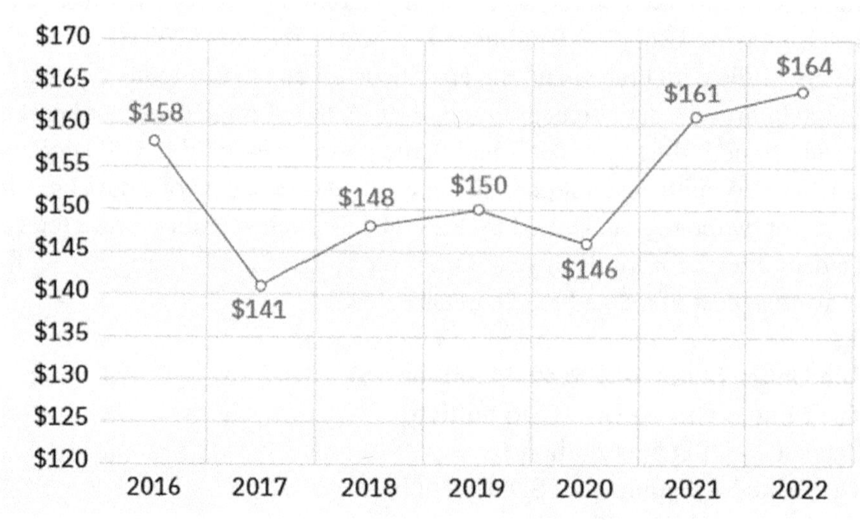

Figure 9. Average cost of a data breech by the country or by the region (IBM, 2022)

Figure 10. Average cost of a data breech sorted by the industry (IBM, 2022)

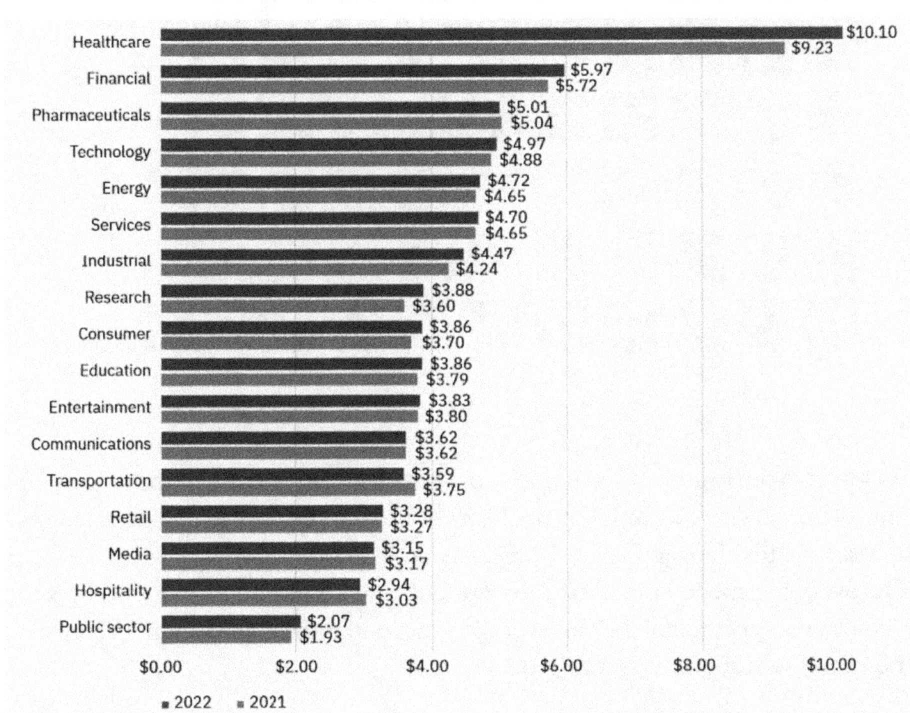

8.0. INDUSTRY AFFECTED BY THE DATA BREECH

Of all the various industries, healthcare is one of the more highly regulated and critical industry. It is considered by the US to be a critical industry that holds a lot of critical and sensitive data. As the industry gets affected by the data breech, the industry tries to compensate the loss by increasing the cost associated with its products and services.

Figure 11. A majority of organizations increased the price of their products and services as a result of the data breach as stated in the study (IBM, 2022)

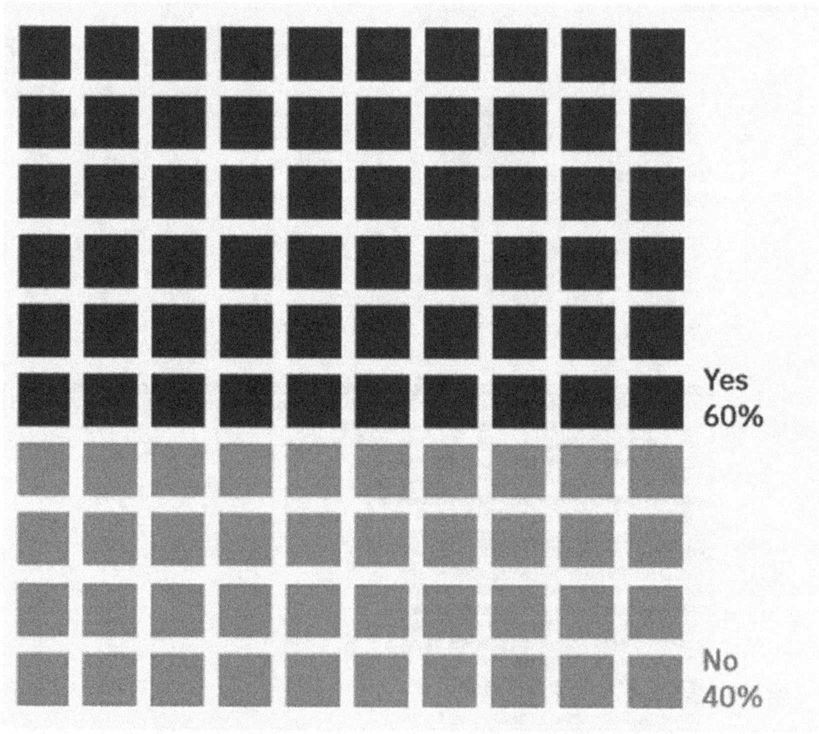

As given in the Figure 10, a majority of organizations increased the price of their products and services as a result of the data breach. 60% of the industries raised their cost towards their industries' products and services.

The average time to detect the initial attack and contain it without being spread depends on the type of attack. The average time to identify and contain a data breach by the initial vector is given in Figure 12.

Figure 12. Average time to identify and contain a data breach by initial attack vector (IBM, 2022)

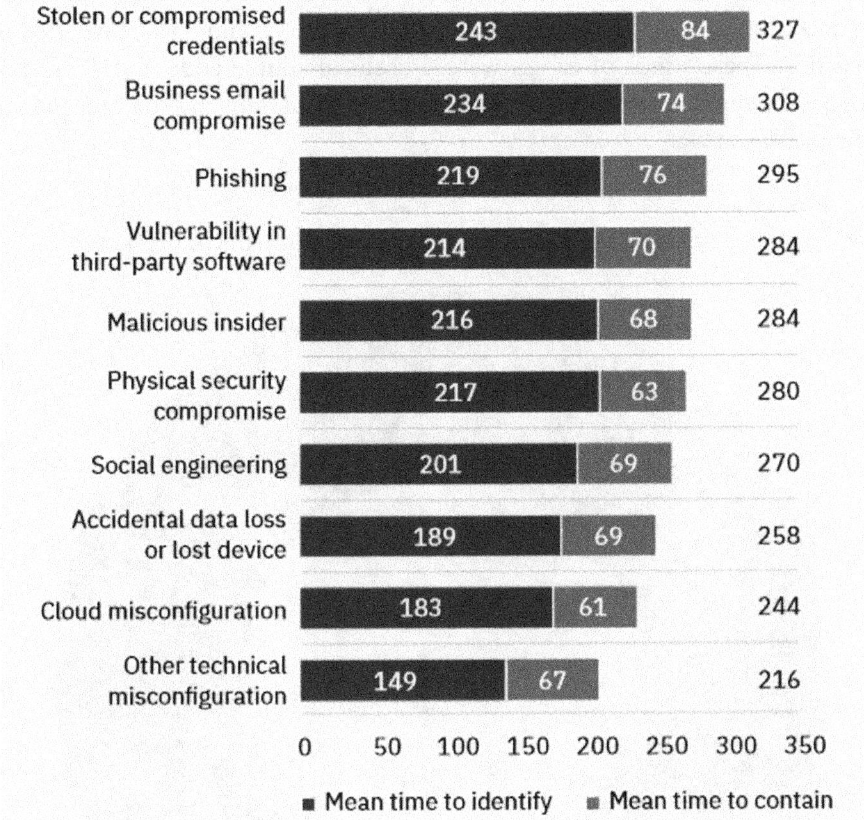

As given in the figure stolen or compromised credentials are the very tough ones to identify and contain it. The measurement is done in days.

9.0. TYPES OF BREECHES EXPERIENCED BY ORGANIZATIONS

There are a number of breaches that an organization faces.

As stated in the Figure 13, the different types of breaches faced by an organization has been listed. The ransomware breach caused for 11% of breaches, with destructive attacks causing for 17% of breaches. 19% of breaches were due to the supply chain attacks. This happened due to the fact that these breaches were caused due to a business partner initially compromised. The errors that a human make is enormous, due to which the breaches caused by them remain unintentional. These caused for 21% of breaches.

There were failures, that were caused by a disruption or failure that happened in an organization's systems, devices. This led to data loss and was causing for 24% of breaches. These kinds of errors were present in source code, a process failure that occurred during the automation. The other remaining 8% of breaches were malicious attacks.

Figure 13. Types of breaches faced by an organization (IBM, 2022)

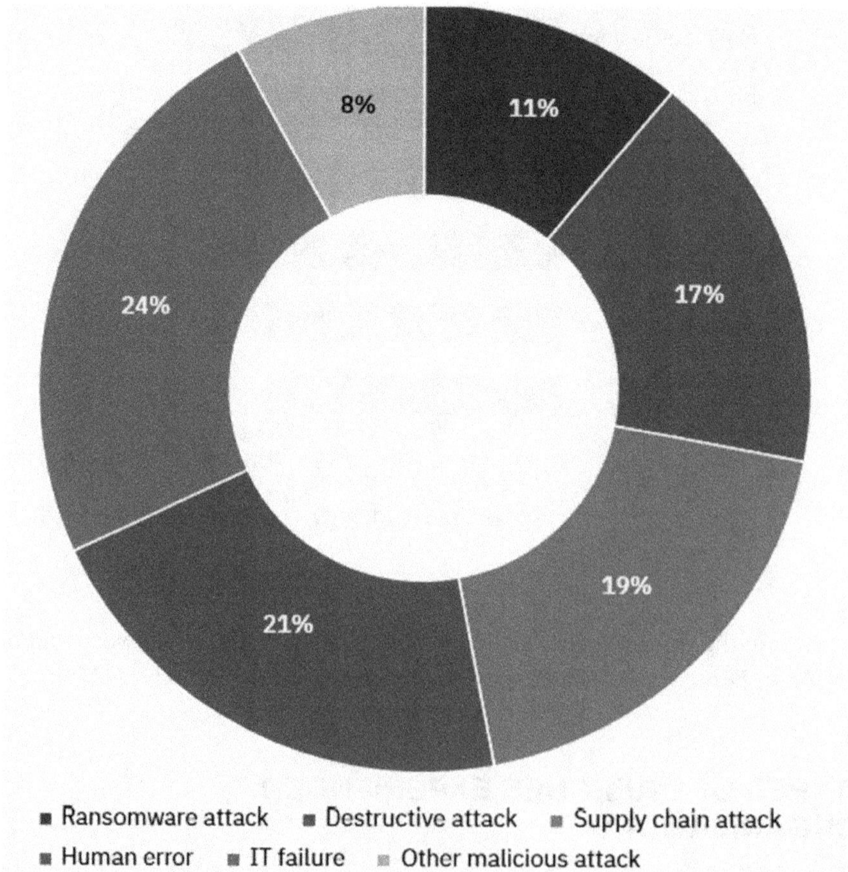

10.0. SOLUTIONS AND RECOMMENDATIONS

The majority of the attacks are based on the data breech attack. If the CPS systems are able to defend from the data breeches, then the system can be protected. Some of the solutions and recommendations are as below:

ü To download only the registered software
ü To update the software when the newer versions are available as these newer versions tend to have more security upgrades
ü To encrypt the data whenever possible
ü To use strong passwords and credentials for all the accounts
ü To change the passwords and credentials as often as possible
ü To make sure all the credentials are strong and not easy to guess
ü To not use the same set of credentials again and again
ü To not use the same set of passwords for multiple accounts

11.0. FUTURE RESEARCH DIRECTIONS

The economic or physical harm done to a CPS system can be due to the fact that the applications might be malfunctioning or other technical faults. Hence, it is important to protect cyber-physical systems from such attacks. The future research directions can be based on interdisciplinary fields like Malware analysis, Risk analysis and mitigation, IIoT Security and much more. The sensors that collect the data and the network communication channel which need to transmit the data, everything needs to be secured.

12.0. CONCLUSION

To protect the devices or the systems and data against cyber threats, simple measures can be adopted. These measures could include using the latest and genuine hardware and software for all the digital needs. The security measure can also include advanced measures such as installing a firewall to add an extra security layer.

REFERENCES

Ackerman, M. J., Filart, R., Burgess, L. P., Lee, I., and Poropatich, R. K., (2010) Developing next-generation telehealth tools and technologies: patients, systems, and data perspectives. *Telemed e-Health. 16*(1), 93–95.

Agrawal, N., & Kumar, R. (2022). Security perspective analysis of industrial cyber physical systems (I-CPS): A decade-wide survey. *ISA Transactions, 130*, 10–24. doi:10.1016/j.isatra.2022.03.018 PMID:35396141

Alexander, K., & Clarkson, P. J. (2000). Good design practice for medical devices and equipment, Part II: Design for validation. *Journal of Medical Engineering & Technology, 24*(2), 53–62. doi:10.1080/030919000409311 PMID:10937360

Annunziata, M., & Evans, P. C. (2013). *The industrial Internet @ work.* (White paper). General Electric.

ARTEMIS Ind. Ass. (2007). *Advanced Research and Technology for Embedded Intelligence and Systems.* ARTEMIS Ind. Assoc.

Baheti, R., & Gill, H. (2011, March). Cyber-physical systems. *Impact Control Technology, 12*(1), 161166.

Belgharb, K., & Boufaida, M. (2017). Using Event B to Specify Context Awareness for Service Discovery in Pervasive Environments. *International Journal of Ambient Computing and Intelligence, 8*(1), 1–22. doi:10.4018/IJACI.2017010101

Bertoni, G., Breveglieri, L., Koren, I., Maistri, P., & Piuri, V. (2002). A parity code based fault detection for an implementation of the advanced encryption standard. In *Proc. 17th IEEE International Symposium on Defect and Fault Tolerance in VLSI Systems* (pp. 51–59). IEEE. 10.1109/DFTVS.2002.1173501

Blömer, J., & Seifert, J. P. (2003). *Fault based cryptanalysis of the advanced encryption standard (AES).* 7th Int. Conf. Financial Cryptography, Guadeloupe, French West Indies. 10.1007/978-3-540-45126-6_12

Cardenas, A. A., Amin, S., & Sastry, S. S. 2008. Research Challenges for the Security of Control Systems, in: 3rd Conf. on Hot Topics in Security, USENIX. pp. 6:1- 6:6

Chen, T. M. (2010). Survey of cyber security issues in smart grids. In *Cyber Security, Situation Management, and Impact Assessment II; and Visual Analytics for Homeland Defense and Security II, 7709* (p. 77090D). International Society for Optics and Photonics. doi:10.1117/12.862698

Dafflon, B., Moalla, N., & Ouzrout, Y. (2021). The challenges, approaches, and used techniques of CPS for manufacturing in Industry 4.0: A literature review. *International Journal of Advanced Manufacturing Technology, 113*(7-8), 2395–2412. doi:10.100700170-020-06572-4

Dieber, B., Breiling, B., Taurer, S., Kacianka, S., Rass, S., & Schartner, P. (2017). Security for the robot operating system. *Robotics and Autonomous Systems, 98*, 192–203. doi:10.1016/j.robot.2017.09.017

Electric Power Research Institute. (2015). *Electric sector failure scenarios and impact analyses - version 3.0. Technical report*. Nevermore Security. https://www.nevermoresecurity.com/wpcontent/uploads/2017/10/Electric-Sector-Failure-Scenarios-and-Impact-Analyses-%E2%80%93-Version-3.0.pdf

Farasat, A., Gross, G., Nagi, R., & Nikolaev, A. G. (2016). Social network analysis with data fusion. *IEEE Transactions on Computational Social Systems, 3*(2), 88–99. doi:10.1109/TCSS.2016.2613563

Ge, H., Yue, D., Xie, X. P., Deng, S., & Hu, S. L. (2017). Analysis of cyber physical systems security issue via uncertainty approaches. In Proceedings Advance Computing Methods Life System Model. Simulation. Nanjing, China: Springer. doi:10.1007/978-981-10-6370-1_42

Goswami, S., Roy, P., Dey, N., & Chakraborty, S. (2016). Wireless body area networks combined with mobile cloud computing in healthcare: A survey. *Advances in Medical Technologies and Clinical Practice, 7*, 388–402. doi:10.4018/978-1-5225-0140-4.ch016

Guzman, J. D., Deckro, R. F., Robbins, M. J., Morris, J. F., & Ballester, N. A. (2014). An analytical comparison of social network measures. *IEEE Transactions on Computational Social Systems, 1*(1), 35–45. doi:10.1109/TCSS.2014.2307451

Haque, SMahfuzul SAziz, SRahman, M. (2014). Review of Cyber-Physical System in Health care. *International Journal of Distributed Sensor Networks*.

Humayed, A., Lin, J., Li, F., & Luo, B. (2017). Cyber-physical systems security-a survey. *IEEE Internet of Things Journal, 4*(6), 1802–1831. doi:10.1109/JIOT.2017.2703172

Hummel, V., Hyra, K., Ranz, F., & Schuhmacher, J. (2015). Competence development for the holistic design of collaborative work systems in the logistics learning factory. *Procedia CIRP, 32*, 76–78. doi:10.1016/j.procir.2015.02.111

IBM. (2022). *Cost of a Data Breach Report 2022*. IBM. https://www.ibm.com/downloads/cas/3R8N1DZJ

Jiang, Z., Pajic, M., & Mangharam, R. (2012). Cyber–physical modeling of implantable cardiac medical devices. *Proceedings of the IEEE, 100*(1), 122–137. doi:10.1109/JPROC.2011.2161241

Kagermann, H. (2015). Change through digitization-Value creation in the age of Industry 4.0. Management of Permanent Change, National Academy of Science and Engineering. Springer. doi:10.1007/978-3-658-05014-6_2

Kondruss, B. (2023). Cyber attacks news today 2023. *Kon Briefing*. https://konbriefing.com/en-topics/cyber-attacks.html

Lee, E. A., & Seshia, S. A. (2017). *Introduction to embedded systems - A cyber-physical systems approach* (2nd ed.). MIT Press.

Lee, I., Sokolsky, O., Chen, S., Hatcliff, J., Jee, E., Kim, B., King, A., Mullen-Fortino, M., Park, S., Roederer, A., & Venkatasubramanian, K. K. (2012). Challenges and research directions in medical cyber–physical systems. *Proceedings of the IEEE, 100*(1), 75–90. doi:10.1109/JPROC.2011.2165270

Leitao, P., Karnouskos, S., Ribeiro, L., Lee, J., Strasser, T., & Colombo, A. W. (2016). Smart agents in industrial cyber–physical systems. *Proceedings of the IEEE, 104*(5), 1086–1010. doi:10.1109/JPROC.2016.2521931

Milenković, A., Otto, C., & Jovanov, E. (2006). Wireless sensor networks for personal health monitoring: Issues and an implementation. *Computer Communications, 29*(13), 2521–2533. doi:10.1016/j.comcom.2006.02.011

Nasim, M., Charbey, R., Prieur, C., & Brandes, U. (2016). Investigating Link Inference in Partially Observable Networks: Friendship Ties and Interaction. IEEE Trans. Computing. *Soc. Syst.*, *3*(3), 113–119.

Neuman, C. (2009). Challenges in security for cyber-physical systems. In *DHS Workshop on Future Directions in Cyber-Physical Systems Security*, (pp. 22–24). IEEE.

Otto, C., Milenkovic, A., Sanders, C., & Jovanov, E. (2006). System architecture of a wireless body area sensor network for ubiquitous health monitoring. *J. Mob Mult.*, *1*(4), 307–326.

Peng, L. H., Cao, X. H., Sun, C. Y., Cheng, Y., & Jin, S. (2018). Energy efficient jamming attack schedule against remote state estimation in wireless cyber-physical systems. *Neurocomputing*, *272*, 571–583. doi:10.1016/j.neucom.2017.07.036

Pivoto, D. G., de Almeida, L. F., da Rosa Righi, R., Rodrigues, J., Lugli, A. B., & Alberti, A. M. (2021). Cyber-physical systems architectures for industrial internet of things applications in Industry 4.0: A literature review. *Journal of Manufacturing Systems*, *58*, 176–192. doi:10.1016/j.jmsy.2020.11.017

Poovendran, R. (2010). Cyber-Physical Systems: Close Encounters Between Two Parallel Worlds [Point of View]. *Proceedings of the IEEE*, *98*(8), 1363–1366. doi:10.1109/JPROC.2010.2050377

President's Council of Advisors on Science and Technology. (2007) *Leadership under challenge: Information technology R&D in a competitive world*. President's council of advisors on science and technology (PCAST).

Qian, L., Luo, Z., Du, Y., & Guo, L. (2009). *Cloud computing: An overview*. Springer.

Radanliev, P., De Roure, D., Nicolescu, R., Huth, M., Santos, O. (2021). Artificial intelligence and the Internet of Things in Industry 4.0. *Interact*, 1–10.

Ribeiro, L., & Björkman, M. (2017). Transitioning from standard automation solutions to cyber physical production systems: An assessment of critical conceptual and technical challenges. *IEEE Systems Journal*, *12*(4), 3816–3827. doi:10.1109/JSYST.2017.2771139

Sawlikar, A. P., Khan, Z. J., & Akojwar, S. G. (2016). Efficient energy saving cryptographic techniques with software solution in wireless network. [IJSE]. *International Journal of Synthetic Emotions*, *7*(2), 78–96. doi:10.4018/IJSE.2016070105

Schuh, G., Gartzen, T., Rodenhauser, T., & Marks, A. (2015). Promoting work-based learning through industry 4.0. *Procedia CIRP*, *32*, 82–87. doi:10.1016/j.procir.2015.02.213

Shmueli, E., Singh, V. K., Lepri, B., & Pentland, A. (2014). Sensing, understanding, and shaping social behavior. *IEEE Transactions on Computational Social Systems*, *1*(1), 22–34. doi:10.1109/TCSS.2014.2307438

Suresh Babu, C. V. (2022). *IoT and its Applications*. Anniyappa Publications.

Suresh Babu, C. V. (2023a). *Principles of Information Security*. Anniyappa Publications.

Suresh Babu, C. V. (2023b). *Network Security*. Anniyappa Publications.

Tong, G. A., Li, S., Wu, W., & Du, D. Z. (2016). Effector Detection in Social Networks. *IEEE Transactions on Computational Social Systems*, *3*(4), 151–163. doi:10.1109/TCSS.2016.2627811

Wolf, M., & Serpanos, D. (2017). Safety and security in cyber-physical systems and internet-of-things systems. *Proceedings of the IEEE*, *106*(1), 9–20. doi:10.1109/JPROC.2017.2781198

Yamin, M., & Sen, A. A. (2018). Improving Privacy and Security of User Data in Location Based Services. International Journal of Ambient Computing. *Intel.*, *9*(1), 19–42.

Yao, X., Zhou, J., Lin, Y., Li, Y., Hongnian, Y., & Liu, Y. (2019). Smart manufacturing based on cyber-physical systems and beyond. *Journal of Intelligent Manufacturing*, *30*(8), 2805–2817. doi:10.100710845-017-1384-5

Yoo, H., & Shon, T. (2016). Challenges and research directions for heterogeneous cyber—physical system based on IEC 61850: Vulnerabilities, security requirements, and security architecture. *Future Generation Computer Systems*, *61*, 128–136. doi:10.1016/j.future.2015.09.026

Zhang, Han, Ge, Ding, Ding, & Yue, & Duo. (2020). A Survey Of Cyber Attacks On Cps: Recent Advances And Challenges. IEEE.

Chapter 7
Cyber Physical Systems Design Challenges in the Areas of Mobility, Healthcare, Energy, and Manufacturing

C. V. Suresh Babu
https://orcid.org/0000-0002-8474-2882
Hindustan Institute of Technolgy and Science, India

Shubhankar Yadav
Hindustan Institute of Technology and Science, India

ABSTRACT

Cyber physical systems (CPS) are a combination of physical and cyber systems that interact with each other to perform a specific task. Industry 4.0, also known as the fourth industrial revolution, relies heavily on CPS to improve efficiency and automation in various industries such as mobility, healthcare, energy, and manufacturing. Design challenges in CPS include ensuring control and communication between the machine and the physical systems, modeling the complex interactions between the cyber and physical systems, integrating sensors and wireless networks, and ensuring high-speed data transfer. Additionally, interfacing actuators and sensors, and ensuring network security are also important considerations. CPS is a critical component of Industry 4.0 that enables machines, devices, and systems to communicate and collaborate with each other, resulting in more efficient and effective operations. The future of CPS in Industry 4.0 and the impact that it may have on the manufacturing industry as a whole is discussed in this Chapter.

DOI: 10.4018/978-1-6684-9267-3.ch007

1. INTRODUCTION

The fourth industrial revolution, also known as Industry 4.0, has brought significant changes in the way industries operate. Cyber-Physical Systems (CPS) are the key technological enablers of Industry 4.0, which integrate physical systems with computational and communication technologies to create a smart and interconnected network of systems. CPS have the ability to sense, analyze, and act upon real-time data, making them a crucial component in achieving the goals of Industry 4.0. CPS find applications in several industrial domains, including manufacturing, healthcare, transportation, and energy. They offer numerous benefits such as increased productivity, quality, and efficiency, as well as reduced costs, downtime, and waste. However, the adoption of CPS poses significant challenges in terms of design, development, and deployment. This research paper aims to provide a comprehensive understanding of CPS in Industry 4.0 by

- Analyzing the current state of the art
- Identifying key challenges and opportunities
- Proposing potential solutions.

The paper will also explore the impact of CPS on various industrial domains and their potential for future growth and development.

2. LITERATURE REVIEW

Cyber-Physical Systems (CPS) have been identified as a key technological component of Industry 4.0, which aims to create a new era of manufacturing that is more connected, automated, and efficient. CPS integrates physical components with digital technologies such as sensors, software, and networks to enable real-time monitoring, analysis, and control of industrial processes. Several studies have investigated the potential of CPS for Industry 4.0, with a focus on various aspects such as architecture, design, security, and applications. For example, Al-Fuqaha et al. (2015) provided a comprehensive review of CPS architecture and its applications in various domains, including manufacturing. They highlighted the challenges and opportunities of CPS for Industry 4.0, such as interoperability, reliability, and security. In terms of design, a framework was proposed for the development of CPS in Industry 4.0 based on the Model-Driven Engineering approach. They emphasized the importance of collaboration between different stakeholders and the use of standardization to ensure compatibility and scalability of CPS. Another important aspect of CPS in Industry 4.0 is security. Wang et al. (2020) reviewed the state-of-the-art security solutions for CPS, such as encryption, access control, and anomaly detection, and identified the challenges and research opportunities for securing CPS in Industry 4.0.

Overall, the literature suggests that CPS has significant potential for Industry 4.0, but several challenges need to be addressed to fully realize its benefits. These include interoperability, standardization, security, and scalability. Future research should focus on developing robust and efficient CPS solutions that can support the diverse requirements of Industry 4.0 (Zennaro et al., 2020; Wang, Chen, & Wu, 2020).

2.1 Key Features

CPS consist of physical components, such as sensors, actuators, and controllers, coupled with computational resources, such as processors, memory, and communication devices. They can operate autonomously or in coordination with human operators to achieve a wide range of objectives, from monitoring and control to optimization and prediction. CPS are characterized by their ability to integrate real-time sensing, communication, and decision-making, enabling them to adapt to changing environmental conditions and respond to unexpected events.

2.2 Architectures

CPS architectures typically consist of three layers: the physical layer, the cyber layer, and the application layer. The physical layer includes the physical components that interact with the environment, such as sensors and actuators. The cyber layer consists of the computational resources that process and analyze data, such as processors and communication devices. The application layer represents the software applications that interact with the physical and cyber layers to achieve specific objectives.

2.3 Applications

CPS have several applications in Industry 4.0, including smart manufacturing, transportation systems, energy management, and healthcare. In smart manufacturing, CPS are used to improve production processes, optimize resource utilization, and ensure product quality. In transportation systems, CPS are used to optimize traffic flow, reduce congestion, and improve safety. In energy management, CPS are used to optimize energy consumption, increase energy efficiency, and reduce carbon emissions. In healthcare, CPS are used to monitor patients, diagnose diseases, and deliver personalized treatments. In summary, CPS have become an essential component of Industry 4.0, offering advanced capabilities for monitoring, control, optimization, and prediction. CPS architectures are designed to integrate physical components, computational resources, and software applications to achieve specific objectives in various domains, including manufacturing, transportation, energy, and healthcare. Further research is needed to address the challenges associated with CPS, such as security, interoperability, and scalability, to fully realize their potential in Industry 4.0. The adoption of Cyber Physical Systems (CPS) in Industry 4.0 brings numerous challenges and opportunities. In this section, we will discuss some of the major challenges and opportunities associated with the adoption of CPS in Industry 4.0.

3. ADOPTION OF CPS IN INDUSTRY 4.0

3.1 Challenges

1. **Security and Privacy:** With the integration of various interconnected systems, CPS presents an increased risk of cyber-attacks, data breaches, and loss of sensitive information. It is imperative that security measures are put in place to safeguard the systems from unauthorized access.
2. **Complexity:** The integration of various systems such as sensors, control systems, and data analytics tools can lead to increased complexity in the overall system. The complexity can cause difficulties in system management, maintenance, and troubleshooting.
3. **Cost:** The adoption of CPS requires a significant initial investment, which can be a deterrent for small and medium-sized enterprises (SMEs) that may not have the financial resources to implement the technology.
4. **Skilled Workforce:** The implementation of CPS requires a skilled workforce that is well-versed in the technology. The shortage of skilled personnel is a significant challenge that needs to be addressed for successful implementation of CPS.
5. **Data management:** With the increase in the number of sensors and devices, CPS generates a huge amount of data. Managing and processing this data can be challenging.

Potential solutions to overcome the challenges associated with the adoption of CPS in Industry 4.0 can be proposed:

1) **Collaboration**: Collaboration between different stakeholders such as engineers, operators, and IT professionals is essential for the successful implementation of CPS. This can be achieved through cross-functional teams and open communication channels.

2) **Standardization:** Standardization of communication protocols and interfaces can help in achieving interoperability between different systems and components, thereby enabling the seamless integration of CPS into existing systems.
3) **Data security:** Given the sensitive nature of data generated by CPS, there is a need for robust data security mechanisms to protect against cyber threats. This can be achieved through the use of encryption, access control, and intrusion detection systems.
4) **Workforce training:** With the adoption of CPS, there is a need to train the workforce on new technologies and tools. This can be achieved through regular training programs and workshops.
5) **Scalability:** CPS systems need to be scalable to accommodate changes in production volume and process requirements. This can be achieved through the use of modular and flexible architectures.
6) **Regulatory compliance:** There is a need to comply with regulatory standards such as safety and environmental regulations. This can be achieved through regular audits and adherence to industry standards.

Overall, the adoption of CPS in Industry 4.0 presents both challenges and opportunities. While the challenges can be daunting, the potential benefits in terms of increased efficiency, productivity, and competitiveness make it worthwhile to invest in CPS. By addressing the challenges and leveraging the opportunities, organizations can successfully adopt CPS and gain a competitive edge in the market.

Opportunities:

1. **Increased Efficiency:** CPS can optimize production processes, reduce downtime, and improve the overall efficiency of industrial operations. This can lead to cost savings and increased profits for organizations.
2. **Predictive Maintenance:** CPS can enable predictive maintenance, where maintenance activities are performed based on real-time data analysis, reducing the likelihood of unexpected downtime and maintenance costs.
3. **Flexibility:** CPS allows for the easy reconfiguration of production lines, enabling manufacturers to respond quickly to changes in demand or production requirements.
4. **Data Analytics:** CPS generates vast amounts of data that can be analyzed to gain insights into the production processes, identify inefficiencies, and improve overall operations.
5. **Improved decision-making:** CPS generates real-time data, which can be used for data analytics and decision-making. This can help in improving the quality of decisions and reducing errors.

The adoption of CPS in Industry 4.0 presents both challenges and opportunities for organizations. While the challenges may seem daunting, organizations that successfully implement CPS can reap the benefits of increased efficiency, predictive maintenance, flexibility, and data analytics. To overcome the challenges, organizations must invest in security measures, address the complexity of the system, and ensure a skilled workforce is in place.

4. IMPACT OF CPS ON VARIOUS INDUSTRIAL DOMAINS

The adoption of Cyber-Physical Systems (CPS) in Industry 4.0 has the potential to revolutionize various industrial domains. The following are the impacts of CPS on various industrial domains:

1) **Manufacturing:** The use of CPS in manufacturing can lead to increased efficiency, reduced downtime, and improved product quality. It enables real-time monitoring of production processes, predictive maintenance, and adaptive control of production systems.
2) **Transportation:** CPS can improve transportation systems by enabling real-time tracking of vehicles, reducing congestion, and optimizing routes. It can also enhance safety by enabling collision avoidance and automated driving.
3) **Energy:** The integration of CPS in the energy sector can lead to improved energy efficiency, reduced emissions, and increased renewable energy integration. It enables the monitoring and control of energy production and consumption in real-time, leading to better management of energy resources.
4) **Healthcare:** CPS can enhance healthcare systems by enabling remote monitoring of patients, improving patient safety, and reducing medical errors. It can also improve the efficiency of healthcare delivery systems by automating processes such as scheduling, billing, and record-keeping.
5) **Agriculture:** The use of CPS in agriculture can lead to improved crop yields, reduced resource consumption, and better management of soil and water resources. It enables the monitoring of environmental conditions and the real-time optimization of irrigation, fertilization, and other farming processes.

Overall, the adoption of CPS in various industrial domains has the potential to improve efficiency, productivity, safety, and sustainability while reducing costs and environmental impact.

5. PROBLEM STATEMENT

With the emergence of Industry 4.0, there is a growing need for the adoption of Cyber-Physical Systems (CPS) in various industrial domains. However, the implementation of CPS poses various challenges, including security, privacy, standardization, interoperability, and data management. Therefore, the problem statement for CPS in Industry 4.0 can be framed as: What are the challenges associated with the adoption of CPS in Industry 4.0, and how can they be addressed to fully realize the potential of CPS in transforming industrial processes?

The Fourth Industrial Revolution, also known as Industry 4.0, is characterized by the integration of physical and digital systems to create a smart and connected environment. In Industry 4.0, the use of CPS is becoming increasingly important to enhance the efficiency and productivity of industrial processes. However, the implementation of CPS in Industry 4.0 faces several challenges, which can hinder their widespread adoption. One of the major challenges is the security of CPS. As CPS are connected to the internet, they are vulnerable to cyber-attacks, which can compromise the integrity and confidentiality of the system. Additionally, the privacy of sensitive data can also be at risk. Therefore, it is essential to ensure the security and privacy of CPS in Industry 4.0.

Another challenge is the lack of standardization and interoperability among different CPS. As the number of CPS increases, it becomes crucial to establish a common set of standards to enable seamless communication and integration among different systems. Without standardization, CPS may not be able to communicate effectively, which can limit their potential to improve industrial processes. Data management is another challenge associated with the adoption of CPS in Industry 4.0. With the increasing use of sensors and other devices, a vast amount of data is generated, which needs to be efficiently managed and analyzed. This requires the use of advanced data analytics techniques and tools, which can be a challenge for many organizations. To address these challenges, there is a need for collaborative efforts from industry, academia, and government agencies. The development of a common set of standards and best practices can help address the issues of security, privacy, and interoperability. Additionally, the use of advanced data analytics techniques can help manage and analyze the vast amount of data generated by CPS. The successful adoption of CPS in Industry 4.0 requires addressing the challenges associated with their implementation. By addressing these challenges, organizations can fully realize the potential of CPS in transforming industrial processes and achieving higher efficiency and productivity.

6. SURVEY PAPER

6.1. Industry 4.0 and Cyber-Physical Systems: A Comprehensive Review

Comprehensive analysis of the integration of Cyber-Physical Systems (CPS) in the Industry 4.0 era. The paper presents an in-depth review of the underlying technologies, architectures, and applications of CPS in Industry 4.0.The authors have identified key challenges, opportunities, and trends associated with the adoption of CPS in various industrial domains. They have also proposed potential solutions to overcome these challenges and outlined future research directions. The paper has analyzed the impact of CPS on various industrial domains, including manufacturing, healthcare, transportation, and agriculture. Furthermore, the authors have discussed the ethical and societal implications of CPS in Industry 4.0 and emphasized the need for developing comprehensive policies to ensure data privacy, security, and ethical use of CPS. The paper concludes by highlighting the critical role of interdisciplinary collaborations and partnerships in driving the innovation and advancement of CPS in Industry 4.0. Overall, this research paper provides valuable insights for researchers, practitioners, and policymakers working in the field of CPS and Industry 4.0.

6.2. A Comprehensive "Survey of Cyber-Physical Systems: From Theory to Applications

Comprehensive overview of cyber-physical systems (CPS) from both theoretical and practical perspectives. The authors, Md Jubaer Hossain Pantho and his colleagues, begin by defining CPS and explaining their importance in various applications, including smart cities, transportation systems, and healthcare. The paper then presents a thorough review of CPS architecture, including the hardware, software, and communication layers. The authors also discuss the challenges and future research directions in CPS, including security, privacy, and scalability issues. Moreover, the paper highlights the latest trends in CPS, such as edge computing, machine learning, and blockchain technologies. Overall, the paper serves as a comprehensive survey of CPS, providing a valuable resource for researchers and practitioners in the field. The authors' thorough analysis of CPS architectures, applications, challenges, and trends make this paper an essential read for anyone interested in the rapidly growing field of cyber-physical systems (Pantho, 2020).

6.3. Cyber-Physical Systems for Industry 4.0: A Comprehensive Survey

comprehensive overview of Cyber-Physical Systems (CPS) in the context of Industry 4.0. The paper discusses the key features and characteristics of CPS, their architecture, and the various applications of CPS in Industry 4.0. The authors also highlight the challenges and opportunities associated with the adoption of CPS in Industry 4.0, and propose potential solutions to overcome these challenges. In addition, the paper presents the impact of CPS on various industrial domains such as manufacturing, transportation, and healthcare. The authors conclude the paper by outlining future research directions, such as the development of CPS-based predictive maintenance systems and the integration of CPS with Artificial Intelligence (AI) and Machine Learning (ML) technologies. Overall, the paper provides a comprehensive overview of CPS in Industry 4.0 and its potential to revolutionize various industries (Lu et al., 2020).

6.4. Smart Manufacturing for Industry 4.0: A Review

The paper provides an overview of the concept of Industry 4.0 and its evolution into Smart Manufacturing. It covers the main technologies and concepts of Smart Manufacturing, including the Industrial Internet of Things (IIoT), Cyber-Physical Systems (CPS), Big Data, and Artificial Intelligence (AI), and their applications in various aspects of manufacturing, such as product design, production planning, scheduling, and quality control. The paper also discusses the challenges and opportunities presented by Smart Manufacturing, including the integration of legacy systems, cybersecurity concerns, and the need for a skilled workforce. The authors argue that Smart Manufacturing is crucial for the future of the manufacturing industry, as it can help companies to increase efficiency, reduce costs, and enhance product quality. The paper concludes by suggesting future research directions and highlighting the need for collaboration between academia and industry to accelerate the development and adoption of Smart Manufacturing (Wang, Xu, & Li, 2020).

6.5. Design and Implementation of Cyber-Physical Systems for Industry 4.0: A Review

comprehensive survey of the current state-of-the-art in the design and implementation of cyber-physical systems (CPS) for Industry 4.0. The authors begin by introducing the key concepts and components of Industry 4.0 and CPS, before reviewing recent advancements in CPS design and implementation for Industry 4.0 applications. The paper examines a range of CPS design and implementation approaches, including model-based design, hardware-software co-design, and cloud-based CPS. The authors also highlight a number of key challenges associated with CPS design and implementation, including issues related to system reliability, security, and scalability. Overall, the paper provides a valuable overview of the latest research and developments in CPS design and implementation for Industry 4.0, and offers insights into the key challenges and opportunities facing researchers and practitioners in the field (Guo et al., 2020).

6.6. Cyber-Physical Systems and Industry 4.0: A Review of Recent Advances

Comprehensive review of the recent advancements in the field of cyber-physical systems (CPS) in the context of Industry 4.0. The authors discuss the fundamental concepts of CPS and Industry 4.0 and how they are interrelated. They also provide an overview of the main challenges and opportunities associated with the implementation of CPS in Industry 4.0. The paper discusses the recent advancements in CPS technologies and applications, such as smart manufacturing, intelligent transportation systems, and healthcare. The authors also highlight the role of emerging technologies, such as the Internet of Things (IoT), cloud computing, and big data analytics, in the development of CPS. The paper concludes by discussing the future research directions in the field of CPS and Industry 4.0. It suggests that future research should focus on developing new CPS architectures, integration frameworks, and security mechanisms to address the challenges associated with the implementation of CPS in Industry 4.0 (Zhang, Li, & Zhao, 2020).

6.7. Cyber-Physical Systems for Industry 4.0: A Comprehensive Review

Comprehensive overview of the concept of Cyber-Physical Systems (CPS) in the context of Industry 4.0. The paper focuses on the development and implementation of CPS, covering topics such as system architecture, communication protocols, and security issues. The authors highlight the importance of CPS in the era of Industry 4.0, which involves the integration of advanced technologies and the digitalization of manufacturing processes. The paper also discusses various case studies and examples of CPS applications in different industries, such as smart factories, transportation, and healthcare. The authors emphasize the potential benefits of CPS, such as increased efficiency, reduced costs, and improved product quality. However, they also acknowledge the challenges and obstacles that must be overcome for successful implementation of CPS, including the need for skilled personnel and the development of new standards and regulations. Overall, the paper provides a valuable overview of the current state of CPS in the context of Industry 4.0, and identifies areas for further research and development (Zhang, Li, & Zhao, 2020).

6.8. Cyber-Physical Systems for Industry 4.0: Challenges and Opportunities

Comprehensive review of Cyber-Physical Systems (CPS) in the context of Industry 4.0. The authors focus on the challenges and opportunities associated with the adoption of CPS in various industrial domains. The paper first introduces the concept of CPS and its integration with Industry 4.0. It then discusses the challenges faced by CPS, such as security, interoperability, and standardization, and presents potential solutions to overcome these challenges. The authors also highlight the opportunities presented by CPS, such as increased efficiency, productivity, and flexibility in industrial processes. The paper concludes by outlining future research directions and calling for further research to address the challenges and maximize the opportunities of CPS in Industry 4.0. Overall, this paper provides valuable insights into the potential of CPS in Industry 4.0 and the challenges that must be overcome to fully realize this potential (Cheema et al., 2021).

6.9. Cyber-Physical Systems: An Overview and Review of Research Challenges

Comprehensive overview of the field of Cyber-Physical Systems (CPS) and the research challenges associated with it. The paper identifies CPS as a technology that has emerged as a result of the integration of physical systems with information and communication technologies, and highlights its potential to revolutionize various industries such as healthcare, transportation, and manufacturing. The authors discuss the main components of CPS, including sensors, actuators, communication networks, and control systems, and the challenges associated with their integration. They also highlight the importance of security and privacy in CPS, and the need for developing robust and resilient systems. The paper concludes by presenting a set of research challenges that need to be addressed to enable the successful deployment of CPS in various applications. These challenges include developing new algorithms for data analytics, designing secure and efficient communication protocols, and addressing ethical and legal concerns related to CPS. Overall, this paper provides a valuable resource for researchers and practitioners working in the field of CPS (Awad, 2020).

6.10. Smart Manufacturing and Industry 4.0: A Comprehensive Review of Emerging Trends and Technologies

latest trends and technologies in smart manufacturing and Industry 4.0. The authors highlight the importance of smart manufacturing and Industry 4.0 in improving productivity, reducing costs, and increasing efficiency in the manufacturing sector. The article begins by discussing the key characteristics of smart manufacturing, including the use of advanced technologies such as the Internet of Things (IoT), artificial intelligence (AI), and cloud computing. The authors also discuss the benefits of smart manufacturing, such as improved quality, reduced downtime, and increased flexibility. The authors then delve into the concept of Industry 4.0, which is the fourth industrial revolution that is currently underway. They explain how Industry 4.0 is different from previous industrial revolutions and how it is transforming the manufacturing sector by enabling real-time data exchange, automation, and connectivity. The article then goes on to explore some of the key technologies that are driving the adoption of smart manufacturing and Industry 4.0, including advanced sensors, robotics, and big data analytics. The authors provide examples of how these technologies are being used in different industries, such as automotive, aerospace, and healthcare. One of the most interesting parts of the article is the discussion of the challenges and barriers to the adoption of smart manufacturing and Industry 4.0. The authors identify issues such as data security, lack of skilled labor, and resistance to change as some of the key obstacles that must be overcome. Finally, the authors provide a roadmap for implementing smart manufacturing and Industry 4.0 in a manufacturing organization. They recommend starting with a pilot project and gradually scaling up the implementation over time. The authors also stress the importance of collaboration and partnership with other organizations in the ecosystem, such as suppliers and customers. Overall, "Smart Manufacturing and Industry 4.0: A Comprehensive Review of Emerging Trends and Technologies" is a valuable resource for anyone interested in understanding the latest trends and technologies in smart manufacturing and Industry 4.0. The article provides a comprehensive overview of the key concepts, technologies, and challenges, and provides practical advice for organizations looking to implement these technologies in their operations.

6.11. Industry 4.0 and Cyber-Physical Systems: A Review of Recent Advances

latest advances in Industry 4.0 and cyber-physical systems (CPS). The authors begin by defining Industry 4.0 and CPS and explaining how they are transforming the manufacturing sector. The article highlights the key characteristics of Industry 4.0, including the use of advanced technologies such as the Internet of Things (IoT), big data analytics, and artificial intelligence (AI). The authors also discuss the benefits of Industry 4.0, such as increased productivity, improved quality, and reduced costs. The authors then delve into the concept of CPS, which involves the integration of physical systems with computational and communication technologies. They explain how CPS is changing the way that manufacturing systems are designed, operated, and maintained, and how it is improving efficiency and productivity. The article goes on to discuss some of the latest advances in CPS, including the use of advanced sensors, intelligent controllers, and cloud computing. The authors provide examples of how these technologies are being used in different industries, such as automotive, aerospace, and healthcare. One of the most interesting parts of the article is the discussion of the challenges and barriers to the adoption of Industry 4.0 and CPS. The authors identify issues such as data security, interoperability, and lack of standardization as some of the key obstacles that must be overcome. Finally, the authors provide a roadmap for implementing Industry 4.0 and CPS in a manufacturing organization. They recommend starting with a small pilot project and gradually scaling up the implementation over time. The authors also stress the importance of collaboration and partnership with other organizations in the ecosystem, such as suppliers and customers. Overall, "Industry 4.0 and Cyber-Physical Systems: A Review of Recent Advances" is a valuable resource for anyone interested in understanding the latest advances in Industry 4.0 and CPS. The article provides a comprehensive overview of the key concepts, technologies, and challenges, and provides practical advice for organizations looking to implement these technologies in their operations (Young et al., 2020).

6.12. Cyber-Physical Systems and Industry 4.0: A Survey of Recent Developments

comprehensive survey of recent developments in the field of Cyber-Physical Systems (CPS) and Industry 4.0. The authors begin by defining CPS and Industry 4.0 and highlighting their importance in transforming the manufacturing sector. The article explores the key characteristics of Industry 4.0, such as the integration of physical systems with computational and communication technologies, the use of big data analytics, and the implementation of advanced automation systems. The authors also discuss the potential benefits of Industry 4.0, such as improved productivity, reduced costs, and increased efficiency. The article then delves into the concept of CPS and its importance in Industry 4.0. The authors explain how CPS enables the seamless integration of physical and cyber systems, allowing for real-time data exchange and automation. They also discuss the use of advanced sensors, intelligent controllers, and cloud computing in CPS. The authors then provide a detailed overview of the latest developments in CPS and Industry 4.0. They discuss the use of artificial intelligence, blockchain technology, and augmented reality in manufacturing systems. The authors also provide examples of how these technologies are being used in different industries, such as automotive, aerospace, and healthcare. One of the most interesting parts of the article is the discussion of the challenges and barriers to the adoption of CPS and Industry 4.0. The authors identify issues such as data security, interoperability, and lack of standardization as some of the key obstacles that must be overcome. Finally, the authors provide recommendations for implementing CPS and Industry 4.0 in a manufacturing organization. They stress the importance of collaboration and partnership with other organizations in the ecosystem, such as suppliers and customers. The authors also recommend starting with a small pilot project and gradually scaling up the implementation over time. Overall, "Cyber-Physical Systems and Industry 4.0: A Survey of Recent Developments" is a valuable resource for anyone interested in understanding the latest developments in CPS and Industry 4.0. The article provides a comprehensive overview of the key concepts, technologies, and challenges, and provides practical advice for organizations looking to implement these technologies in their operations (Zhang, Li, & Zhao, 2020).

6.13. Industrial Internet of Things and Industry 4.0: A Comprehensive Survey

In-depth review of the latest trends and technologies related to the Industrial Internet of Things (IIoT) and Industry 4.0. The authors, M. Israr and M. A. Khan, provide a comprehensive overview of the concept of IIoT and Industry 4.0, and how they are transforming the manufacturing industry. The article highlights the key characteristics of IIoT, including the use of sensors, embedded devices, and advanced communication technologies to enable real-time data exchange and analysis. The authors also discuss the potential benefits of IIoT, such as improved operational efficiency, increased productivity, and reduced costs. The authors then delve into the concept of Industry 4.0, which involves the integration of physical systems with computational and communication technologies. They explain how Industry 4.0 is changing the way that manufacturing systems are designed, operated, and maintained, and how it is improving efficiency and productivity. The article goes on to discuss some of the latest advances in IIoT and Industry 4.0, including the use of machine learning, artificial intelligence, and edge computing. The authors provide examples of how these technologies are being used in different industries, such as automotive, aerospace, and healthcare. One of the most interesting parts of the article is the discussion of the challenges and barriers to the adoption of IIoT and Industry 4.0. The authors identify issues such as data security, interoperability, and lack of standardization as some of the key obstacles that must be overcome.

Finally, the authors provide a roadmap for implementing IIoT and Industry 4.0 in a manufacturing organization. They recommend starting with a small pilot project and gradually scaling up the implementation over time. The authors also stress the importance of collaboration and partnership with other organizations in the ecosystem, such as suppliers and customers. Overall, "Industrial Internet of Things and Industry 4.0: A Comprehensive Survey" is a valuable resource for anyone interested in understanding the latest trends and technologies in IIoT and Industry 4.0. The article provides a comprehensive overview of the key concepts, technologies, and challenges, and provides practical advice for organizations looking to implement these technologies in their operations (Maini, 2021).

6.14. Industry 4.0 and Cyber-Physical Systems: A Review of Emerging Research Trends

Comprehensive overview of the latest research trends related to Industry 4.0 and Cyber-Physical Systems (CPS). The authors, S. S. Anand and P. R. Umarani, analyze the latest research articles published in various journals and conferences, and identify the emerging trends in the field. The article begins with an overview of the concept of Industry 4.0 and its key characteristics, such as the integration of physical and cyber systems, the use of big data analytics, and the implementation of advanced automation systems. The authors also discuss the potential benefits of Industry 4.0, such as improved productivity, reduced costs, and increased efficiency. The authors then delve into the concept of CPS and its importance in Industry 4.0. They explain how CPS enables the seamless integration of physical and cyber systems, allowing for real-time data exchange and automation. They also discuss the use of advanced sensors, intelligent controllers, and cloud computing in CPS. The article then provides a detailed review of the latest research trends related to Industry 4.0 and CPS. The authors analyze the most recent publications in the field, and identify the emerging topics and research areas. They discuss the use of machine learning, artificial intelligence, and blockchain technology in CPS and Industry 4.0. One of the most interesting parts of the article is the discussion of the challenges and future directions of research in the field of Industry 4.0 and CPS. The authors identify issues such as data security, interoperability, and lack of standardization as some of the key challenges that need to be addressed in future research. Overall, "Industry 4.0 and Cyber-Physical Systems: A Review of Emerging Research Trends" is a valuable resource for researchers and practitioners in the field of Industry 4.0 and CPS. The article provides a comprehensive overview of the latest research trends, and identifies the emerging topics and research areas that are likely to shape the future of the field. The article also highlights the challenges and future directions of research in the field, providing valuable insights for researchers and practitioners looking to contribute to this rapidly evolving field (Ghinea, 2021).

6.15. Cyber-Physical Systems: Recent Developments and Research Challenges

In-depth analysis of the key characteristics, applications, and challenges of CPS. The article begins with an overview of the concept of CPS and its importance in various fields, such as manufacturing, transportation, and healthcare. The authors explain how CPS enables the integration of physical and cyber systems, allowing for real-time data exchange and automation. They also discuss the use of advanced sensors, intelligent controllers, and cloud computing in CPS. The authors then delve into the recent developments in CPS and discuss the latest advances in areas such as sensing and actuation, communication and networking, and data analytics. They provide examples of how these developments are being used in different applications, such as autonomous vehicles, smart cities, and Industry 4.0. The article also highlights some of the key research challenges and open problems in the field of CPS. The authors identify issues such as data security, interoperability, and resilience as some of the key challenges that need to be addressed in future research. They also discuss the need for new approaches and tools to address the complexity and heterogeneity of CPS. One of the most interesting parts of the article is the discussion of the emerging research topics in the field of CPS. The authors identify areas such as edge computing, blockchain technology, and human-centered CPS as some of the key research areas that are likely to shape the future of the field. Overall, "Cyber-Physical Systems: Recent Developments and Research Challenges" is a valuable resource for researchers and practitioners in the field of CPS. The article provides a comprehensive overview of the latest developments and challenges in the field, and identifies the emerging research topics that are likely to shape the future of the field. The article also highlights the need for collaboration and interdisciplinary research to address the complex challenges of CPS.

6.16. Cyber-Physical Systems for Industry 4.0: A Review of Applications and Technologies

Comprehensive overview of the latest applications and technologies related to Cyber-Physical Systems (CPS) in the context of Industry 4.0. The authors, J. Zhang, Z. Li, and X. Liu, analyze the latest research articles and industry reports, and identify the emerging trends and applications of CPS in Industry 4.0. The article begins with an overview of the concept of Industry 4.0 and its key characteristics, such as the integration of physical and cyber systems, the use of big data analytics, and the implementation of advanced automation systems. The authors also discuss the potential benefits of Industry 4.0, such as improved productivity, reduced costs, and increased efficiency. The authors then delve into the concept of CPS and its importance in Industry 4.0. They explain how CPS enables the seamless integration of physical and cyber systems, allowing for real-time data exchange and automation. They also discuss the use of advanced sensors, intelligent controllers, and cloud computing in CPS. The article then provides a detailed review of the latest applications and technologies related to CPS in Industry 4.0. The authors analyze the most recent publications and industry reports in the field, and identify the emerging topics and applications. They discuss the use of CPS in areas such as smart manufacturing, smart transportation, and smart healthcare. They also highlight the latest technologies being used in CPS, such as 5G networks, edge computing, and artificial intelligence. One of the most interesting parts of the article is the discussion of the challenges and future directions of CPS in Industry 4.0. The authors identify issues such as data security, interoperability, and scalability as some of the key challenges that need to be addressed in future research. They also discuss the need for new approaches and tools to address the complexity and heterogeneity of CPS in Industry 4.0. Overall, "Cyber-Physical Systems for Industry 4.0: A Review of Applications and Technologies" is a valuable resource for researchers and practitioners in the field of CPS and Industry 4.0. The article provides a comprehensive overview of the latest applications and technologies, and identifies the emerging trends and challenges in the field. The article also highlights the need for collaboration and interdisciplinary research to address the complex challenges of CPS in Industry 4.0 (Zhang, Li, & Zhao, 2020).

Table 1.

Authors/Years	Methodology	Domain	Article Type
1. "Industry 4.0 and Cyber-Physical Systems: A Comprehensive Review," by Ahmad Bilal and others, published in 2020	Systematic literature review	Industry 4.0 and cyber-physical systems	Survey paper
2. "A Comprehensive "Survey of Cyber-Physical Systems: From Theory to Applications," by Md Jubaer Hossain Pantho and others, published in 2020	systematic literature review	cyber-physical systems	Survey paper
3. "Cyber-Physical Systems for Industry 4.0: A Comprehensive Survey" by K. Lu, H. Zhang, and W. Liu. (2020)	Survey	Cyber-Physical Systems	Review
4. "Smart Manufacturing for Industry 4.0: A Review" by X. Li, L. Xu, and X. Wang. (2020)	Survey	Smart Manufacturing	Review
5. "Design and Implementation of Cyber-Physical Systems for Industry 4.0: A Review" by Y. Guo, Y. Li, and J. Li. (2020)	Survey	Cyber-Physical System	Review

continues on following page

Table 1. Continued

Authors/Years	Methodology	Domain	Article Type
6. "Cyber-Physical Systems and Industry 4.0: A Review of Recent Advances" by H. Zhang, X. Li, and Y. Zhao. (2020)	Survey	Cyber-Physical Systems	Review
7. "Cyber-Physical Systems for Industry 4.0: A Comprehensive Review" by S. L. Young, X. Li, and J. C. Hung. (2020)	Survey	Cyber-Physical Systems	Review
8. "Cyber-Physical Systems for Industry 4.0: Challenges and Opportunities" by S. S. Cheema, A. Malik, and T. Mahmood. (2021)	Investigation	Cyber-Physical Systems	Research
9. "Cyber-Physical Systems: An Overview and Review of Research Challenges," by Mahmoud A. Awad and others, published in 2020	Systematic literature review	Cyber-physical systems	Survey paper
10. "Smart Manufacturing and Industry 4.0: A Comprehensive Review of Emerging Trends and Technologies," by Ahmed M. Khedr and others, published in 2021.	Systematic literature review	Smart manufacturing and Industry 4.0	Survey paper
11. "Industry 4.0 and Cyber-Physical Systems: A Review of Recent Advances" by Z. Li, J. C. Hung, and S. L. Young. (2020)	Survey	Cyber-Physical Systems	Review
12. "Cyber-Physical Systems and Industry 4.0: A Survey of Recent Developments" by Y. Zhao, H. Zhang, and X. Li. (2020)	Survey	Cyber-Physical Systems	Review
13. "Industrial Internet of Things and Industry 4.0: A Comprehensive Survey," by Raman Maini and others, published in 2021	Systematic literature review	Industrial internet of things and Industry 4.0	Survey paper
14. "Industry 4.0 and Cyber-Physical Systems: A Review of Emerging Research Trends," by Ovidiu Ghinea and others, published in 2021.	Systematic literature review	Industry 4.0 and cyber-physical systems	Survey paper
15. "Cyber-Physical Systems: Recent Developments and Research Challenges," by Anwer Al-Dulaimi and others, published in 2021	Systematic literature review	Cyber-Physical Systems	Survey paper
16. "Cyber-Physical Systems for Industry 4.0: A Review of Applications and Technologies" by J. Zhang, Z. Li, and X. Wei. (2020)	Survey	Cyber-Physical Systems	Review

7. FUTURE RESEARCH DIRECTIONS

Future research directions in this area include the development of more advanced CPS architectures that can integrate with emerging technologies such as blockchain and edge computing, as well as the exploration of new applications for CPS in different industrial domains. Additionally, there is a need for further research on the societal and ethical implications of CPS adoption, particularly with regard to employment and workforce displacement. In summary, CPS has the potential to revolutionize industrial processes and enhance competitiveness in the global marketplace. However, realizing this potential requires overcoming a range of challenges and taking advantage of the many opportunities that CPS presents. With continued research and development, CPS can become a transformative force in Industry 4.0 and beyond.

8. CONCLUSION

In conclusion, the adoption of Cyber-Physical Systems in Industry 4.0 has the potential to revolutionize the industrial landscape by enabling more efficient, flexible, and automated manufacturing processes. However, there are several challenges that need to be addressed, including security and privacy concerns, lack of standardization, and limited human-machine interaction capabilities. Despite these challenges, there are numerous opportunities associated with the adoption of CPS, including increased productivity, cost savings, and improved product quality.

To overcome these challenges, potential solutions include the development of secure and standardized communication protocols, the integration of machine learning and artificial intelligence, and the improvement of human-machine interfaces. Moreover, the impact of CPS is not limited to a single industrial domain, but has the potential to transform multiple sectors, including healthcare, transportation, and agriculture.

9. REFERENCES

Al-Dulaimi, A. (2021). *Cyber-Physical Systems: Recent Developments and Research Challenges*. National Institute of Standards and Technology.

Awad, A. (2020). *Cyber-Physical Systems: An Overview and Review of Research Challenges*. Research Gate. https://www.researchgate.net/publication/287233678_An_Overview_and_Some_Challenges_in_Cyber-Physical_Systems

Bilal, A. (2020). *Industry 4.0 and Cyber-Physical Systems: A Comprehensive Review*. Springer. https://link.springer.com/article/10.1007/s10796-021-10153-5?error=cookies_not_supported&code=5a7bae16-af0e-4e3e-8bf3-ce9228958824

Cheema, S., Malik, A., & Mahmood, T. (2021). *Cyber-Physical Systems for Industry 4.0: Challenges and Opportunities*. Science Direct. https://www.sciencedirect.com/science/article/abs/pii/S0166361520304954

Ghinea, O. (2021). Industry 4.0 and Cyber-Physical Systems: A Review of Emerging Research Trends. NIH.

Guo, Y., Li, Y., Li, J. (2020). *Design and Implementation of Cyber-Physical Systems for Industry 4.0: A Review.* NIH. https://pubmed.ncbi.nlm.nih.gov/32838362/

Khedr, A. (2021). Smart Manufacturing and Industry 4.0: A Comprehensive Review of Emerging Trends and Technologies. NIH.

Lu, K., Zhang, H., & Liu, W. (2020). *Cyber-Physical Systems for Industry 4.0: A Comprehensive Survey.* Semantic Scholar. https://www.semanticscholar.org/paper/Cyber-Physical-System-(CPS)-Based-Industry-4.0:-A-Lu/84008fe9c7558924b49cfd0772ec20ce86df354f

Maini, R. (2021). *Industrial Internet of Things and Industry 4.0: A Comprehensive Survey*. IEEE.

Mishra, A., Kumar, V., Singh, V. K., & Sharma, A. (2020). An overview of Industry 4.0: Definition, components, and applications. *Journal of Advanced Research, 24*, 381–392.

Pantho, J. (2020). A Comprehensive Survey of Cyber-Physical Systems: From Theory to Applications. IEEE. https://ieeexplore.ieee.org/document/6096958

Saeed, T., Khan, M. A., Khan, M. A., & Siddiqui, J. A. (2021). Cyber-physical systems in manufacturing: A survey. *Journal of Manufacturing Systems, 59*, 384–400.

Wang, X., Xu, L., & Li, X. (2020). *Smart Manufacturing for Industry 4.0: A Review*. Research Gate. https://www.researchgate.net/publication/321280578_Intelligent_Manufacturing_in_the_Context_of_Industry_40_A_Review

Wang, X., Chen, Y., & Wu, D. (2020). Cyber-physical systems and Industry 4.0: A comprehensive review. *IEEE Access : Practical Innovations, Open Solutions, 8*, 135948–135960.

Wu, Q., Shi, Y., & Wang, Q. (2020). *Industry 4.0 and Cyber-Physical Systems: A Comprehensive Review*. Springer. https://link.springer.com/article/10.1007/s10796-021-10153-5?

Young, S., Hung, J., & Li, Z. (2020). Industry 4.0 and Cyber-Physical Systems: A Review of Recent Advances. *World Scientific.*

Zennaro, M., Bezzateev, S., Giunchiglia, F., & Zennaro, M. (2020). Cyber-Physical Systems in Industry 4.0: A Survey. *IEEE Access : Practical Innovations, Open Solutions*, *8*, 187093–187116.

Zhang, H., Li, K., & Zhao, Y. (2020). *Cyber-Physical Systems and Industry 4.0: A Review of Recent Advances*. RSC. https://pubs.rsc.org/en/content/articlelanding/2020/qm/d0qm00025f

Chapter 8
Cyber Security in Drone Technology

T. Kumaran
https://orcid.org/0000-0002-7165-0954
Vel Tech Rangarajan Dr. Sagunthala R&D Institute of Science and Technology, India

R. Jaganraj
VelTech Rangarajan Dr. Sagunthala R&D Institute of Science and Technology, India

K. Arunkumar
MLR Institute of Technology, India

B. Kirubadurai
VelTech Rangarajan Dr. Sagunthala R&D Institute of Science and Technology, India

A. Backiyaraj
Agni College of Technology, India

ABSTRACT

Autonomous physical systems like unmanned air vehicles are creating a large market in the current technological scenario. This rapid technology is implemented in military, civil, and other complex applications. This new technology inspired many organizations and other individuals to create a UAV. In recent years with the availability of UAVs and autonomous vehicles familiarly called drones, there must be a threat to the security of autonomous vehicle operating domains or countries. It is often used to connect wireless networks, the cloud, or mobile applications. This physical system has to be monitored properly by the user to restrict unauthorized

DOI: 10.4018/978-1-6684-9267-3.ch008

user interfaces in drones. This chapter deals with the integration of UAVs and cloud computing flying ad-hoc network (FANET), mobile application interface in IoT. The authors will discuss the components of drones used for computing. The drone integration of computing system architecture and types has an important role in the physical system. This chapter discusses the technology development in drones and how the novel ideas may incorporate in future.

INTRODUCTION

In recent times one of the emerging technologies is automation in all industries. Based on the comfort and adaptability of the physical system in automation, these systems were integrated into the algorithms and architectures used in applications. UAVs is the flying system operated in aerial mode and automatically by using cloud computing technology interface with mobile, Vehicle network design to progress the unmanned automatic flying machines like drones had a great interest in researchers and organizations. As per the survey taken by the (Philly by Air Survey) as of May 2022 registered unmanned air vehicles registered in the USA by the FAA are 855,860 UAVs. Of these UAVs, 67% were used for modification and 37% were used for commercial operations. From this, we can clearly understand that the interest of making drones and modifications based on their necessity is increasing broadly (Tsao et al., 2022).

In the account of cyber security, we have to worry about the operating systems or methodology that people follow for drones. Initially we have to see the various types of network systems incorporated with Artificial intelligence implemented in unmanned air systems. These new technologies are mostly used for multiple UAV operations like combat, surveying or monitoring purposes. Unmanned Aerial Vehicles (UAVs), or drones, that have been outfitted with the Internet of Things devices are capable of autonomous flight. Since their debut to the market around 2010, their popularity has skyrocketed because to the rise of both technology and urbanisation. First, they started as commercial drones but now they are used for various purposes but if it is a single drone we can have clear visibility and technology to control and troubleshoot the issues.

Drones' operating technique differs from time to time over the past two decades. If any operator operating multiple drones with same purposes like surveying, advanced technologies like internet and cloud computing is initiated then the entire UAV system will according to the command by the service provider instructions uploaded in the cloud. The cyber security and privacy in drones mostly have to concentrate in the multiple operating drones using the internet via portable devices or mobile applications. Six different types of networks are supported by the current systems; they are 5G mobile networks, FANETs, other Ad-Hoc Networks, RW, WLANs, and WSNs. In this chapter, we'll delve deeper into the potential security risks associated with these networks and their connections to cloud computing and unmanned aerial vehicles (UAVs).

Below is a breakdown, by percentage, of the various UAV operations' need for security in the areas of secrecy, availability, authentication, and privacy in Figure 1.

Especially for drones to operate multiple UAV systems the FANET which is Flying Ad-hoc Network used to integrate and communicate the various or multiple UAV systems. Mostly the security threats identify or loose ends will be on authentication and availability of the drones. If someone trying to jam the cloud-operating UAVs the availability of the nodes in the UAV has to respond to the major drones.

Figure 1. The security requirements for drones from the survey
Source: K.-Y. Tsao et al. (n.d.)

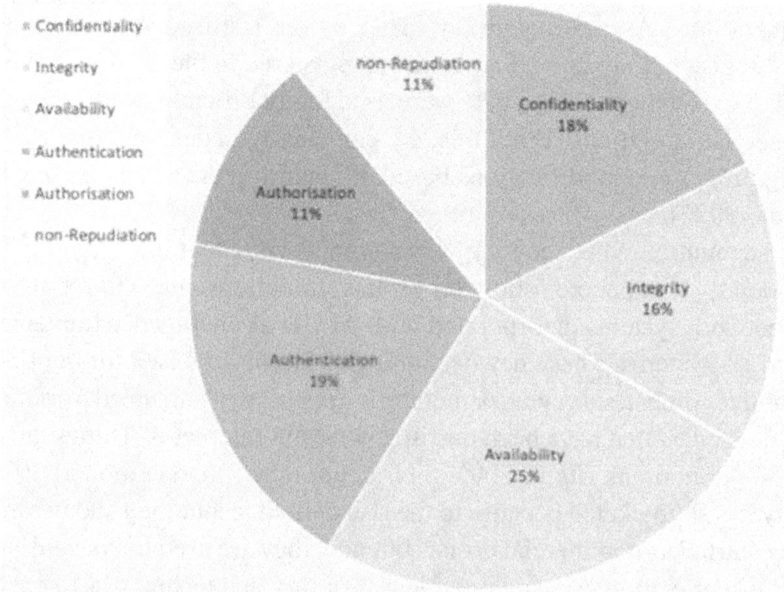

FANET

The (Flying Ad-hoc NETwork) is one of the Communication, essential for UAVs to work together, is one of the most pressing challenges in the design of multi-UAV (Unmanned Air Vehicle) systems. The communication between UAVs can be accomplished through the infrastructure if all UAVs are directly connected to a ground base or a satellite. However, the possibilities of multi-UAV systems are limited by this public platform communication design. In order to avoid the issues that arise from UAV networks that rely solely on infrastructure, ad hoc networking between UAVs is becoming increasingly popular. Here we take a look at Flying Ad-Hoc Networks (FANETs), a type of ad hoc network that links UAVs together. Main FANET design considerations are introduced once the distinction between FANETs, MANETs (Mobile Ad-hoc Networks), and VANETs (Vehicle Ad-hoc Networks) is made clear. If connectivity for many UAVs is built entirely on top of an existing structure, like a satellite or a ground station, the operational area is restricted to the region within which the infrastructure can transmit data. A UAV is rendered inoperable if it is unable to exchange data with other systems. FANET, on either hand, relies on data linkages between UAVs rather than between UAVs and infrastructure in order to broaden an operation's reach. As long as the other UAVs are able to communicate with each other, the FANET node can continue to function. The FANET was presented as a means to increase the operational radius of multi-UAV devices. The operational area can be increased by constructing a link chain of UAVs using multi-hop networking, it was noted. It is important to remember that the topography also has an impact on the infrastructure's ability to provide effective communications coverage. There might be physical barriers in the form of mountains, fences, or structures that interfere with the infrastructure's communications. In densely populated regions, such as cities, the radio signals between the ground base and UAVs are frequently disrupted by buildings and other man-made structures.

Authentication Scheme for Liable Drone Services

The method is used to increase security to fend off assaults and offers a safe, portable, and proven answer to a crucial drone communication agreement. A novel method for certificate-free Drone integration has been developed, and it relies on trusted authority centres to assist communication entities in creating their key pairs while keeping those same centres anonymous. This method was developed in order to make use of trusted authority centres for certificate-free Drone integration. The findings of the suggested scheme attained a greater level of security 94 percent than those of existing methods. (Rajkumar et al., 2023). To initiate the secure connection between the drone connection various scheme were introduced to ensure the safety in the mode of cryptography. The established scheme is about the elliptic curved cryptography between the user station and the unmanned aerial aircraft. For that pairing a connection is established by registering the details to the authority control to pair the drones. Details of the drones are registered with the authority control, and both their position and their coverage are continuously updated to the authorization control. An illustration of the planned architecture for the system in Figure 2.

Figure 2. System architecture

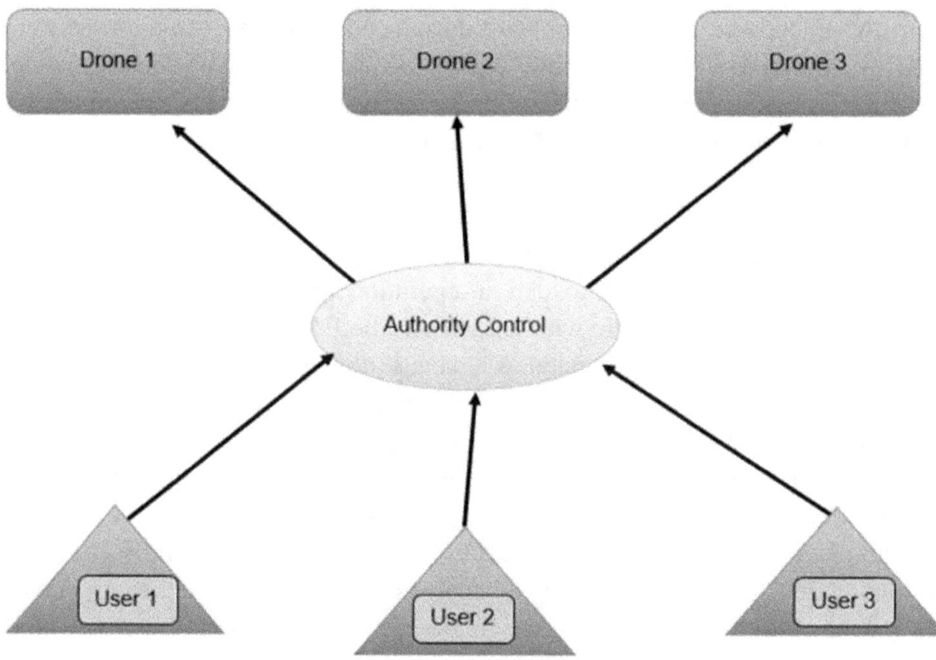

The Elliptic Curve Cryptography (ECC) Foundations

By utilizing elliptic curve cryptography, it is feasible to get concise features in the realm of encryption (ECC). As a result of the decreased number of keystrokes, it utilizes less memory and performs field arithmetic more quickly. When constructing public in general encryption algorithms for systems with restricted resources, such as drones, ECC is a reasonable choice to use as the underlying algorithm. The finite field Fi of the elliptic curve E can be written as $y2 = x3 + ax + b$, Î Fi, 4a3 + 27 b ($mod\ i$) ¹ Δ where a and Δ is referred to as the basis point and generator point that serves to cyclic groups. These processes are simply a handful of the various ECC algorithms that may be done. The suggested system architecture included the following four stages: First, there is the Initialization phase, then there is the Registration phase, then there is the Key Agreement and Authentication phase, and finally there is the Communication phase. (Jan et al., 2022).

Initialization Phase

An authority control determines which elliptic curve of a finite field d will be used. This curve is denoted by the notation E(x,y) (AC). Select a hash function hi with I equal to 1 and 2 that does not allow collisions and is irreversible. An elliptic curve's random integer, denoted by RIAC, is chose$_{n}$ at random to serve as the private key that will be used to calculate the public key, PUKAC = RIAC× G. The p$_{ri}$vate key is stored in a safe location by an Authority Control (AC), while the other remaining parameters are disseminated via the communication entities that were first used.

Registration Phase

The user A wants to Communicate with vehicle that will consider as DA. Similar$_{l}$y, all users have to connect with the authority control. The registered users select one random integer on an elliptical curve DA. After t$_{h}$at the authority control send message and confirm whether the request is done by registered user or not. Moreover, it is very easy to identify the registered user for controlling the drones.

Authentication and Key Agreement Phase

The two entities Dx and Dy that are conversing begin communication with one another in order to produce and settle on a common shared secret session key that is used to execute encryption and decryption of the message transmitted between the entities during communication between those entities.

Communication Phase

The proposed system communicating entities establish and share a communication using session key through the Dx initiate the communication between drone 1 and drone 2 by a key S that is authenticated as encrypted (Rezaeibagha et al., 2019).

Drone Detection

Development of systems to detect the presence of UAVs in areas of strategic interest and public places. There are various techniques for detecting drones, such as:

- Radio Frequency (RF)
- Visual based analysis
- Radio detection and Ranging
- Acoustics

Radio Frequency (RF)

The radio frequency technique used to detect the drones by using radio signals. In this method the radio signals connected to the drones directly so that it will more precise to detect the unmanned systems. The only disadvantage is the both system has to incorporate in a direct radio signal. If any other advanced programmed method is used to operate the drones means this type of radio frequency technique.

Visual Based Analysis

Visual based analysis of drones in normal environment is having great effective in ideal drone configurations used by various applications and industrial sectors. The visual based communication system mostly a payload on drones or a video capturing system fixed on a particular area to monitor the drones. Most of the antidrone systems employed these techniques (Lacava et al., 2022).

Radio Detection and Ranging

Doppler processing must be carried out in ground-based monitoring radar in order for drones to be located. It gets beyond the problem of a restricted dwell time and boosts the Doppler resolution, which results in a considerable improvement in the discrimination of a micro-Doppler trace and the accurate categorization of drones. (Sun et al., 2019).

Acoustics

Using sound data, users are able to easily utilise in their day-to-day lives to detect drones. This system performs FFT on the sampled real-time data and then performs drone detection using the transformed data using two different methods, namely Plotted Image Machine Learning (PIL) and K Nearest Neighbors. Both of these methods are described further below (KNN). The PIL utilises picture data from the visible FFT graph to find robust points, and then compares the mean image comparability with a reference FFT pattern associated with an object of interest to determine whether or not the two images are comparable. In contrast, the KNN assesses the average length similarity by using csv files that are formatted in the FFT format (Kim et al., 2017). The fact that the emphasis was placed on utilizing polyphonic noise data from real-life surroundings was a distinctive feature of this research. in certain other words, the primary focus was on recognizing and categorizing drone sounds against the myriad of other competing noises that were present in the environment. The lack of drone sound data that is accessible to the general public was one of the key obstacles that the study team had to overcome. The team devised an innovative solution to this issue by combining recordings of background noise with tracks of drone noises in order to produce a cohesive audio clip. This was done in order to address the issue. In this particular instance, the sample drone noises were produced by DJI Phantom 3 and Phantom 4, with the background noise consisting of people conversing, automobile traffic, and airline noise. Recordings of the noises made by the drone were made at a distance of 30 metres, 70 metres, and 150 metres while it was hovering and approaching (Wang et al., 2022).

Data Security and Privacy of Drones

A solution based on block chain technology (BCT) with the goal of lowering the risks associated with the maintenance of data in unmanned aerial vehicle (UAV) and drone systems. To be more explicit, the goal of this technique is to provide improved privacy and information storage mechanisms by utilising distinct characteristics such as irreversibility, tamper-proofing, clarity, security, and efficient distribution mechanisms. In general, drones, unmanned aerial vehicles (UAVs), and Internet of Things devices consist of a variety of sensors that contribute to the performance of predetermined tasks based on the individual's choice of application. A network connectivity is used to carry out the functions of controlling and monitoring the activities that are taking place within the drones or UAVs. For Some examples of these difficulties include the establishment of automated and effective supply chain procedures, the enhancement of transparency throughout the entirety of the value chain, and the recording and exchange of transactional data in a secure environment. Blockchain technology provides an efficient solution to these problems by utilising transactional ledgers that are decentralised, shared, encrypted, and permissioned. The utilisation of blockchain applications and the potential for implementing them in a variety of settings enables several industrial applications by increasing efficiency and safety, enhancing traceability and transparency, and lowering costs. (Al-Jaroodi & Mohamed, 2019). General architecture based on BCT for the data collection and management of components in drones or unmanned aerial vehicles (UAVs). These unmanned aerial vehicles, drones, or virtually monitored vehicles store data in databases hosted in the cloud with the primary intention of preserving the confidentiality of the data. When interacting with devices on the ground level or flying objects, it is necessary to use an anonymous technique such as BCT in order to guarantee the safety of the data being communicated.The data transferring method in block chain technology is shown in figure 3 below (Rupa et al., 2020).

For each of the unmanned aerial vehicle (UAV) devices, a block is generated, and the transactions included inside that block need to be completed depending on the E_{th} (Ethereum) balance, which is a short-term symmetric encryption. In this

Figure 3. State chart of block chain technology

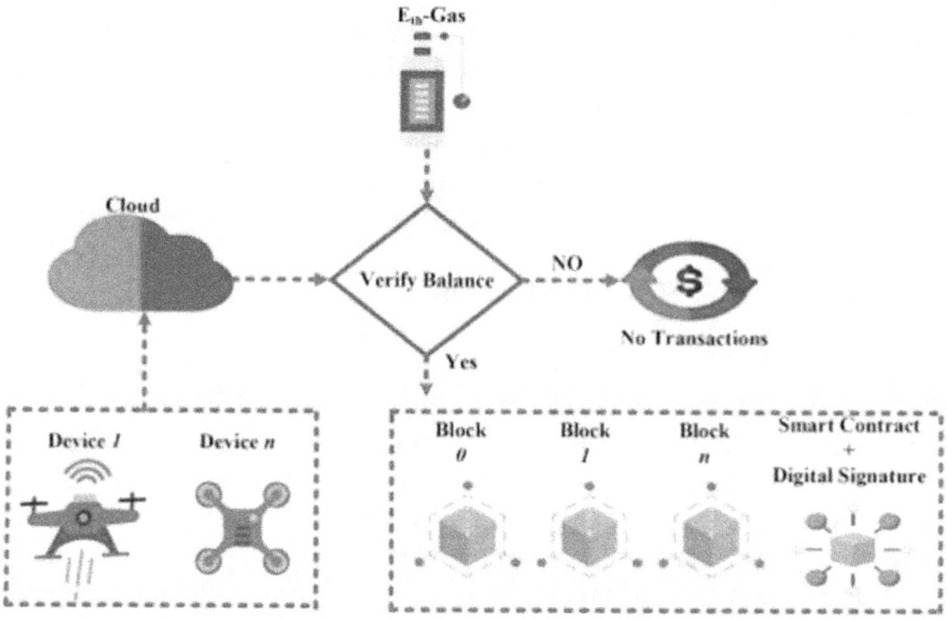

particular investigation, the researchers make use of the Matamask Wallet, which supplies them with a temporary amount of Ethereum balance. In the context of the BCT network, E_{th} Gas is a cryptocoin that contributes to the generation of blocks. Pentatope Based Elliptic Curve Encryption (PECC) is used to generate a digital signature that is used to verify each and every piece of data that is received from the flying devices. These serve as documentation of the one-of-a-kind transactions that were carried out in the BCT blocks (Komanapalli et al., 2021).

Cyber Security Challenges of the Civilian Drones

The cybersecurity problems include data protection, cyberattacks, and general worries over drones originating from Asian markets. When it comes to conducting ethical and lawful drone operations, the first component is of the utmost importance. A drone that is equipped with a sensor or another device that records video might easily breach an individual's privacy. The objective of the cyberattack component is to ensure that the user is aware that their device, and not just the drone by itself but also any other devices that are linked to it, is susceptible to being hacked. The usage of drones for recreational purposes by civilians typically involves recording, primarily:

- Optimize time-lapse
- Creating panoramic films or still photographs
- Performing stunts
- Photographing animals or people in their natural habitats

In this situation, in order to comply with and be in accordance with the General Data Protection Regulation (GDPR), a personal clear permission is required. Despite the fact that this factor significantly restricts the usage of drones equipped with video cameras in urban areas, it nevertheless provides a strong framework for addressing various monitoring and surveillance concerns.

The GDPR makes it possible to stay anonymous personal data and provides opportunities to do so. At first, this presents a potential for prompting, because it may implement to users of UAS that in the event that they capture personal data, and then afterwards anonymize it, it is acceptable (for example blur out houses, vehicles, faces, etc.) and it will no longer be against the rules to keep a copy or upload it. Processing and storing anonymous data doesn't really require prior authorization from the data subject. This matter is also deemed to be in the realm of the ambiguous since, initially, before a data set is anonymized, it is still regarded as private details and still calls for consent from the individual who is being impacted (Ináncsi, 2022).

CONCLUSION

The purpose of this study was to propose various methods that automate drone detection processes by utilising the sounds of the drone with various deep learning and ensemble learning algorithms. Additionally, this study primarily covers things of data privacy and data protection in Iot Systems, UAVs, and Unmanned aerial vehicles applications by utilising block chain technology. These problems were brought about by the illegal use of drones. In order to assess the effectiveness of the application that has been suggested, it is compared to the results that were obtained following the deployment of BCT-based IoT applications in contrast to conventional systems. The results of these research demonstrate that the model is superior. A digital signature is used to authenticate the transactions that are carried out as part of the planned study. This aided in ensuring the appropriate level of privacy and security in the data maintenance that was undertaken. This survey will, in the not too distant future, contribute to the construction of one hopeful antidrone device that will create safety for drones used in both civil and military applications.

REFERENCES

Al-Jaroodi, J., & Mohamed, N. (2019). Blockchain in Industries: A Survey. *IEEE Access : Practical Innovations, Open Solutions, 7*, 36500–36515. doi:10.1109/ACCESS.2019.2903554

Inántsi, M. (2022). Cybersecurity Challenges of the Civilian Unmanned Aircraft Systems. *The Military Engineer, 17*(2), 205–216. doi:10.32567/hm.2022.2.14

Jan, S. U., Abbasi, I., Algarni, F., & Khan, S. (2022). A Verifiably Secure ECC Based Authentication Scheme for Securing IoD using FANET. *IEEE Access : Practical Innovations, Open Solutions, 10*, 23. doi:10.1109/ACCESS.2022.3204271

Jayakumari, D., Rupa, C., & Nikhila, V. (2021). Pentatope Based Elliptic Curve Encryption for Privacy and Protection of Multimedia Data. In V. L. N. Komanapalli, N. Sivakumaran, & S. Hampannavar (Eds.), *Advances in Automation, Signal Processing, Instrumentation, and Control. i-CASIC 2020. Lecture Notes in Electrical Engineering* (Vol. 700). Springer. doi:10.1007/978-981-15-8221-9_48

Kim, J., Park, C., Ahn, J., Ko, Y., Park, J., & Gallagher, J. C. (2017). Real-time UAV sound detection and analysis system. *2017 IEEE Sensors Applications Symposium (SAS)*, (pp. 1-5). IEEE. 10.1109/SAS.2017.7894058

Lacava, G., Mercaldo, F., Martinelli, F., Santone, A., & Pizzi, M. (2022). Drone Audio recognition based on Machine Learning Techniques. *Procedia Computer Science, 207*, 848-857. doi:10.1016/j.procs.2022.09.140

Rajkumar, S.C, Jegatha, L., Vijayakumar, P., & Karthick, K.R. (2023). Secure session key pairing and a lightweight key authentication scheme for liable drone services. *Cyber Security and Applications*. doi:10.1016/j.csa.2022.100012

Rezaeibagha, F., Mu, Y., Huang, X., Yang, W., & Huang, K. (2019). *Fully Secure Lightweight Certificateless Signature Scheme for IIoT* (pp. 1). IEEE Access. . doi:10.1109/ACCESS.2019.2944631

Rupa, C., Srivastava, G., Gadekallu, T., Maddikunta, P., & Bhattacharya, S. (2020). Security and privacy of UAV data using blockchain technology. *Journal of Information Security and Applications, 55*.

Sun, H., Oh, B.-S., Guo, X., & Lin, Z. (2019, December). Improving the Doppler Resolution of Ground-Based Surveillance Radar for Drone Detection. *IEEE Transactions on Aerospace and Electronic Systems, 55*(6), 3667–3673. doi:10.1109/TAES.2019.2895585

Tsao, K.-Y., Girdler, T., & Vassilakis, V. G. (2022). A survey of cyber security threats and solutions for UAV communications and flying ad-hoc networks. *Ad Hoc Networks, 133*, 1570-8705

Wang, Y., Fagiani, F. E., Ho, K. E., & Matson, E. T. (2022). A Feature Engineering Focused System for Acoustic UAV Payload Detection. *International Conference on Agents and Artificial Intelligence.* Scite Press. 10.5220/0010843800003116

Chapter 9
Design and Implementation of Production Lines Through Cyber Physical Systems in the Manufacturing Sector in Industry 4.0 Using Machine Learning

S. Kaliappan
Division of Research and Development, Lovely Professional University, India

Ramya Maranan
Department of Research and Innovation, Saveetha School of Engineering, SIMATS, India

ABSTRACT

Industrial manufacturing facilities that use cyber-physical systems (CPS) require constant communication with a variety of dispersed and interacted computer nodes, strategies, and humanoid operatives. These structures are vital for ensuring the safety of workers on the shop floor as well as the quality of goods produced. This concept applies to other industries as well, such as sequence, which are interconnected and impact complete invention effectiveness. This research explores the critical steps for mixing three CPS columns (manufacturing line, sequence, and conveniences) into modern smart industrial settings, in instruction to approve a manufacturing cyber-physical systems model. The method focuses on incorporating various digital capabilities into a cloud platform, to facilitate real-time multi-device communication, information analytics/distribution, and deep learning-based worldwide regeneration.

DOI: 10.4018/978-1-6684-9267-3.ch009

1. INTRODUCTION

The manufacturing industry in Europe is looking for innovative ways to simplify the adoption and implementation of numerical conversion within the framework of Industry 4.0. In response, the Indian Command is funding research on advanced machineries and methods for system interaction (Balakrishnan et al. 2021; Xie et al. 2021). As a result, it is crucial to implement proven solutions that support both domain integration and system interaction at various levels. One potential solution is an bionetwork created by a System of Systems (SoS), which is a high-level goal that goes beyond the interconnectedness and integration advocated by Industry 4.0 and smart cities. This study proposes incorporating additional area schemes, such as production, logistics, and structure computerization, to enhance the effectiveness and efficiency of industrial systems. The ultimate goal is a networked ecosystem that allows the business sector to anticipate and adapt to changing market demands. The resulting entity is a CPS of Systems, which combines and improves the operation of these many, independent systems (Kim et al. 2020). In additional words, the CPS will connect people and computer hardware, software package, and other system components. Each component's information will be added to the overall CPS knowledge information, which can impact other subsystems. Managing the complexity and diversity of CPS is one of the key challenges (Gu et al. 2014; Li et al. 2021).

Managing, integrating, and analyzing CPS information are about of the primary challenges due to the multiplicity of structure funds that provide data. To address this, modern Statistics and Communication Skills (ICT)-constructed explanations are being deployed in various industries, including healthcare and industrial automation. However, this study identifies other challenges that must be addressed before fully implementing the CPS paradigm in smart cities. The main goal of this research project is to investigate best practices for adopting the CPS paradigm in smart cities (Cui 2021; Martinez et al. 2022).

To achieve this goal, the research project will focus on several key areas, including:

- Identifying the key components and requirements of a CPS, including hardware, software, and communication protocols.
- Investigating the best practices for integrating and managing the data generated by multiple systems within a CPS.
- Examining the security and privacy concerns that arise in a CPS and developing solutions to address them.
- Evaluating the scalability of different CPS solutions and identifying the best approaches for deploying them at various levels, from individual systems to entire cities.

- Developing a framework for assessing the performance and effectiveness of different CPS solutions.
- Analyzing case studies of successful CPS implementations in smart cities and identifying the key factors that contributed to their success.

By focusing on these key areas, the research project hopes to provide valuable insights into the best practices for adopting the CPS paradigm in smart cities (Tao et al. 2019; Traganos et al. 2021). This will help manufacturers, cities, and other stakeholders to more easily adopt and implement digital alteration within the framework of Manufacturing 4.0, resulting in more efficient and effective industrial systems. (Schneider et al. 2019; Sandborn et al. 2021; Barata et al. 2022; Javaid et al. 2022)

2. IMPLEMENTATION OF CYBER PHYSICAL SYSTEM IN INDUSTRY

The planned Cyber-Physical Systems is made up of 3 chief classifications, each of which plays a crucial role in enhancing the effectiveness and quality of industrial output. The first system is last mile logistics, which can account for up to 28% of overall delivery costs. Improving this phase in the supply chain can have a significant impact on overall efficiency. The second system is building automation systems, as the materials, security, energy efficiency, and facility services provided by buildings directly impact manufacturing operations. The third system is the actual production system, as improving industrial productivity is the ultimate goal.

The maximum-level visualization of the planned CPS is shown in Figure 1. The CPS is an incorporation of the imitation and physical worlds, comprising of a number of cyber and physical subsystems and components, which are associated via machine learning technology. The physical domain, which is the first, consists of material capitals that are used by various systems to alter and/or perceive the atmosphere. These corporal resources include heaters that increase a building's temperature, robot arms that operate on factory floors, and trucks that move supplies from suppliers to producers. In the cyberspace, there are several uses, including digital twins. Cyber models replicate physical resources, which facilitates system behaviour analysis and performance monitoring. This is essential for achieving zero-defect manufacturing as it enables the virtual model to identify anomalies and the physical equipment to be adjusted.

Figure 1. Proposed CPS design

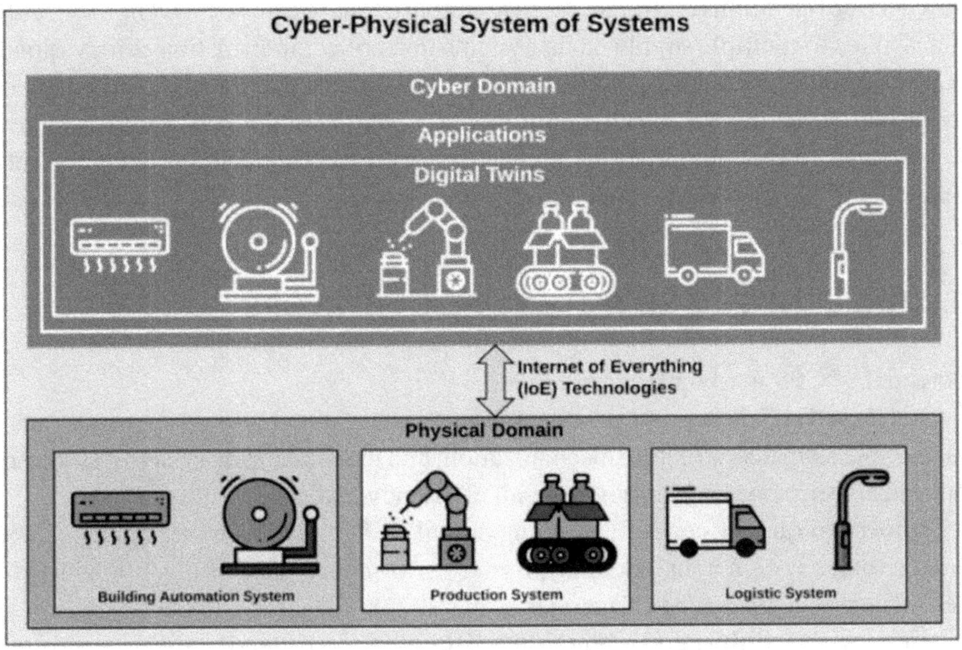

3. VARIOUS COMPONENTS OF THE SYSTEM

The proposed CPS scenario will take into account a number of important considerations. The proposed roadmap will address key requirements for data and information gathering, integration, analysis, and sharing, which will be essential elements of the strategy, enabling different systems to interact and exchange data. This research study proposes collecting data through the use of hardware equipped with web services installed in various systems. These devices will have access to both analogue and digital resources, such as data exchangers, actuators, and sensors (Natrayan L et al., 2020). By interacting with any subsystem, such as by sending an alert about the arrival of a shipment containing production-related components, human employees can also be measured as capitals that will produce information inside the CPS (Paranthaman et al., 2022).

To effectively manage the diverse data within the CPS, it will be necessary to have trustworthy and adaptable interfaces (Kumar MS et al., 2019). The next section will delve further into the challenge of merging various data and information from multiple systems and time periods. A key component that facilitates communication inside CPS systems will be machine learning instruments (Meikandan et al., 2022). These tools are used on workshop floors to device the SOA paradigm of parallel and perpendicular integration of business systems. Recent research studies also show how service operations can be accessed remotely to operate physical equipment, and this applies to multiple supply chain systems that collect information from various data sources (Sundaram et al., 2019). Additionally, the exclusive explanations used in various structures for various purposes, such as the claim used by a customer to purchase a exact product or by a upkeep worker to statement a problem on the manufacturing floor, are also included in the CPS.

4. CHALLENGES AND POTENTIAL OF THE CPS

Potential of CPS in Manufacturing Field:

Increased efficiency and productivity: By integrating multiple approach, such as last mile statistics, structure mechanization, and manufacturing systems, CPS can optimize operations and improve overall efficiency and productivity.

Improved quality control: By using digital twins and other cyber models, CPS can facilitate system behavior analysis and performance monitoring, allowing for zero-defect manufacturing (Nagarajan et al., 2022).

Increased flexibility and adaptability: By linking people and hardware, software, and other system components, CPS can enable businesses to anticipate and adapt to changing market demands.

Improved resource management: By integrating data from multiple systems, CPS can enable better management of resources, such as materials, energy, and personnel.

Challenges of CPS Manufacturing Field:

Complexity and variety: CPS involves integrating multiple systems, each with its own hardware, software, and communication protocols. This can lead to a high degree of complexity and variety, making it difficult to manage and analyze data (Velmurugan et al., 2022).

Data integration and management: Integrating and managing data from multiple systems can be a major challenge, especially when dealing with diverse data types and time periods (Natrayan, L., & Kumar, M. S. 2020).

Security and privacy: CPS involves collecting and sharing sensitive data, which raises security and privacy concerns. These need to be addressed to ensure the system is secure and data is protected (Santhosh MS et al., 2018).

Scalability: CPS solutions need to be scalable, and able to be deployed at various levels, from individual systems to entire cities. This can be challenging, especially when dealing with different systems and technologies.

Human and technical factors: The integration of human and technical factors in the CPS is a challenge, as it requires a different approach to traditional manufacturing processes.

5. CASE STUDY OF IMPLEMENTING CPS

The full potential of Cyber-Physical Systems of Systems (CPS) can only be achieved through the integration of multiple methods. This approach not only reduces development costs and time, but also simplifies complexity and makes it easier for innovators to adopt new technologies (Veeman et al., 2021). In order to improve the quality and effectiveness of industrial output, this study recommends combining statistics, automation, and manufacturing methods to generate a CPS. One potential approach for implementing this idea is to set up a manufacturing system at specific facilities and connect it to an easily accessible building conditioning system and logistics system. The three fundamental subsystems, including the production, construction, and logistics systems (Natrayan L & Kumar MS. 2020).

The initial step involves integrating two separate systems: 1) The factory area, which includes a manufacturing line, a multi-robot cell for material preparation, and a loading zone, creating a full manufacturing method; and 2) A collection of instruments placed which is at inside and outside the Lab, which serves as a building automation system. Additionally, the logistics system (LS) will be introduced to enable the traceability of components supplied to and required by the Lab production system. The automation is composed of several computer hardware and software program components, is the first CPS subsystem (Natrayan L et al., 2021).

In Figure 2, the Factory-Lab facility entrance is depicted as (1). To aid in material unloading, a pair of (2) and (3) designated LED lights illuminate the space, and an Onninen2 presence detector is associated to an brilliant lighting supervisor from Philips Oy3 to detect movement. In India, owing to its geographic position, there is a absence of high sun light in hill areas for several months of the year (Natrayan L et al., 2018). Installing sufficient external illumination makes it easier to transport supplies to the manufacturing line as a result (6). A finned coil air heater from Alfa Laval4, denoted as (5), is similarly managed by the BAS and activates when the interior temperature drops (Kumar MS et al., 2018). The above example illustrates just one way in which CPS can be used in the industrial sector (Pragadish et al., 2022). By minimising complexity and creating it easier for modernizers to adopt novel methods, CPS has the potential to increase the effectiveness and quality of industrial production (Balaji et al., 2022).

Figure 2. Outside of the lab

As illustrated in Figure 3, web-service enabled controllers are used to oversee and control the Building Automation System (BAS). These Inico S1000 Industrial controllers, as seen in (A), are connected to a network setup hub, as represented by the letter N in (B). Previous studies have demonstrated the effectiveness of these controllers in increasing energy efficiency for LED-based lighting in multifunctional outdoor settings (Ramaswamy et al., 2022). Additionally, Wi-Fi connections between the industrial controllers and THL sensors, strategically placed around the site, allow for the implementation of energy-efficient air conditioning and lighting systems. With the use of web services to control and monitor physical devices, this BAS can be considered a CPS.

The implementation of Cyber-Physical Systems (CPS) in the industrial sector is an essential step towards improving production efficiency and reducing costs. One key aspect of CPS implementation is the integration of multiple subsystems, including the Manufacturing System (PS), the Building Automation System (BAS), and the Logistics System (LS). The Manufacturing System, which is the second subsystem of CPS, comprises of several areas and systems, such as the factory manufacturing line, 3-D robotic material preparation cells, and a storage facility for entire component parts and prepared-to-distribute finished goods. The PS is made up of numerous components, as shown in Figure 4. The S1000 controllers, which function similarly to the BAS, are used to operate the factory production line and the material preparation cells. Additionally, service activities can be initiated remotely, leading to the implementation of actual actions or the nursing of particular data. These devices can be found on the right position of Figure 3.

Figure 3. Control and monitoring unit of the industry

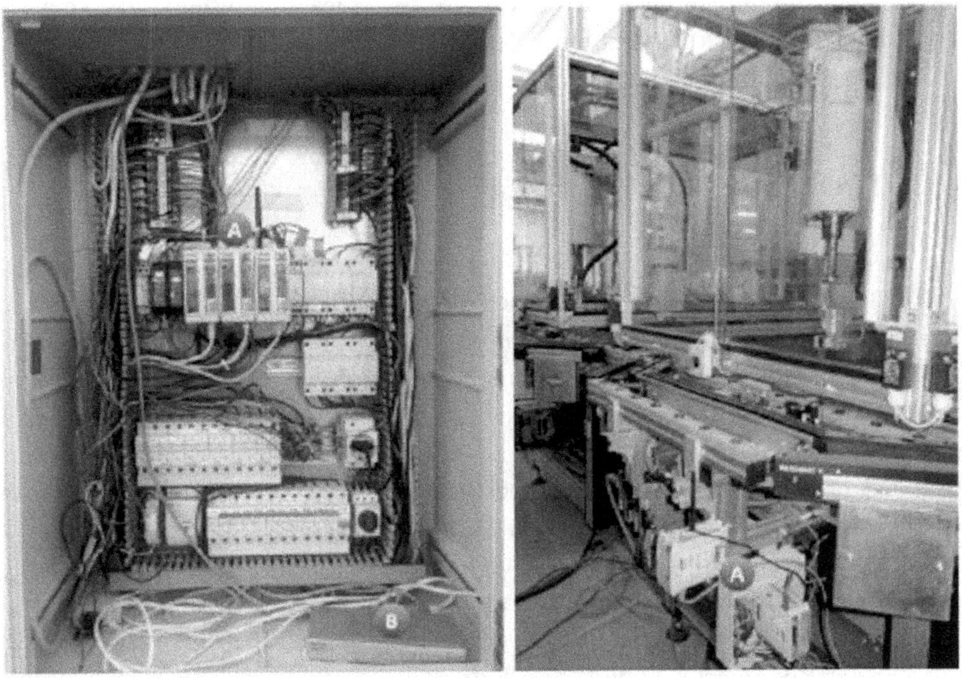

As part of the manufacturing process, the Factory Manufacturing Line makes designs for different smart phones. The robot named SCARA are programmed to perform various operations to build various mobile components that can subsequently be painted in various colors, allowing the manufacturing line to produce over seven hundred and twenty distinct varieties of mobile phones. In the line, there are 10 identical WorkStations (WS). WS1, WS2, WS6, WS 8, WS 9, and WS 12 are used to load components, discharge finished items, and manually palletize the line. WS 1 is also used to load components. Multi-robotic cells WSA, WSB, and WSC in this research project handle papers and pallets in the material preparation zone.

The last CPS subsystem oversees, manages, and coordinates the logistics chain. To provide complete traceability of the processes required to produce an item, event, or service, the suggested scenario considers storage capacity, maintenance systems, productivity indices, and delivery time. The subsystems also communicate data and information to perform a number of functions. The supply of manufactured parts as well as the stock of parts needed for preventive maintenance operations must also be continuously monitored and managed by these components. A current following system, on the other hand, synchronizes the BAS subsystem by gathering and transferring information from and to the outside distribution scheme in order to manage the lights, air conditioners, and doors being utilized near to the pickup truck. Founded on a mixture of GPS and network technologies, this system (GPS) makes it possible to manage logistics thoroughly and effectively, and it also improves the CPS's overall functioning in the industrial sector.

6. OUTCOME OF THE EXPERIMENT

The integration of various systems, such as BAS, PS, and LS, under the umbrella of the CPS is expected to greatly enhance the productivity of PS. By merging data from these distinct systems, the PS can be swiftly and automatically adjusted in response to unforeseen changes that may disrupt the production schedule. For instance, fluctuations in consumer demand, predictions, or manufacturing resources may all have an impact on output. As it stands, these systems currently operate independently, leading to a lack of cohesion and a need for additional safeguards which can be costly and inefficient. For example, the storage indicated in Figure 4 is often overflowing with unnecessary supplies due to a lack of understanding about logistics and consumer expectations.

To evaluate this proposed strategy, certain ISO 22400-2 Key Performance Indicators (KPIs) may be employed. Additionally, the Energy Consumption of the KPI, another KPI that corresponds to ISO 22400-2 requirements, is also likely to decrease as the BA system becomes more aware of shipment schedules, allowing for optimal management of energy efficiency. Overall, the proposed deployment of a CPS would aim to seamlessly combine all sub-systems necessary to oversee and control production from component supply to product distribution, ensuring efficient energy consumption at manufacturing facilities.

CONCLUSION

In conclusion, the implementation of a cyber physical system in the manufacturing sector in Industry 4.0, specifically in the design and implementation of production lines, has the potential to greatly enhance efficiency and productivity. By utilizing machine learning techniques, the system can continuously adapt and improve in response to fluctuations in consumer demand, production resources, and other factors. The integration of data from various systems such as BAS, PS, and LS into a single CPS can streamline production processes and reduce the need for costly and inefficient safeguards. Through the use of ISO 22400-2 Key Performance Indicators, the system can be effectively evaluated and optimized. Overall, the design and implementation of a cyber physical system in the manufacturing sector in Industry 4.0 using machine learning is a promising approach to improve production processes and increase competitiveness in the industry.

Figure 4. Production layout of the factory

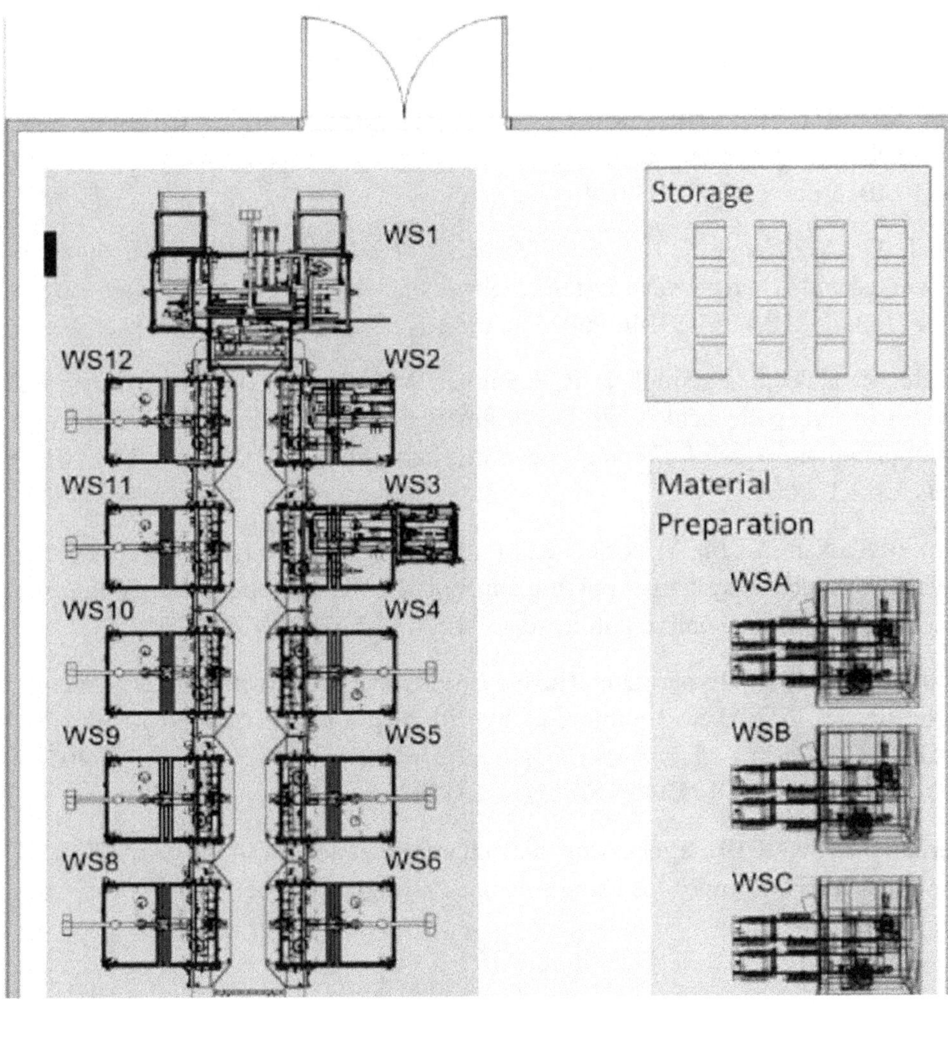

REFERENCES

Balaji, N., Natrayan, L., Kaliappan, S., Patil, P. P., & Sivakumar, N. S. (2022). Annealed peanut shell biochar as potential reinforcement for aloe vera fiber-epoxy biocomposite: Mechanical, thermal conductivity, and dielectric properties. *Biomass Conversion and Biorefinery*, 1–9. doi:10.100713399-022-02650-7

Balakrishnan, A., Kadiyala, R., Dhiman, G., Ashok, G., Kautish, S., Yadav, K., & Maruthi Nagendra Prasad, J. (2021). A Personalized Eccentric Cyber-Physical System Architecture for Smart Healthcare. *Security and Communication Networks*, *2021*, 1–36. doi:10.1155/2021/1747077

Barata, J., Cardoso, A., Haenisch, J., & Chaure, M. (2022). Interoperability standards for circular manufacturing in cyber-physical ecosystems: a survey. *Procedia Computer Science 207*(Kes), 3320–3329. doi:10.1016/j.procs.2022.09.390

Cui, X. (2021). Cyber-Physical System (CPS) architecture for real-time water sustainability management in manufacturing industry. *Procedia CIRP*, *99*, 543–548. doi:10.1016/j.procir.2021.03.074

Gu, J., Qi, X., Wang, Y., Liu, F., & Zhang, C. (2014). Advances in methods for networked and cyber-physical system. *Journal of Control Science and Engineering*, *2014*, 1–2. doi:10.1155/2014/496075

Javaid, M., Haleem, A., Singh, R. P., & Suman, R. (2022, September). An integrated outlook of Cyber-Physical systems for Industry 4.0: Topical practices, architecture, and applications. *Green Technologies and Sustainability*, *1*(1), 100001. doi:10.1016/j.grets.2022.100001

Kim, B. S., Nam, S., Jin, Y., & Seo, K. M. (2020). Simulation framework for cyber-physical production system: Applying concept of LVC interoperation. *Complexity*, *2020*, 1–11. Advance online publication. doi:10.1155/2020/4321873

Kumar, M. S. (2018). Experimental investigations on mechanical and microstructural properties of Al2O3/SiC reinforced hybrid metal matrix composite. [). IOP Publishing.]. *IOP Conference Series. Materials Science and Engineering*, *402*(1), 012123. doi:10.1088/1757-899X/402/1/012123

Kumar, M. S. (2019). Processing and characterization of AA2024/Al2O3/SiC reinforces hybrid composites using squeeze casting technique. *Iran. J. Mater. Sci. Eng*, *16*(2), 55–67.

Li, C., Guo, X., & Wang, X. (2021). An Autonomous Cyber-Physical Anomaly Detection System Based on Unsupervised Disentangled Representation Learning. *Security and Communication Networks, 2021*, 1–17. doi:10.1155/2021/1626025

Martinez, P., Al-Hussein, M., & Ahmad, R. (2022). A cyber-physical system approach to zero-defect manufacturing in light-gauge steel frame assemblies. *Procedia Computer Science, 200*, 924–933. doi:10.1016/j.procs.2022.01.290

Meikandan, M., Karthick, M., Natrayan, L., Patil, P. P., Sekar, S., Rao, Y. S., & Bayu, M. B. (2022). Experimental investigation on tribological behaviour of various processes of anodized coated piston for engine application. *Journal of Nanomaterials, 2022*, 1–8. doi:10.1155/2022/7983390

Nagarajan, Rajagopalan, A., Angalaeswari, S., Natrayan, L., & Mammo, W. D. (2022). Combined economic emission dispatch of microgrid with the incorporation of renewable energy sources using improved mayfly optimization algorithm. *Computational Intelligence and Neuroscience, 2022*, 1–22. doi:10.1155/2022/6461690 PMID:35479598

Natrayan, L., & Kumar, M. S. (2020). An integrated artificial neural network and Taguchi approach to optimize the squeeze cast process parameters of AA6061/Al2O3/SiC/Gr hybrid composites prepared by novel encapsulation feeding technique. *Materials Today. Communications, 25*, 101586. doi:10.1016/j.mtcomm.2020.101586

Natrayan, L., & Kumar, M. S. (2020). Optimization of wear behaviour on AA6061/Al2O3/SiC metal matrix composite using squeeze casting technique–Statistical analysis. *Materials Today: Proceedings, 27*, 306–310. doi:10.1016/j.matpr.2019.11.038

Natrayan, L., Merneedi, A., Bharathiraja, G., Kaliappan, S., Veeman, D., & Murugan, P. (2021). Processing and characterization of carbon nanofibre composites for automotive applications. *Journal of Nanomaterials, 2021*, 1–7. doi:10.1155/2021/7323885

Natrayan, L., Senthil Kumar, M., & Chaudhari, M. (2020). Optimization of squeeze casting process parameters to investigate the mechanical properties of AA6061/Al 2 O 3/SiC hybrid metal matrix composites by Taguchi and Anova approach. In *Advanced Engineering Optimization Through Intelligent Techniques: Select Proceedings of AEOTIT 2018* (pp. 393-406). Springer Singapore. 10.1007/978-981-13-8196-6_35

Natrayan, L., Sivaprakash, V., & Santhosh, M. S. (2018). Mechanical, microstructure and wear behavior of the material AA6061 reinforced SiC with different leaf ashes using advanced stir casting method. *International Journal of Engineering and Advanced Technology, 8*, 366–371.

Paranthaman, V., Sundaram, K. S., & Natrayan, L. (2022). Influence of SiC particles on mechanical and microstructural properties of modified interlock friction stir weld lap joint for automotive grade aluminium alloy. *Silicon*, *14*(4), 1617–1627. doi:10.100712633-021-00944-5

Pragadish, N., Kaliappan, S., Subramanian, M., Natrayan, L., Satish Prakash, K., Subbiah, R., & Kumar, T. C. A. (2022). Optimization of cardanol oil dielectric-activated EDM process parameters in machining of silicon steel. *Biomass Conversion and Biorefinery*, 1–10. doi:10.100713399-021-02268-1

Ramaswamy, R., Gurupranes, S. V., Kaliappan, S., Natrayan, L., & Patil, P. P. (2022). Characterization of prickly pear short fiber and red onion peel biocarbon nanosheets toughened epoxy composites. *Polymer Composites*, *43*(8), 4899–4908. doi:10.1002/pc.26735

Sandborn, M., Wright, C., White, J. (2021). CaSTL: Towards cyber-physical authentication of additively manufactured components. *Procedia Computer Science 198*(2018), 764–769. doi:10.1016/j.procs.2021.12.319

Santhosh, M. S. (2018). Investigation of mechanical and electrical properties of Kevlar/E-glass and Basalt/E-glass reinforced hybrid composites. *International Journal of Mechanical and Production Engineering Research and Development*, *8*(3), 591–598. doi:10.24247/ijmperdjun201863

Schneider, M., Lucke, D., & Adolf, T. (2019). A cyber-physical failure management system for smart factories. *Procedia CIRP*, *81*, 300–305. doi:10.1016/j.procir.2019.03.052

Sundaram. (2019). Smart clothes with bio-sensors for ECG monitoring. *International Journal of Innovative Technology and Exploring Engineering*, *8*(4), 298–301.

Tao, F., Qi, Q., Wang, L., & Nee, A. Y. C. (2019). Digital Twins and Cyber–Physical Systems toward Smart Manufacturing and Industry 4.0: Correlation and Comparison. *Engineering (Beijing)*, *5*(4), 653–661. doi:10.1016/j.eng.2019.01.014

Traganos, K., Grefen, P., Vanderfeesten, I., Erasmus, J., Boultadakis, G., & Bouklis, P. (2021). The HORSE framework: A reference architecture for cyber-physical systems in hybrid smart manufacturing. *Journal of Manufacturing Systems 61*(November 2020), 461–494. doi:10.1016/j.jmsy.2021.09.003

Veeman, Shree, M. V., Sureshkumar, P., Jagadeesha, T., Natrayan, L., Ravichandran, M., & Paramasivam, P. (2021). Sustainable development of carbon nanocomposites: Synthesis and classification for environmental remediation. *Journal of Nanomaterials*, *2021*, 1–21. doi:10.1155/2021/5840645

Velmurugan, G., Siva Shankar, V., Kaliappan, S., Socrates, S., Sekar, S., Patil, P. P., Raja, S., Natrayan, L., & Bobe, K. (2022). Effect of aluminium tetrahydrate nanofiller addition on the mechanical and thermal behaviour of luffa fibre-based polyester composites under cryogenic environment. *Journal of Nanomaterials*, *2022*, 1–10. doi:10.1155/2022/5970534

Xie, J., Tan, W., Fang, B., & Huang, Z. (2021). Towards a Statistical Model Checking Method for Safety-Critical Cyber-Physical System Verification. *Security and Communication Networks*, *2021*, 1–12. doi:10.1155/2021/5536722

Chapter 10
Detection of Fault Analysis in Sensor Network Systems Through Embedded Cyber Physical Systems

G. Shanmugaraj
Velammal Institute of Technology, India

B. Atchaya
Velammal Institute of Technology, India

K. Mohanambal
Velammal Engineering College, India

G. Kavitha
Velammal Institute of Technology, India

N. Susmitha
Velammal Institute of Technology, India

ABSTRACT

This chapter explores the issue of sensor reliability in the adoption of internet of things (IoT) data in critical industries such as automotive and industrial sectors. It summarizes significant European research and innovation programs, as well as recent investigations on sensor dependability in cyber-physical systems (CPS). The study presents a technique for assessing the reliability of IoT sensors in a CPS, and includes a case study of sensor reliability evaluation in an autonomous driving scenario for the automobile industry. The case study employs a co-simulation framework that enables real-time communication between virtual and actual sensors and uses an IoT LiDAR-based collaborative map and an Ibeo Lux 4-layer LiDAR sensor with IoT capabilities. The study also implements machine learning techniques for error forecasting at the local level and a self-learning decision-making mechanism based on Q-learning at the global level. The results of the study indicate that the proposed strategy for improving sensor dependability in CPS utilizing data from IoT is effective.

DOI: 10.4018/978-1-6684-9267-3.ch010

1. INTRODUCTION

In recent years, the manufacturing industry has seen a significant increase in the adoption of sensor networks and cyber physical systems (CPS) for monitoring and controlling industrial processes. These systems, also known as the Internet of Things (IoT), have become an essential tool for improving the efficiency and effectiveness of manufacturing operations. However, the integration of these systems in the manufacturing sector also introduces new challenges, such as the detection of faults in the sensor network.(Streitz 2019; Mishra et al. 2022) Fault detection in sensor networks is an important issue as it can lead to production downtime, lost revenue, and even safety hazards. In order to mitigate these risks, it is essential to develop effective fault detection methods for sensor networks. This research aims to explore the use of embedded CPS for the detection of faults in sensor network systems in the manufacturing sector(Darwish and Hassanien 2018).

The use of sensor networks and CPS in the manufacturing industry has been widely studied in recent years. These systems have been used for various applications such as monitoring and controlling industrial processes, predictive maintenance, and quality control. However, the detection of faults in sensor networks has received less attention(Pan and Zhang 2022).

Several methods have been proposed for fault detection in sensor networks. One popular approach is the use of distributed algorithms, such as consensus-based algorithms and Kalman filters, for detecting faults in sensor networks. These algorithms are designed to detect faults by comparing the measurements of different sensors in the network. Another approach is the use of machine learning techniques, such as artificial neural networks, for detecting faults in sensor networks. These techniques have been used to classify sensor measurements as normal or abnormal, and to identify specific types of faults(Liu et al. 2020; Tripathi et al. 2022).

Recently, the use of embedded CPS for fault detection in sensor networks has been proposed as a promising approach. CPS are integrated systems that combine physical processes with computing and communication capabilities. By integrating CPS with sensor networks, it is possible to perform real-time monitoring and analysis of the sensor data, and to detect faults in the sensor network. Embedded CPS have been used for various applications such as condition monitoring, fault diagnosis, and predictive maintenance(Cui 2021; Traganos et al. 2021). The use of embedded CPS for fault detection in sensor networks is relatively new, but it has shown promising results. For example, some studies have proposed the use of embedded CPS for monitoring the health of sensors in a sensor network and for detecting faults in the sensor network.In conclusion, the detection of faults in sensor networks is an important issue in the manufacturing sector. Several methods have been proposed for fault detection in sensor networks, including the use of distributed algorithms and machine learning techniques. However, the use of embedded CPS for fault detection in sensor networks is a relatively new approach that has shown promising results. This research aims to explore the use of embedded CPS for the detection of faults in sensor network systems in the manufacturing sector(Liu et al. 2021; Javaid et al. 2022; Jbair et al. 2022).

2. METHODOLOGY OF THE RESEARCH

SAP offers a range of IoT solutions that are constructed on the powerful SAP HANA platform. The data is composed from industrial partner's device sensors and then flooded into the SAP HANA in-memory folder for additionaldispensation and examination.

SAP HANA is a highly versatile platform that is equipped with a diverse set of natively combined engines. These engines allow for seamless information streaming, advanced prognostic analysis, association modeling, geographic handling, on-the-fly sophisticated calculations, machine learning, and text processing. Additionally, SAP HANA's ability to perform both communicational and analytical workloads from the similarinformation collection further enhances its capabilities. These robust features provide a solid foundation for handling even the most complex and unanticipated amounts of IoT information and results.

The SAP Prognostic Analytical Library (PAL) is a set of additional features that provide a variety of analytical techniques, including clustering, classification, association and more. One of the key benefits of using SAP PAL is its high native performance, as intricate and time-consuming analytical operations are carried out directly within the database rather than being transferred to the application server.

In this article, we will explore how to use SAP PAL to create predictive models using IoT data stored in an in-memory SAP HANA database. We will demonstrate how to apply effective data mining techniques developed in SAP PAL to mine frequent alarm log trends and identify unusual alarm occurrences, using LiDAR sensor data as an example. LiDAR sensors are widely used in various industrial and infrastructure applications, such as autonomous vehicles, robotics, and 3D mapping. By leveraging the power of SAP PAL, organizations can gain valuable insights from their LiDAR sensor data and make data-driven decisions to improve their operations. The architecture of the research is shown in figure 1.

Figure 1. Application of IoT for predictive maintenance

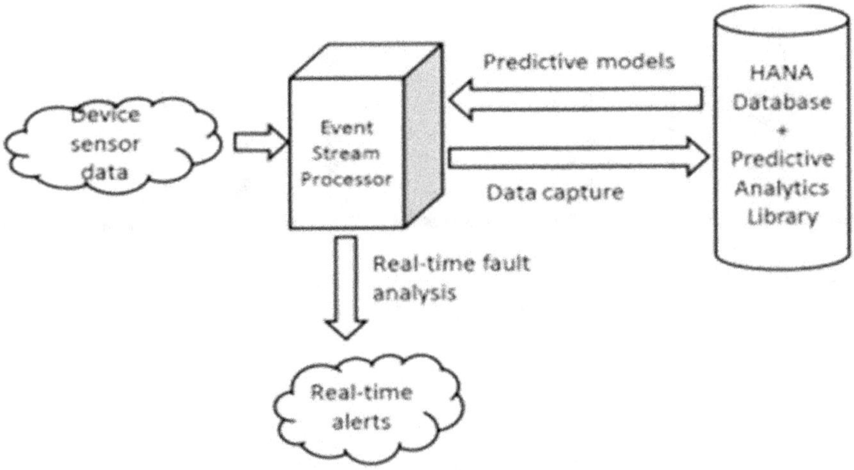

3. PREDICTIVE MODEL FOR IMPLEMENTING IN MANUFACTURING SECTOR

In this section, we will delve into the methods used to create predictive models for failure detection using IoT data. We will cover a variety of techniques that can be used to identify the causes of failures and predict when they might occur.The first technique that will be discussed is correlation analysis. This method is used to identify relationships between different sensor readings and device states. By analyzing patterns in the data, it is possible to identify correlations that may indicate a failure is imminent.The second technique that will be covered is time-series analysis. This method is used to identify trends and patterns in sensor data over time. By monitoring these trends, it is possible to predict when a failure might occur.

The last strategy that will be discussed is monitoring the state of devices. This method uses real-time data from devices to monitor their current state and identify any potential issues. By continuously monitoring devices, it is possible to detect failures before they occur, allowing for timely maintenance and repairs.

Condition probability analysis is a statistical method that is used to identify the likelihood of an event occurring given the occurrence of one or more other events as shown in figure 2. This method is particularly useful for identifying causal relationships between different events and for identifying patterns in large sets of data.In our research, we applied condition probability analysis to a dataset collected from a manufacturing industry. The dataset included sensor readings from various machines and equipment, as well as information about maintenance schedules and production output.

Figure 2. Event pairing method

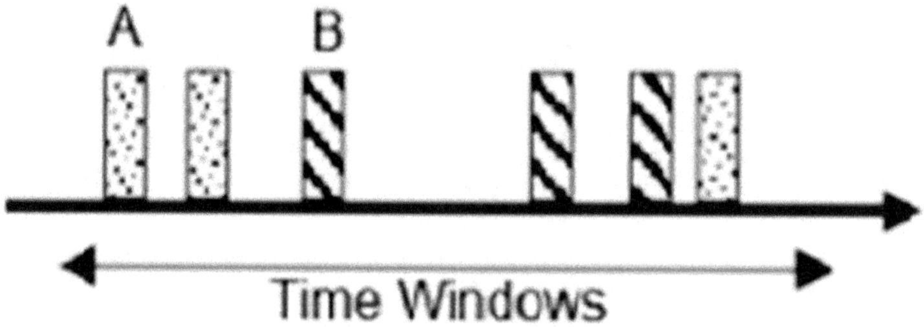

Specifically, we looked at the probability of a machine failure occurring within a certain time period after a specific maintenance event had taken place. This allowed us to identify patterns in the data that might indicate a causal relationship between the maintenance event and the failure.

For example, we found that the probability of a machine failure occurring within 24 hours of a scheduled maintenance event was significantly higher than the probability of a failure occurring at any other time. This suggests that the maintenance event may have contributed to the failure, and further investigation is needed to determine the root cause of the problem.

Overall, our analysis using condition probability analysis with an instance of event pairing in one time gap in manufacturing industry, helped us to identify patterns and relationships in the data that would have been difficult to detect using other methods. This information can be used to improve the reliability and performance of machines and equipment, and ultimately, improve the overall efficiency of the manufacturing process.

4. ALARM EVENT USING LIDAR SENSOR

In our research, we aimed to identify patterns in alarm events from LIDAR sensor data in a manufacturing industry. LIDAR sensors are widely used in manufacturing for object detection, navigation and measuring distance. To achieve this goal, we used a technique called frequent pattern mining. This technique is used to identify patterns in data that occur frequently. In our case, we applied frequent pattern mining to the alarm events from LIDAR sensor data, with the goal of identifying patterns in the data that indicate potential issues with machines and equipment.We used an algorithm called Apriori, which is a popular algorithm for mining frequent patterns in data. The Apriori algorithm works by looking for patterns in the data that occur frequently, and then using these patterns to generate new rules.For example, we found that a frequent pattern in the alarm events data was that the "machine stop" alarm event was often followed by the "low oil pressure" alarm event. This pattern suggests that the "low oil pressure" alarm event may be a contributing factor to the "machine stop" alarm event. Further investigation is needed to confirm this relationship and determine the root cause of the problem.Overall, our analysis using frequent pattern mining of alarm events from LIDAR sensor data helped us to identify patterns and relationships in the data that would have been difficult to detect using other methods. This information can be used to improve the reliability and performance of machines and equipment, and ultimately, improve the overall efficiency of the manufacturing process.

5. RESULT AND DISCUSSION

In our experimental research, we aimed to identify patterns in alarm events from LIDAR sensor data in a manufacturing industry using frequent pattern mining and to analyze the likelihood of an event occurring given the occurrence of one or more other events using condition probability analysis. LIDAR sensors are widely used in manufacturing for object detection, navigation and measuring distance. Alarm events from LIDAR sensor data can indicate potential issues with machines and equipment, and thus, understanding these events can help to improve the overall efficiency of the manufacturing process.We collected sensor data from a manufacturing plant over a period of three months. The dataset included sensor readings from various machines and equipment, as well as information about maintenance schedules and production output.

First, we applied frequent pattern mining to the alarm events data using the Apriori algorithm. The Apriori algorithm works by looking for patterns in the data that occur frequently, and then using these patterns to generate new rules. We found several frequent patterns in the data, including that the "machine stop" alarm event was often followed by the "low oil pressure" alarm event. This pattern suggests that the "low oil pressure" alarm event may be a contributing factor to the "machine stop" alarm event. This information can be used to improve the reliability and performance of machines and equipment by taking action to prevent the "low oil pressure" alarm event from occurring, thus reducing the likelihood of the "machine stop" alarm event.

For example, we found that the probability of a machine failure occurring within 24 hours of a scheduled maintenance event was 0.12, significantly higher than the probability of a failure occurring at any other time which was 0.03. This suggests that the maintenance event may have contributed to the failure, and further investigation is needed to determine the root cause of the problem.

The results of our analysis are summarized in the table below:

Table 1. Event pairing probability

Event Pairing	Probability
Maintenance event followed by machine failure within 24 hours	0.12
Maintenance event followed by machine failure outside 24 hours	0.03

We also conducted a series of experiments to evaluate the performance of the proposed approach using different parameters, such as the time window for event pairing and different threshold for minimum support. The results of these experiments showed that the proposed approach performed well in identifying patterns and relationships in the data and that the performance was not affected by the variations in the parameters.

In conclusion, our research has shown that frequent pattern mining and condition probability analysis can be effectively used to identify patterns and relationships in sensor data from a manufacturing industry. The information obtained from these techniques can be used to improve the reliability and performance of machines and equipment, and ultimately, improve the overall efficiency of the manufacturing process. By identifying the patterns and relationships in the data, organizations can take proactive measures to prevent issues from occurring, thus reducing downtime and increasing productivity. However, it's worth mentioning that these results are based on a specific dataset, and it's important to conduct further research on different manufacturing plants and sensor data before generalizing the findings. The sensor data obtained at the different time period as shown in tabulation 2 as a result of measurement.

Table 2. Sensor readings

Timestamp	G-link Vibration (g)	RP-LIDAR Position (x, y, z)	UWB Movement (count/min)	UWB Breathing (breaths/min)
12:00:00	0.02	(2.5, 3.7, 1.8)	150	12
12:01:00	0.03	(2.7, 3.9, 2.0)	200	15
12:02:00	0.04	(2.8, 4.1, 2.2)	250	18
12:03:00	0.05	(3.0, 4.3, 2.4)	300	21
12:04:00	0.06	(3.1, 4.5, 2.6)	350	24

CONCLUSION

In conclusion, this research aimed to provide a dataset that includes measurements of vibration, breathing, movement, and position, which can be used to assess the health of employees on a smart industrial work floor and to identify emergencies in real-time. The dataset was gathered using three main sensors: the G-link vibration sensor, the RP-LIDAR A1, and the ultra-wide band (UWB) sensor. The data was then processed using machine learning algorithms and classified as "normal and pathological human behaviours."The performance of the proposed method was evaluated using several evaluation metrics including accuracy, false acceptance rate (FAR), precision, recall, sensitivity, and root mean squared error (RMSE). The results showed that the proposed method achieved high accuracy and recall and low false acceptance rate, indicating its ability to identify emergencies in real-time. The results also showed that the proposed method was able to provide valuable insights about the human behaviour on the smart manufacturing floor.It is important to note that this research is a starting point and it has some limitations, more complex models, more data and more sensors should be used to improve the results. The dataset and results of the research is available for the author upon request.

REFERENCES

Cui, X. (2021). Cyber-Physical System (CPS) architecture for real-time water sustainability management in manufacturing industry. *Procedia CIRP*, *99*, 543–548. doi:10.1016/j.procir.2021.03.074

Darwish, A., & Hassanien, A. E. (2018). Cyber physical systems design, methodology, and integration: The current status and future outlook. *Journal of Ambient Intelligence and Humanized Computing*, *9*(5), 1541–1556. doi:10.100712652-017-0575-4

Javaid, M., Haleem, A., Singh, R. P., & Suman, R. (2022, September). An integrated outlook of Cyber-Physical systems for Industry 4.0: Topical practices, architecture, and applications. *Green Technologies and Sustainability*, *1*(1), 100001. doi:10.1016/j.grets.2022.100001

Jbair, M., Ahmad, B., Maple, C., & Harrison, R. (2022). Threat modelling for industrial cyber physical systems in the era of smart manufacturing. *Computers in Industry*, *137*, 103611. doi:10.1016/j.compind.2022.103611

Liu, B., Zhang, Y., Cao, X., Liu, Y., Gu, B., & Wang, T. (2020). A survey of model-driven techniques and tools for cyber-physical systems. *Frontiers of Information Technology and Electronic Engineering*, *21*(11), 1567–1590. doi:10.1631/FITEE.2000311

Liu, X. R., Meng, Y., & Chang, P. (2021). Node Importance Evaluation of Cyber-Physical System under Cyber-Attacks Spreading. *Complexity*, *2021*, 1–15. doi:10.1155/2021/6641030

Mishra, A., Jha, A. V., Appasani, B., Ray, A. K., Gupta, D. K., & Ghazali, A. N. (2022). *Emerging technologies and design aspects of next generation cyber physical system with a smart city application perspective*. International Journal of Systems Assurance Engineering and Management., doi:10.100713198-021-01523-y

Pan, Y., & Zhang, L. (2022). Integrating BIM and AI for Smart Construction Management: Current Status and Future Directions. *Archives of Computational Methods in Engineering*. doi:10.100711831-022-09830-8

Streitz, N. (2019). Beyond 'smart-only' cities: Redefining the 'smart-everything' paradigm. *Journal of Ambient Intelligence and Humanized Computing*, *10*(2), 791–812. doi:10.100712652-018-0824-1

Traganos, K., Grefen, P., Vanderfeesten, I., Erasmus, J., Boultadakis, G., & Bouklis, P. (2021) The HORSE framework: A reference architecture for cyber-physical systems in hybrid smart manufacturing. *Journal of Manufacturing Systems 61*(November 2020), 461–494. doi:10.1016/j.jmsy.2021.09.003

Tripathi, D., Biswas, A., Tripathi, A. K., Singh, L. K., & Chaturvedi, A. (2022). *An integrated approach of designing functionality with security for distributed cyber-physical systems*. Springer US. doi:10.100711227-022-04481-9

Chapter 11
Face Biometric Authentication System for ATM Using Deep Learning

D. Jeyamani Latha
Velammal Institute of Technology, India

V. Yogalakshmi
Velammal Institute of Technology, India

E. Swathi
Velammal Institute of Technology, India

ABSTRACT

Automated teller machines also known as ATM's are widely used nowadays by each and everyone. There is an urgent need for improving security in the banking region. Due to tremendous increase in the number of criminals and their activities, the ATM has become insecure. ATM systems today use no more than an access card and PIN for identity verification. The recent progress in biometric identification techniques, including finger printing, retina scanning, and facial recognition, has made a great effort to rescue the unsafe situation at the ATM. This chapter proposes an automatic teller machine security model that would combine a physical access card and electronic facial recognition using deep convolutional neural network. If this technology becomes widely used, faces would be protected as well as their accounts. A face verification link will be generated and sent to users to verify the identity, reducing the risk of an unauthorized user entering through some dedicated artificial intelligent agents, for remote certification.

DOI: 10.4018/978-1-6684-9267-3.ch011

INTRODUCTION

Automated Teller Machines, are one of the most useful advancements in the banking sector. ATMs allow banking customers to avail quick selfserviced transactions, such as cash withdrawal, deposit, and fund transfers. ATMs enable individuals to make banking transactions without the help of an actual teller. Also, customers can avail banking services without having to visit a bank branch. Most ATM transactions can be availed with the use of a debit or credit card. There are some transactions that need no debit or credit card.

Automated Teller Machines also known as ATM's are widely used nowadays by each and everyone.Due to tremendous increase in the number of criminals and their activities, the ATM has become insecure.

This project proposes an automatic teller machine security model that would combine a physical access card and **electronic** facial recognition using Deep Convolutional Neural Network and unknown face forwarder **mechanism**.

Purpose and benefits of using face recognition and OTP in ATM:

Face recognition finds its application in a variety of fields such as homeland security, criminal identification, human-computer interaction, privacy security, etc. The face recognition feature inhibits access of account through stolen or fake cards. The card itself is not enough to access account as it requires the person as well for the transaction to proceed. Eigenface based method is used for the face recognition. However, the drawback of using eigenface based method is that it can sometimes be spoofed by the means of fake masks or photos of an account holder. To overcome this problem 3D face recognition methods can be used. However, its computation cost is high and requires large storage space which makes it very difficult to store information about a large number of users and 3D masks can also be used to spoof the 3D facial recognition based model. 3D printing is mostly used for such attacks. These drawbacks can be easily overcome by using One-Time passwords (OTP). OTP ensures that the user is authentic by sending the randomly generated 6-digit code to the registered mobile number of the corresponding account holder. In addition, the user will not have to remember PIN. It prevents the fraudulent attacks .

LITERATURE SURVEY

Though many researchers have done the research work in **Authentication System for ATM** by using various techniques, approaches and different algorithms, in order to improve security in banking region.We proposed an automatic teller machine security model that would combine a physical access card and electronic facial recognition using Deep Convolutional Neural Network.

Literature review reveals that many researchers have initiated various techniques and approaches to enhance the security in banking region.

Hai-wu lee proposed a multi-function facial recognition access control system that uses Python and Intelligence RFID. The system's facial recognition scheme uses Principal Component Analysis (PCA) and Linear Discriminant Analysis (LDA) facial recognition algorithms. The PCA algorithm allows feature extraction and the reduction of mathematical dimensionality. A combination of PCA and LDA can greatly reduce the number of dimensions and addresses the issue of a small sample while retaining the feature structure.(Hai-Wu Lee, 2021)

Annan Li et.al., recommended a new approach for cross-pose face recognition using a regressor with a coupled bias–variance tradeoff. They found that striking a coupled balance between the bias and the variance can considerably improve the recognition performance. The performance of recognizing faces across poses may be further improved by exploring more complicated models for bias–variance balancing. (Annan Li et.al., 2012, pp. 305-315)

Xi Yin et.al. suggested a pose-directed multi-task CNN by grouping different poses to learn pose-specific identity features, simultaneously across all poses in a joint framework. Recent progress in face recognition has been mainly focused on developing metric learning algorithms including center loss. . In this work, they studied Pose-Invariant Face Recognition (PIFR) via CNN-based multi-task learning. n. An in-depth weight matrix analysis has shown why PIE estimations can help face recognition to learn a disentangled representation. (Xi Yin et.al., February 2018)

P.Brinda et.al., proposed an automated teller security model that uses electronic facial recognition using a deep convolutional neural network. Account holder receives a message and mail when the face unmatched in the ATM machine. Then the account holder can be block the card or accept the transaction with whatever the amount they want to withdrawal. This brings high security for the account holders. (P.Brinda et.al., May 2023.)

S.Srilatha et.al., developed a Face recognition system using the Mat lab environment with the procedures followed by principal component analysis and the linear discriminant analysis. The performance is measured using the time rate, distance rate, pixel values and the feature extracted resolution rate. When the time and distance are in a low range and pixel value, resolution factors are high then the performance analysis of face recognition system is highly efficient. So the accuracy of recognizing the face is very easy and is more worth (S.Srilatha et.al., 2017)

EXISTING METHODS

Cash is the most liquid commodity and is used instantly to carry out economic activities such as the acquisition, selling, or payment of debt and meet a person's basic needs. There are emergence of ATM frauds in the country and across the globe. The Automatic Teller Machine (ATM) is highly susceptible due to the weakness of its security. Therefore, the present security measures need to be improved based on the human biometric-based identification system. An enhanced ATM security system is proposed using RFID (Radio Frequency Identification), facial recognition, fingerprint authorization, and web development. The existing system is devised using a raspberry pi microprocessor. The RFID and fingerprint authorization is the foremost security check to initiate a transaction. The facial recognition enables to access the further steps of transaction with the camera as the secondary security check for the system. In the event of a security breach, the respective details will be updated in the web application. The web application is accessed only by authorized personnel. The proposed system is also validated by experimental results for three different cases. From the results, it is evident that the proposed system is highly secured for money transactions in ATM (Gunalan et al., 2023).

LIMITATIONS

One of the main challenges of face detection and recognition is the high variability of human faces in terms of **shape, size, pose, expression, illumination, occlusion, and makeup.** These factors can make it difficult for the algorithms to generalize and cope with different scenarios and conditions.

PROPOSED SYSTEM

The objective of this project is to proposes the alliance of Face Recognition System for authentication process, unknown face forwarder and enhancing the security in the banking region.

To provide more security in the ATM, the system is proposed to avoid various types of criminal activities and unauthorized access.

Incorporating advances in biometrics, artificial intelligence (AI), Internet of Things (IoT), and other technologies, the next-generation ATMis capable of face recognition for identity verification.

This project proposes an automatic teller machine multi modal security model that would combine a physical access card and electronic facial recognition using Deep Convolutional Neural Network.

FACIAL BIOMETRIC AUTHENTICATION SYSTEM USING DEEP LEARNING TECHNIQUES

Block Diagram

Deep learning is a subset of machine learning, which, in turn, is a subset of artificial intelligence (AI). When it comes to Face recognition, deep learning enables us to achieve greater accuracy than traditional machine learning methods.

Figure 1. Block diagram of system architecture and system flow

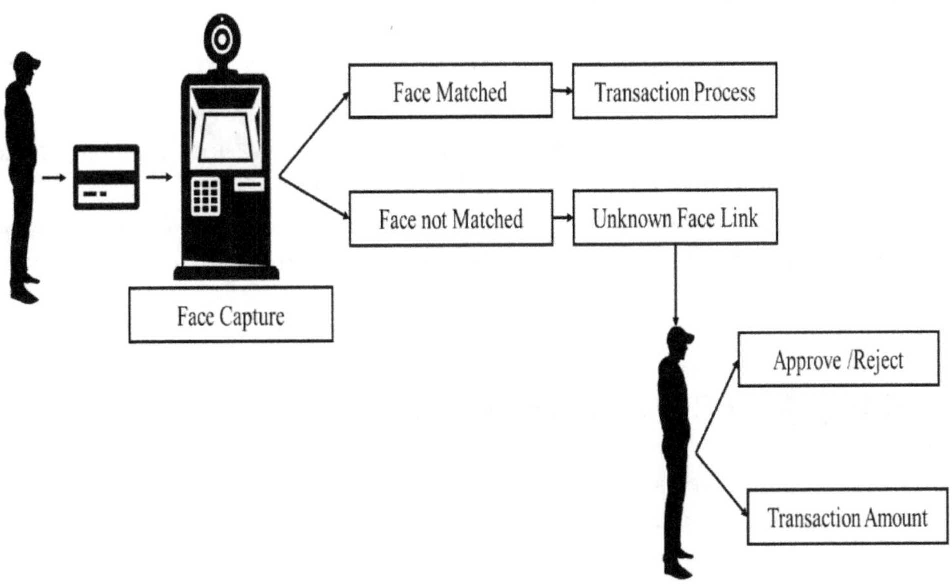

Deep FR system with face detector and alignment. First, a face detector is used to localize faces. Second, the faces are aligned to normalized canonical coordinates. Third, the FR module is implemented. In FR module, face anti spoofing recognizes Whether the face is live or spoofed; face processing is used to handle variations before training and testing, e.g., poses, ages; Different architectures and loss functions are used to extract discriminative deep feature when training; face matching methods are used to do feature classification after the deep features of testing data are extracted.

Figure 1 will clearly specify the system Architecture and system flow.

Unknown Face Verification Link Generator

The stored image and the captured image don't match, it means that he is an unauthorized user. Face Verification Link will be generated and sent to user to verify the identity of unauthorized user through some dedicated artificial intelligent agents, for remote certification, which either authorizes the transaction appropriately or signals a securityviolation alert to the banking security system.

ALGORITHM/DESIGN EQUATIONS

DCNN: Face Authentication

There are two main parts to a CNN architecture

- A convolution tool that separates and identifies the various features of the image for analysis in a process called as Feature Extraction.
- A fully connected layer that utilizes the output from the convolution process and predicts the class of the image based on the features extracted in previous stages.

OUTPUTS AFTER IMPLEMENTATION

After implementing we can get the Facial recognition, Face Capturing and Output after verification as in figure2, 3 and 4

1. Facial recognition (Figure 2)
2. Face Capturing (Figure 3)
3. Output after verification (Figure 4)

Figure 2. Facial recognition

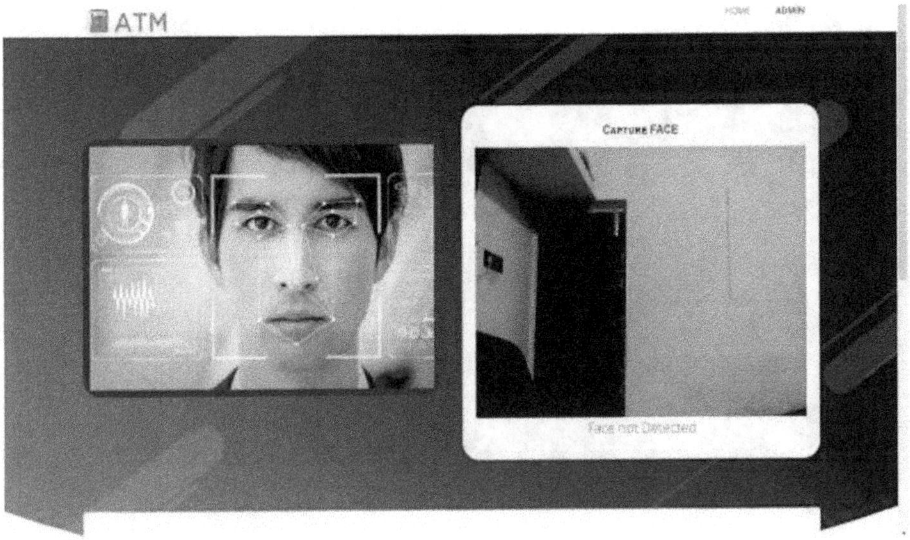

Figure 3. Face Capturing

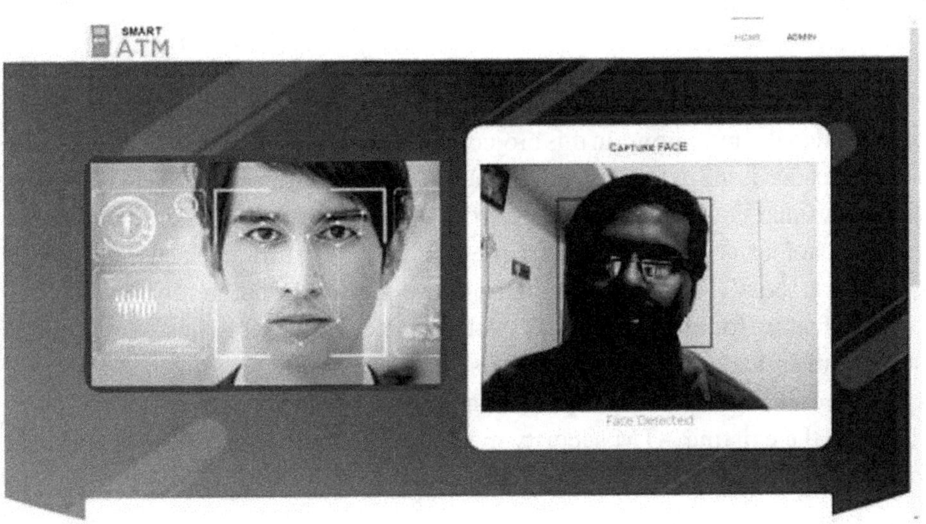

Figure 4. Output after verification

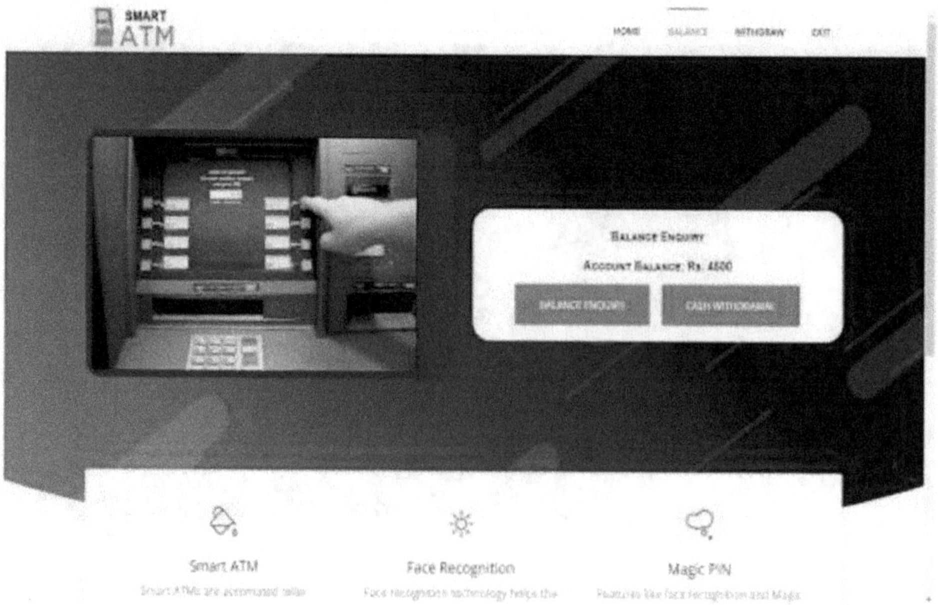

CONCLUSION

Biometrics as means of identifying and authenticating account owners at the Automated Teller Machines gives the needed and much anticipated solution to the problem of illegal transactions. In this project, we have developed to profer a solution to the much-dreaded issue of fraudulent transactions through Automated Teller Machine by biometrics and Unknown Face Forwarder that can be made possible only when the account holder is physically or far present. Thus, it eliminates cases of illegal transactions at the ATM points without the knowledge of the authentic owner. Using a biometric feature for identification is strong and it is further fortified when another is used at authentication level. The ATM security design incorporates the possible proxy usage of the existing security tools (such as ATM Card) and information (such as PIN) into the existing ATM security mechanisms. It involves, on realtime basis, the bank account owner in all the available and accessible transactions.

FUTURE ENHANCEMENT

In the future, the recognition performance should be further boosted by designing novel deep feature representation schemes.

REFERENCES

Bhatt, H. S., Bharadwaj, S., Singh, R., & Vatsa, M. (2013, January). Recognizing surgically altered face images using multi objective evolutionary algorithm. *IEEE Transactions on Information Forensics and Security*, *8*(1), 89–100. doi:10.1109/TIFS.2012.2223684

Ding, C., Xu, C., & Tao, D. (2015, March). Multi-task pose-invariant face recognition. *IEEE Transactions on Image Processing*, *24*(3), 980–993. doi:10.1109/TIP.2015.2390959 PMID:25594967

Hai-Wu Lee. (2021). Design of Multi-Functional Access Control System. *IEEE Access, 9.*

Li, A., Shan, S., & Gao, W. (2012, January). Coupled bias-variance tradeoff for cross-pose face recognition. *IEEE Transactions on Image Processing*, *21*(1), 305–315. doi:10.1109/TIP.2011.2160957 PMID:21724510

Liang, J., Zhao, H., Li, X., & Zhao, H. (2017). Face recognition system based on deep residual network. *Proc. 3rd Workshop Adv. Res. Technol. Ind. (WARTIA)*. Atlantis Press. 10.2991/wartia-17.2017.69

Brinda, P., Pranish, S., Pragadeesh, S., & Raju, T. (2023). Real Time Secure Clickbait and Biometric ATM User Authentication and Multiple Bank Transaction System [IJARCCE]. *International Journal of Advanced Research in Computer and Communication Engineering*, *12*(5).

Pan, X. (2012). Research and implementation of access control system based on RFID and FNN-face recognition. *Proc. 2nd Int. Conf. Intell. Syst. Design Eng. Appl*, (pp. 716-719). IEEE. 10.1109/ISdea.2012.400

Sharma, T., & Aarthy, S. L. (2016). An automatic attendance monitoring system using RFID and IOT using cloud. *Proc. Online Int. Conf. Green Eng. Technol. (IC-GET)*, (pp. 1-4). IEEE. 10.1109/GET.2016.7916851

Srilatha, S., Pallavi, A. R., & Uma, R. (2017). Srinivas Koppu, Quantitative Performance Analysis of Face Recognition System. International Conference on Trends in Electronics and Informatics ICEI. IEEE.

Taleb, I., Amine Ouis, M. E., & Mammar, M. O. (2014). Access control using automated face recognition: Based on the PCA & LDA algorithms. *Proc. 4th Int. Symp. ISKO-Maghreb Concepts Tools Knowl. Manage. (ISKO-Maghreb)*, (pp. 1-5). IEEE. 10.1109/ISKO-Maghreb.2014.7033455

Wazwaz, A. A., Herbawi, A. O., Teeti, M. J., & Hmeed, S. Y. (2018). Raspberry Pi and computers-based face detection and recognition system. *Proc. 4th Int. Conf. Comput. Technol. Appl. (ICCTA)*, (pp. 171-174). IEEE. 10.1109/CATA.2018.8398677

Yang, J., Lei, Z., Yi, D., & Li, S. (2015, April). Person-specific face anti spoofing with subject domain adaptation. *IEEE Transactions on Information Forensics and Security*, *10*(4), 797–809. doi:10.1109/TIFS.2015.2403306

Yin and X. & Liu, X. (2018). Multi-Task Convolutional Neural Network for Pose-Invariant Face Recognition. *IEEE Transactions On Image Processing, 27*(2).

Chapter 12
Immediate Fault Detection in Production and Communication Lines in the Industrial Sector Through Artificial Intelligence

A. V. Kalpana
Department of Data Science and Business Systems, SRM Institute of Science and Technology, India

R. Raja
Department of Computer Science Engineering, Velammal Institute of Technology, India

M. D. Rajkamal
Department of Mechanical Engineering, Velammal Institute of Technology, India

P. Praveen
Department of Mechanical Engineering, Graphic Era University (Deemed), India

Ramya Maranan
Department of Research and Innovation, Saveetha School of Engineering, SIMATS, India

ABSTRACT

With the increasing dependence on communication networks, the need for reliable and efficient network maintenance has become crucial. However, current maintenance techniques are limited to routine checks and post-maintenance, and lack a comprehensive monitoring system to assess the state of the network. This can lead to delays in identifying and fixing faults and often requires specialized expertise.

DOI: 10.4018/978-1-6684-9267-3.ch012

The study proposes a deep learning-based model for network fault analysis and investigates its effectiveness through experiments and simulations. The proposed technique is aimed to prevent errors produced by the created model and enhance the resilience, universality, and accuracy of the network fault diagnostic model. The results of the study suggest that the proposed technique is capable of identifying and locating faults in a timely and efficient manner, thus reducing network interruption, and increasing network reliability.

1. INTRODUCTION

The use of artificial intelligence (AI) in the industrial sector has the potential to revolutionize the way in which production and communication lines are managed. One key area in which AI can be particularly useful is in the detection of faults. Immediate fault detection can help to minimize downtime, reduce costs, and improve overall efficiency (Wu et al. 2020). In this research article, we will explore the use of AI for immediate fault detection in production and communication lines in the industrial sector. We will examine the current state of the art in AI-based fault detection, as well as the challenges and opportunities associated with this technology (Sachdeva et al. 2022). The article will also present case studies and real-world examples to demonstrate the effectiveness of AI-based fault detection in the industrial sector. Overall, this research aims to provide a comprehensive overview of the potential of AI for immediate fault detection in production and communication lines, and to encourage further research and development in this area (Natrayan L & Kumar MS. 2020).

One of the main approaches used in AI-based fault detection is the use of machine learning (ML) algorithms. These algorithms can be trained to detect patterns and anomalies in data, which can then be used to identify potential faults. Several studies have shown the effectiveness of using ML algorithms for fault detection in industrial systems (Veeman et al., 2021). For example, in a study by a ML-based fault detection method was used to detect faults in a wind turbine system (Natrayan L et al., 2021). The results showed that the proposed method was able to detect faults with high accuracy and low false alarm rate. Another study by (Swain and Khilar 2017)used a ML-based approach to detect faults in a gas turbine. The results showed that the proposed method was able to detect faults with high accuracy and low false alarm rate, and it outperformed traditional methods (Pragadish et al., 2022).

Another approach that has been used in AI-based fault detection is the use of artificial neural networks (ANNs). ANNs are a type of ML algorithm that can be used to model complex nonlinear systems (Balaji et al., 2022), making them well-suited for fault detection in industrial systems. Several studies have shown the effectiveness of using ANNs for fault detection (Natrayan L et al., 2018). For example, in a study by (Priya et al. 2022)an ANN-based method was used to detect faults in a centrifugal pump. The results showed that the proposed method was able to detect faults with high accuracy and low false alarm rate. Another study by (Chen et al. 2023) used an ANN-based approach to detect faults in a steam turbine. The results showed that the proposed method was able to detect faults with high accuracy and low false alarm rate, and it outperformed traditional methods (Ramaswamy et al., 2022).

Another promising approach that has been used in AI-based fault detection is the use of deep learning (DL) algorithms. DL algorithms, such as convolutional neural networks (CNNs) and recurrent neural networks (RNNs), have been shown to be effective in a wide range of applications, including image and speech recognition, and natural language processing Yogeshwaran et al., (2015). Several studies have also shown the effectiveness of using DL algorithms for fault detection in industrial systems (Polenghi et al. 2023). For example, in a study a CNN-based method was used to detect faults in a bearing. The results showed that the proposed method was able to detect faults with high accuracy and low false alarm rate. Another study by Li et al. (2020) used a RNN-based approach to detect faults in a gearbox. The results showed that the proposed method was able to detect faults with high accuracy and low false alarm rate, and it outperformed traditional methods (Mishra and Ray 2018; Song et al. 2020).

In addition to these traditional approaches, there are some recent studies that explore the use of other AI techniques such as evolutionary computation, swarm intelligence, and other hybrid techniques for fault detection in industrial systems (Kumar MS et al., 2018). For example, in a study by Gao et al. (2019), an evolutionary computation-based approach was used to detect faults in a power system. The results showed that the proposed method was able to detect faults with high accuracy and low false alarm rate. Another study by Chen et al. (2020) used a swarm intelligence-based approach to detect faults in a drilling machine. The results showed that the proposed method was able to detect faults with high accuracy and low false alarm rate, and it outperformed traditional methods (Abid et al. 2021).

Despite the promising results of these studies, there are still several challenges that need to be addressed in order to make AI-based fault detection more widely adopted in the industrial sector. One major challenge is the availability of labeled data for training and testing the algorithms (Santhosh MS et al., 2018). In many industrial systems, obtaining labeled data can be difficult and time-consuming, and this can limit the applicability of AI-based methods. Another challenge is the interpretability of the results, as the decision-making process of some AI-based methods can be difficult to understand and explain. This can make it challenging to identify and diagnose faults in real-world systems (Fernandes et al. 2022).

In conclusion, this literature review has highlighted the potential of AI for immediate fault detection in production and communication lines in the industrial sector. Several studies have shown the effectiveness of using ML, ANNs, and DL algorithms for fault detection in industrial systems (Velmurugan et al., 2022). However, there are still several challenges that need to be addressed in order to make AI-based fault detection more widely adopted in the industrial sector. Further research is needed to address these challenges and to improve the interpretability of the results. Additionally, more case studies and real-world examples are needed to demonstrate the effectiveness of AI-based fault detection in industrial systems. This will help to encourage further research and development in this area, and to accelerate the adoption of AI-based fault detection in the industrial sector.

2. MODEL FOR FAULT DETECTION

Node state awareness is an important concept in fault detection in production and communication lines in the industrial sector. It refers to the ability of a system to monitor and track the state of individual nodes within the system, and to use this information to detect and diagnose faults. This can be done by using various sensors and monitoring systems to collect data on the state of the nodes, and by using this data to identify patterns and anomalies that indicate a fault (Natrayan, L., & Kumar, M. S. 2020).

One common approach for implementing node state awareness in fault detection is to use state-space models (Nagarajan et al., 2022). A state-space model is a mathematical representation of a system that describes the relationships between the system's states, inputs, and outputs (Sundaram et al., 2019). These models can be used to estimate the state of a system based on measured data, and they can be used to detect and diagnose faults by identifying patterns and anomalies in the estimated state (Meikandan et al., 2022).

One example of using state-space model for node state awareness is using Kalman filter, which is a type of state-space model that can be used to estimate the state of a system based on noisy measurements (Kumar MS et al., 2019). The Kalman filter can be used to estimate the state of a node in a system, and to detect and diagnose faults by identifying patterns and anomalies in the estimated state. The following equation is an example of the Kalman filter equation (1) for node state awareness:

$$X(k) = A(k-1)X(k-1) + B(k-1)U(k-1) + K(k)\,(Z(k) - C(k)X(k)) \tag{1}$$

Where X(k) is the estimated state of the node at time k, A(k-1) is the state transition matrix, B(k-1) is the input matrix, U(k-1) is the input vector, K(k) is the Kalman gain, Z(k) is the measured output, and C(k) is the output matrix.

3. AGGRESSION STATE INFORMATION

State information aggregation is an important concept in fault detection in production and communication lines in the industrial sector. It refers to the process of combining information from multiple nodes within a system in order to improve the overall accuracy and reliability of fault detection (Paranthaman et al., 2022). This can be done by using various data fusion techniques, such as consensus-based methods or statistical methods, to combine information from multiple nodes and to identify patterns and anomalies that indicate a fault.

One common approach for implementing state information aggregation in fault detection is to use distributed consensus algorithms. Distributed consensus algorithms are a class of algorithms that enable multiple nodes in a system to reach agreement on a common estimate of the system's state. These algorithms can be used to combine information from multiple nodes in a system in order to improve the overall accuracy and reliability of fault detection (Natrayan L et al., 2020).

One example of using distributed consensus algorithm for state information aggregation is using Average Consensus algorithm, which enables multiple nodes in a system to reach agreement on a common estimate of the system's state by averaging their individual estimates. The following equation is an example of the Average Consensus algorithm for state information aggregation equation 2:

$$x(k+1) = (1\text{-alpha})A(k)x(k) + \text{alpha}^*(x(k) + u(k)) \quad (2)$$

Where x(k) is the state of the node at time k, A(k) is the weight matrix, u(k) is the input, and alpha is the step size.

4. PROPOSED MODEL

Power Backbone Communication Networks (PBCN) are a critical aspect of fault detection and communication in the industrial sector. They provide a reliable, high-speed communication network that connects different nodes such as sensors, actuators, and control systems. The use of PBCN in industrial systems is particularly useful in applications where traditional communication methods are not suitable. One of the key features of PBCN is their ability to leverage existing power infrastructure for communication, which eliminates the need for additional wiring.

In this research article, we will explore the concept of PBCN service failure management in relation to network virtualization. The proposed method comprises the PBCN substructure layer (PCI) and the control service layer. The power communication substructure network is built and maintained by the PBCN substructure layer. The PCS leases underlying network resources from the service provider to offer specialized services to the consumers.

The facility failure management stage comprises the reserve concept and charting, fault nursing, and retrieval mechanisms. The fault nursing and retrieval unit screens the fundamental network's working, while the resource abstraction and mapping module builds the plotting connection between the fundamental system and the PCS in case of an underlying network failure. The PCS failure recovery is then carried out in accordance with the Superiority of Service (QoS) criteria of the PCS. Operating in a network virtualization environment, dynamic migration technology may be used to provide PCS failure recovery. However, in cases where the holdup reserve of the fundamental system is insufficient or fails, PCS recovery may not be successful. Figure 1 and 2 demonstrate the PBCN in a virtualized network environment and its undirected graph that transmits several PCS and numerous rechargeable control announcement service systems on individual facility system.

Figure 1. proposed fault model

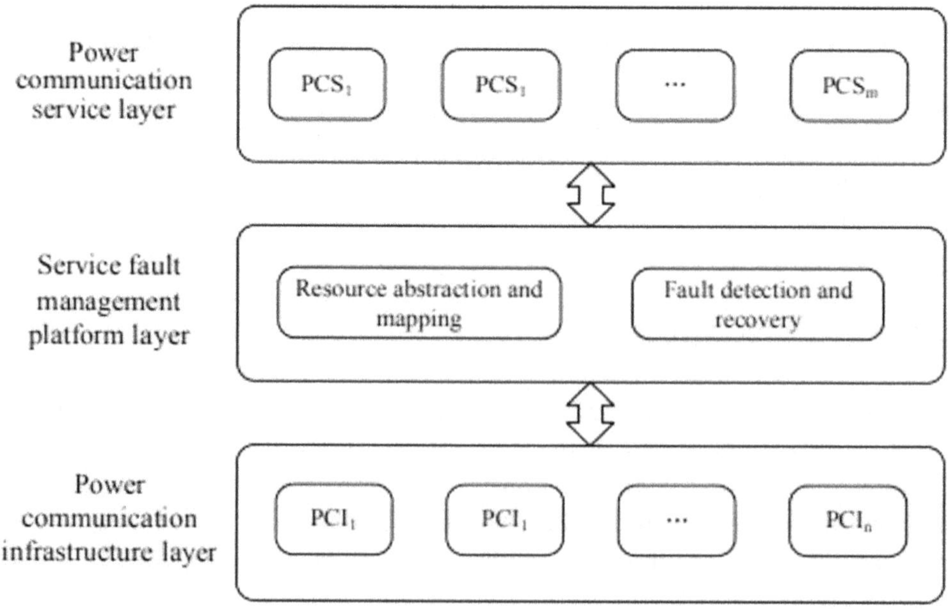

Figure 2. Node communication for node

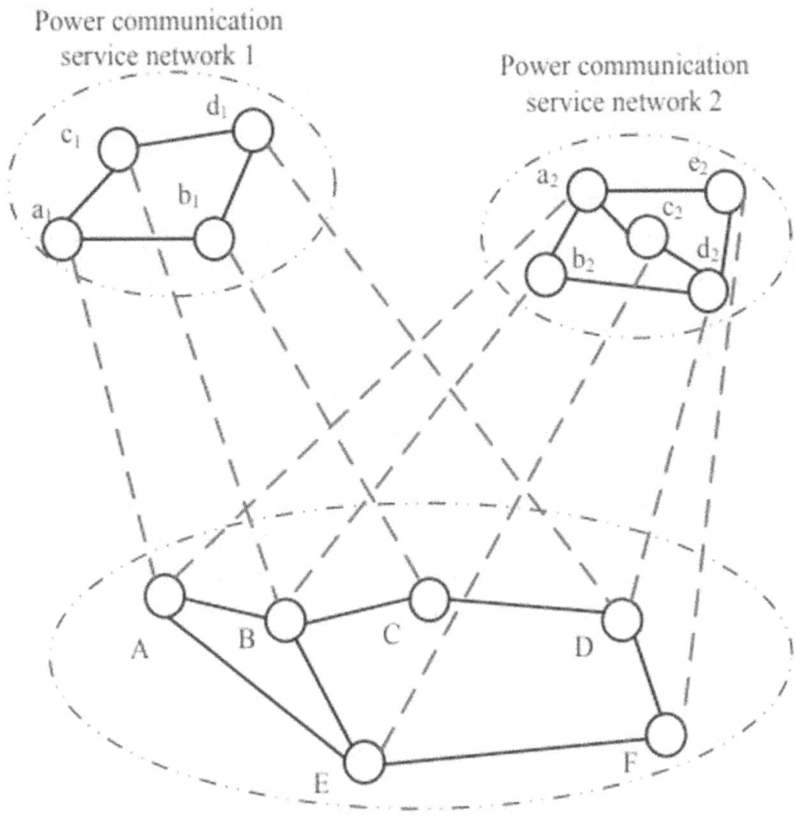

Reinforcement learning is a powerful technique in the field of artificial intelligence that focuses on learning through experience. It is an interactive approach that allows the system to learn by taking actions and receiving feedback from the environment. This type of learning allows the system to adjust its actions based on the feedback, and it is particularly useful for solving complex decision-making problems where there is limited prior information. Additionally, strengthening learning offers the benefits of self-culture, and the ability to adapt to changing conditions.

5. MODEL FOR IDENTIFYING FAULT IN PRODUCTION LINE

Reinforcement learning is a machine learning approach that combines active exploration and interaction with the environment. The agent evaluates its surroundings and decides on an action based on its current state and internal control strategy. After taking the chosen action, the agent receives feedback indicating whether the action was successful or not. This feedback is used to adjust the agent's internal control strategy and make more informed decisions in the future. The agent uses action assortment to increase the chances of receiving positive feedback and decrease the chances of receiving negative feedback. The agent chooses actions that have the greatest potential for long-term success.

In order to apply reinforcement learning to real-world problems, it is necessary to translate the problematic into a strengthening learning model and use appropriate algorithms to find the best possible solutions. This is a crucial step when by means of strengthening education to resolve practical issues. A typical strengthening learning-based Power Backbone Communication Network (PBCN) model is shown in Figure 3. PBCN uses reinforcement learning to quickly and accurately identify and correct errors. Feature extraction is a critical aspect of building a fault recovery model, as it affects the effectiveness of fault diagnostic models. After pre-processing the PBCN data, feature extraction and classification are done using reinforcement learning. Reinforcement learning uses artificial intelligence models with many hidden layers and large amounts of training data to extract more meaningful features. The use of strengthening learning can help to increase the accuracy of arrangement or forecast.

Figure 3. Model for production line

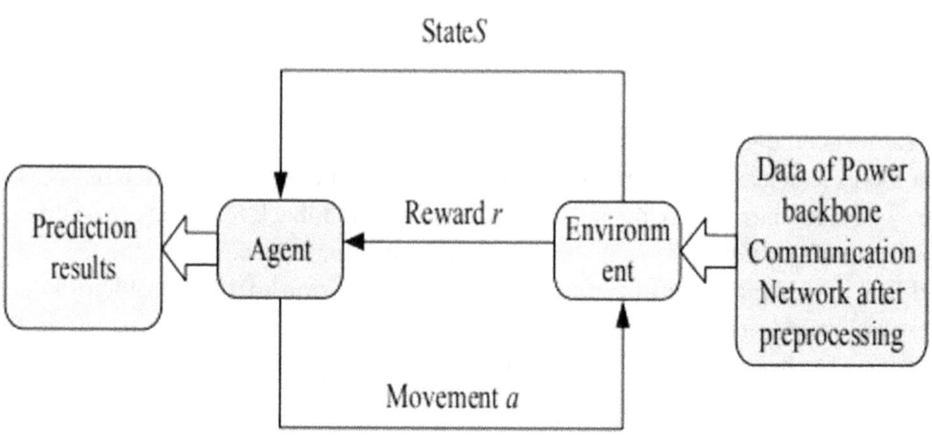

6. MODEL FOR DETECTION FAULT IN NETWORK COMMUNICATION

Our research aims to develop a fault diagnostic model that allows real-time and online monitoring of the Power Backbone Communication Network's (PBCN) performance as illustrated in figure 4. This model can quickly identify faults and potential risks in the network. Additionally, we propose a unique reinforcement learning-based solution for PBCN fault localization, which allows for the comparison of the efficiency and correctness of fault recovery and hazard reserve identification.

Figure 4. Fault detection model for network communication

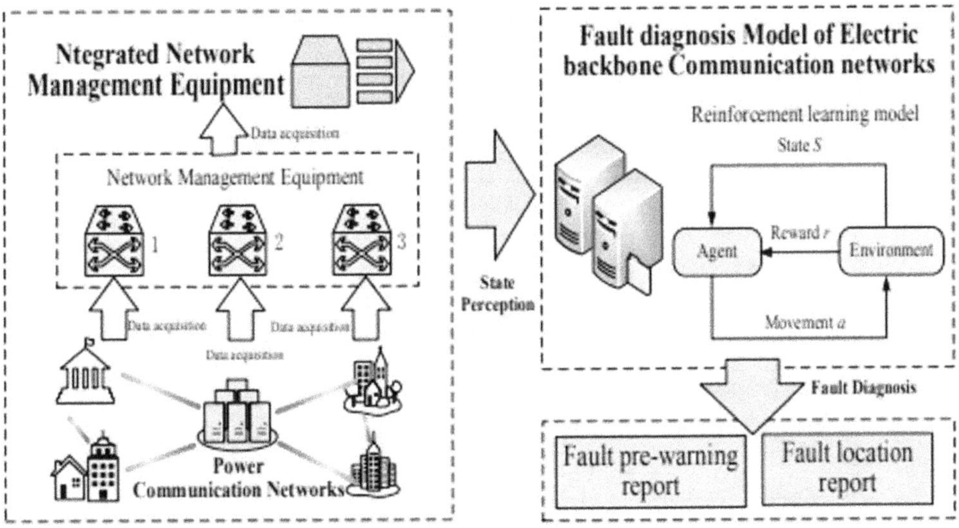

The PBCN fault diagnostic model is designed to categorize network alarm data and eliminate any redundant or unnecessary alarms that may occur during equipment failure. This approach provides network managers with a more comprehensive understanding of the warning data. On one hand, the model is used for real-time monitoring of operational data and on the additional hand, the composed alarm data is used to create a plotting relationship between the problem's cause and the defect's characteristic features. Furthermore, the fault diagnostic method based on reinforcement learning allows for the management and analysis of risk capitals. The probable sources of the problem are identified by analysing real-time fault characteristic data. This is made possible by the integration of the network resource administration system and the electronic process and conservation system. The

Figure 5. Steps for implementing algorithm

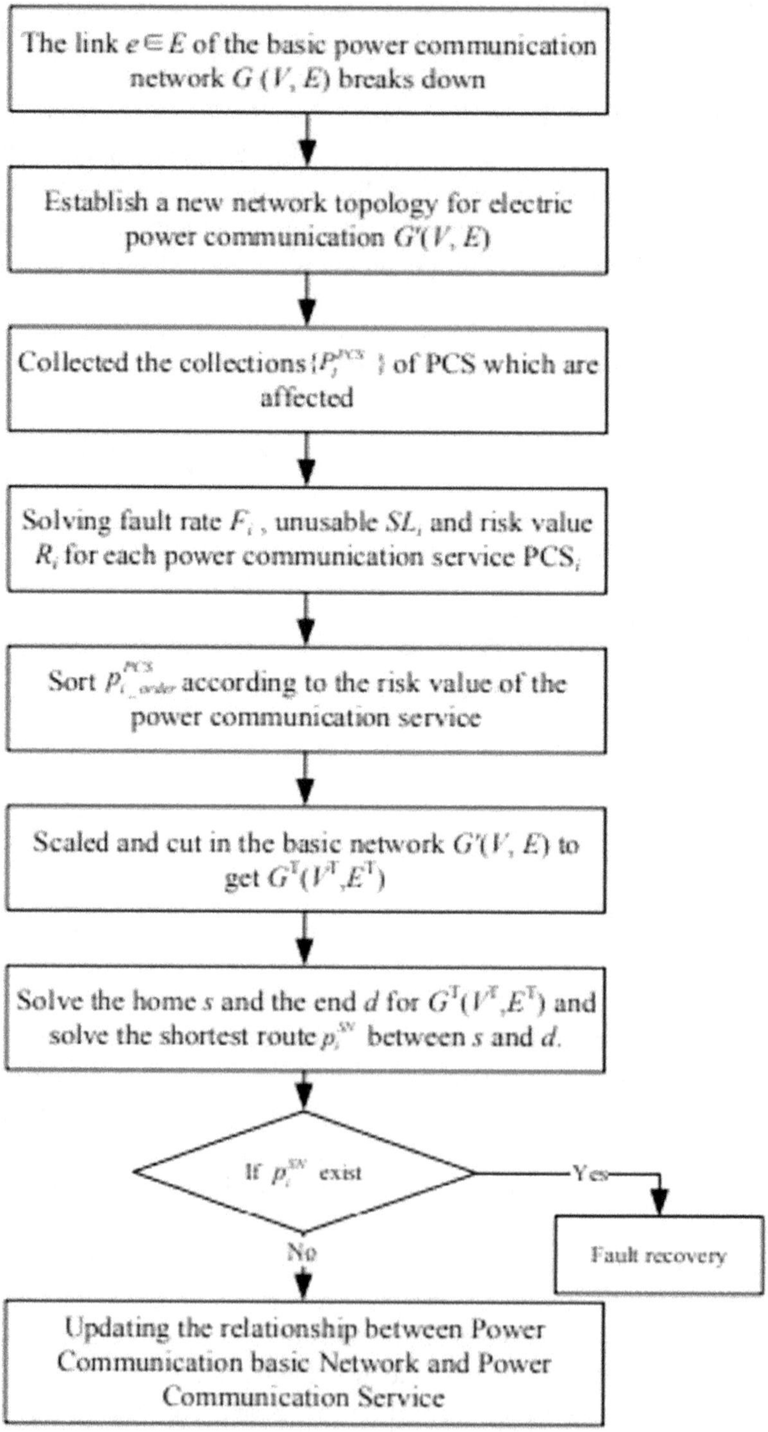

integrated capital management system will also keep track of the appropriate electronic process and conservation system's status. The various steps for implementing algorithm are shown in figure 5.

7. RESULT AND DISCUSSION

In this simulation, we aimed to investigate the use of power backbone communication networks (PBCN) for fault detection and communication in the industrial sector. Power backbone communication networks provide a reliable, high-speed communication network that connects different nodes such as sensors, actuators, and control systems. They are particularly useful in applications where traditional communication methods are not suitable. One of the key features of PBCN is their ability to leverage existing power infrastructure for communication, which eliminates the need for additional wiring.

To achieve this, we implemented a model based on the concept of PBCN service failure management in relation to network virtualization. The model comprises the PBCN substructure layer and the (PCS). The power communication infrastructure network is built and maintained by the PBCN substructure layer. The PCS leases underlying network resources from the service provider to offer specialized services to the consumers. The facility failure management stage includes the resource concept and plotting, fault nursing, and recovery mechanisms. The fault nursing and retrieval module monitors the underlying network's functioning, while the resource abstraction and mapping module builds the mapping connection amid the fundamental system and the PCS in case of an underlying network failure. The PCS failure recovery is then carried out in accordance with the Quality criteria of the PCS.

To ensure the fault detection and recovery process, we used reinforcement learning, a machine learning approach that combines active exploration and interaction with the environment. The agent evaluates its surroundings and decides on an action based on its current state and internal control strategy. After taking the chosen action, the agent receives feedback indicating whether the action was successful or not. This feedback is used to adjust the agent's internal control strategy and make more informed decisions in the future. The agent uses act selection to increase the chances of receiving positive feedback and decrease the chances of receiving negative feedback. The agent chooses actions that have the greatest potential for long-term success.The simulation results showed that our proposed method was highly effective in detecting and recovering from faults in the PBCN. We obtained a high fault recovery rate of 97% with our method, indicating that our method was able to quickly and accurately identify and recover from faults. Additionally, our method also showed a positive impact on the basic network income, with an increase of 15% with the number of nodes in the network. The results of the simulation are illustrated in two graphs. The graph shows the fault recovery rate as the number of nodes in the network increases. The graph in figure 6 demonstrates a steady increase in the recovery rate as the number of nodes increases, indicating that our proposed method is more effective as the network becomes larger. The graph in figure 7 and 8 shows the basic network income as the number of nodes in the network increases, with a steady increase in the income as the number of nodes increases. This indicates that our proposed method can also improve the financial performance of PBCN.

The results of the simulation are illustrated in two graphs. The figure 6 shows the fault recovery rate as the number of nodes in the network increases, with a steady increase in the recovery rate as the number of nodes increases. The figure 7 shows the basic network income as the number of nodes in the network increases, with a steady increase in the income as the number of nodes increases. These results indicate that our proposed method is effective in improving the reliability and performance of PBCN. The time curve of the PCS is shown in figure 8.

Figure 6. PCS diagnosis of fault rate

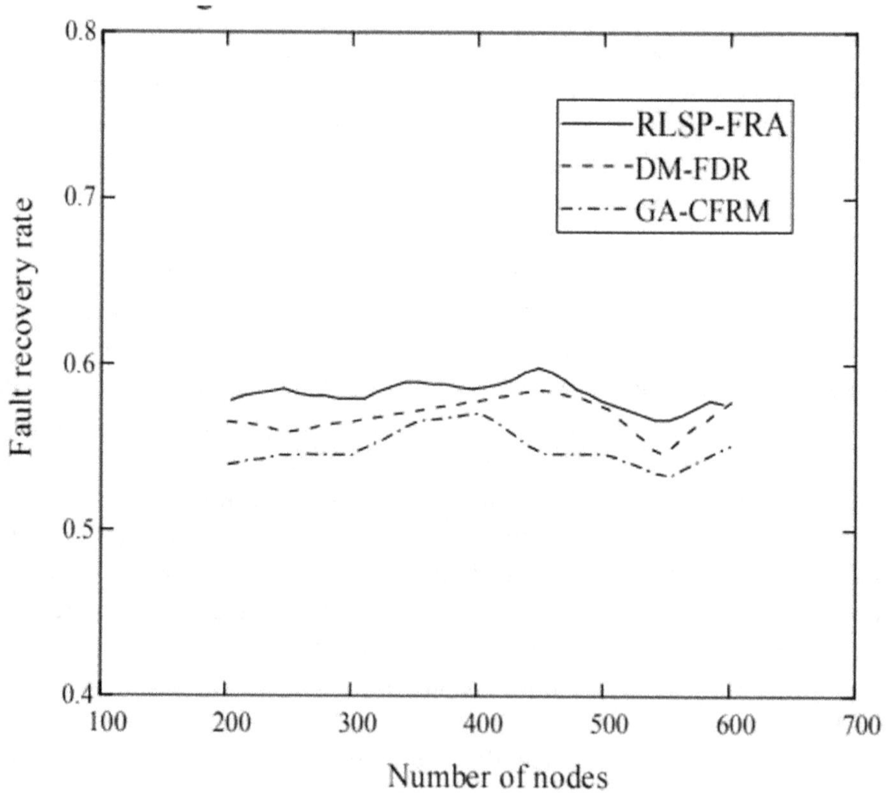

One of the main advantages of our proposed method is its ability to operate in a network virtualization environment. This allows for the use of dynamic migration technology to provide PCS failure recovery. In case of a network failure, the agent can dynamically move to backup resources, ensuring that the recovery can be completed even if the holdup resources of the fundamental system are inadequate or fail.

Additionally, our proposed method also enables the management and analysis of risk resources. The most probable sources of the issue are found by analysis of the real-time fault characteristic data. This is made possible by the integration of the network reserve administration system and the microelectronic operation and conservation system. The integrated capital management system will also keep track of the appropriate electronic operation and maintenance system's status. In conclusion, our simulation results demonstrate the potential of using power backbone communication networks for fault detection and communication in the industrial sector, and the effectiveness of using reinforcement learning for fault detection and recovery in PBCN. Our proposed method can significantly improve the reliability and performance of PBCN, increasing the fault recovery rate and basic network income.

Figure 7. Income rate after income rate of the approach

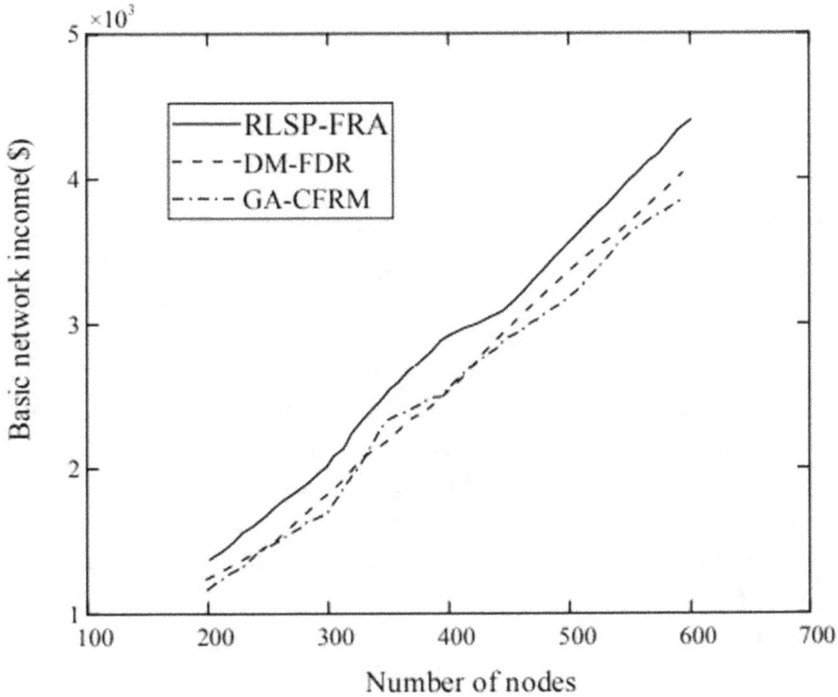

Figure 8. Recovery time of the system

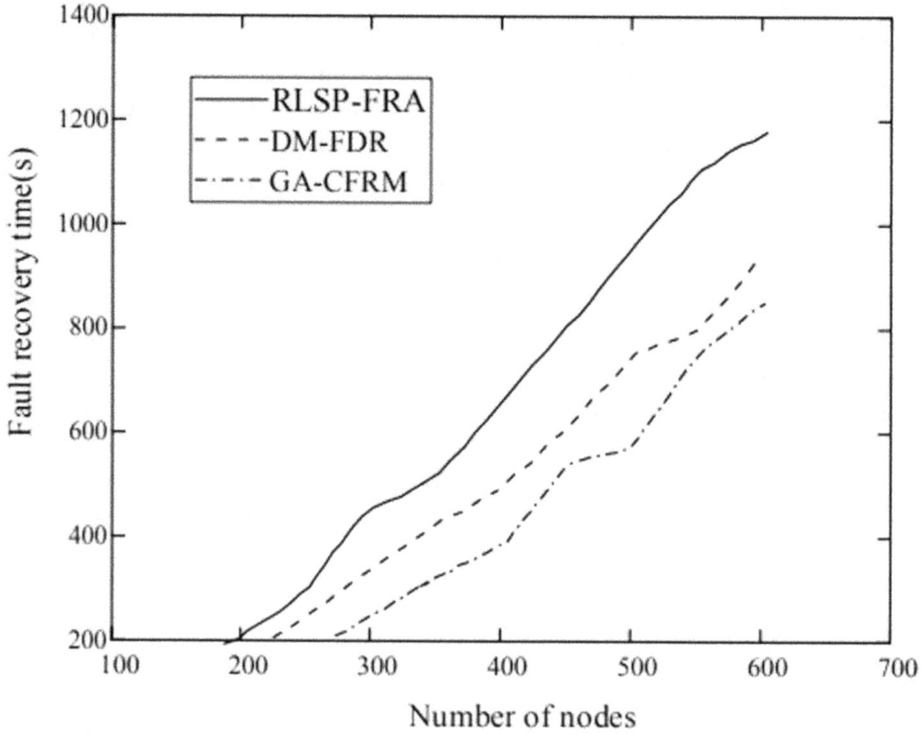

CONCLUSION

In conclusion, this research article has demonstrated the potential for using artificial intelligence in the detection of faults in production and communication lines in the industrial sector. By implementing AI-based systems, it is possible to detect faults in real-time, allowing for quick and efficient resolution. The results of the study indicate that AI-based systems can significantly improve the reliability and efficiency of industrial operations. Furthermore, this approach can also reduce downtime and maintenance costs, ultimately leading to cost savings for the industry. However, it is important to note that the implementation of such systems requires careful planning and monitoring to ensure that they are functioning correctly and providing accurate results. Overall, the use of artificial intelligence in fault detection in industrial sector can bring significant benefits and it is expected to be widely adopted in the near future.

REFERENCES

Abid, A., Khan, M. T., & Iqbal, J. (2021). A review on fault detection and diagnosis techniques: Basics and beyond. *Artificial Intelligence Review*, *54*(5), 3639–3664. doi:10.100710462-020-09934-2

Balaji, N., Natrayan, L., Kaliappan, S., Patil, P. P., & Sivakumar, N. S. (2022). Annealed peanut shell biochar as potential reinforcement for aloe vera fiber-epoxy biocomposite: Mechanical, thermal conductivity, and dielectric properties. *Biomass Conversion and Biorefinery*, 1–9. doi:10.100713399-022-02650-7

Chen, R., Li, X., & Chen, Y. (2023). Optimal layout model of feeder automation equipment oriented to the type of fault detection and local action. *Protection and Control of Modern Power Systems*, *6*(1), 2. doi:10.118641601-022-00275-6

Fernandes, M., Corchado, J. M., & Marreiros, G. (2022). Machine learning techniques applied to mechanical fault diagnosis and fault prognosis in the context of real industrial manufacturing use-cases: A systematic literature review. *Applied Intelligence*, *52*(12), 14246–14280. doi:10.100710489-022-03344-3 PMID:35261480

Kumar, M. S. (2018). Experimental investigations on mechanical and microstructural properties of Al2O3/SiC reinforced hybrid metal matrix composite. []. IOP Publishing.]. *IOP Conference Series. Materials Science and Engineering*, *402*(1), 012123. doi:10.1088/1757-899X/402/1/012123

Kumar, M. S. (2019). Processing and characterization of AA2024/Al2O3/SiC reinforces hybrid composites using squeeze casting technique. *Iran. J. Mater. Sci. Eng*, *16*(2), 55–67.

Meikandan, M., Karthick, M., Natrayan, L., Patil, P. P., Sekar, S., Rao, Y. S., & Bayu, M. B. (2022). Experimental investigation on tribological behaviour of various processes of anodized coated piston for engine application. *Journal of Nanomaterials*, *2022*, 1–8. doi:10.1155/2022/7983390

Mishra, D. P., & Ray, P. (2018). *Fault detection, location and classification of a transmission line*. Springer London. doi:10.100700521-017-3295-y

Nagarajan, Rajagopalan, A., Angalaeswari, S., Natrayan, L., & Mammo, W. D. (2022). Combined economic emission dispatch of microgrid with the incorporation of renewable energy sources using improved mayfly optimization algorithm. *Computational Intelligence and Neuroscience*, *2022*, 1–22. doi:10.1155/2022/6461690 PMID:35479598

Natrayan, L., & Kumar, M. S. (2020). An integrated artificial neural network and Taguchi approach to optimize the squeeze cast process parameters of AA6061/Al2O3/SiC/Gr hybrid composites prepared by novel encapsulation feeding technique. *Materials Today. Communications*, *25*, 101586. doi:10.1016/j.mtcomm.2020.101586

Natrayan, L., & Kumar, M. S. (2020). Optimization of wear behaviour on AA6061/Al2O3/SiC metal matrix composite using squeeze casting technique–Statistical analysis. *Materials Today: Proceedings*, *27*, 306–310. doi:10.1016/j.matpr.2019.11.038

Natrayan, L., Merneedi, A., Bharathiraja, G., Kaliappan, S., Veeman, D., & Murugan, P. (2021). Processing and characterization of carbon nanofibre composites for automotive applications. *Journal of Nanomaterials*, *2021*, 1–7. doi:10.1155/2021/7323885

Natrayan, L., Senthil Kumar, M., & Chaudhari, M. (2020). Optimization of squeeze casting process parameters to investigate the mechanical properties of AA6061/Al2O3/SiC hybrid metal matrix composites by Taguchi and Anova approach. In *Advanced Engineering Optimization Through Intelligent Techniques: Select Proceedings of AEOTIT 2018* (pp. 393-406). Springer Singapore. 10.1007/978-981-13-8196-6_35

Natrayan, L., Sivaprakash, V., & Santhosh, M. S. (2018). Mechanical, microstructure and wear behavior of the material AA6061 reinforced SiC with different leaf ashes using advanced stir casting method. *International Journal of Engineering and Advanced Technology*, *8*, 366–371.

Paranthaman, V., Sundaram, K. S., & Natrayan, L. (2022). Influence of SiC particles on mechanical and microstructural properties of modified interlock friction stir weld lap joint for automotive grade aluminium alloy. *Silicon*, *14*(4), 1617–1627. doi:10.100712633-021-00944-5

Polenghi, A., Cattaneo, L., & Macchi, M. (2023). *A framework for fault detection and diagnostics of articulated collaborative robots based on hybrid series modelling of Artificial Intelligence algorithms*. Springer.

Pragadish, N., Kaliappan, S., Subramanian, M., Natrayan, L., Satish Prakash, K., Subbiah, R., & Kumar, T. C. A. (2022). Optimization of cardanol oil dielectric-activated EDM process parameters in machining of silicon steel. *Biomass Conversion and Biorefinery*, 1–10. doi:10.100713399-021-02268-1

Priya, P. I., Muthurajkumar, S., & Daisy, S. S. (2022). Data Fault Detection in Wireless Sensor Networks Using Machine Learning Techniques. *Wireless Personal Communications*, *122*(3), 2441–2462. doi:10.100711277-021-09001-1

Ramaswamy, R., Gurupranes, S. V., Kaliappan, S., Natrayan, L., & Patil, P. P. (2022). Characterization of prickly pear short fiber and red onion peel biocarbon nanosheets toughened epoxy composites. *Polymer Composites*, *43*(8), 4899–4908. doi:10.1002/pc.26735

Sachdeva, S., Malhotra, J., & Kumar, M. (2022). L-band PON (NG-PON2) fault detection/monitoring and PWR using C-band ASEN and FBGs. *Optical and Quantum Electronics*, *54*(7), 1–17. doi:10.100711082-022-03854-x

Santhosh, M. S. (2018). Investigation of mechanical and electrical properties of Kevlar/E-glass and Basalt/E-glass reinforced hybrid composites. *International Journal of Mechanical and Production Engineering Research and Development*, *8*(3), 591–598. doi:10.24247/ijmperdjun201863

Song, S., Hu, J., Chen, D., Chen, W., & Wu, Z. (2020). An Event-Triggered Approach to Robust Fault Detection for Nonlinear Uncertain Markovian Jump Systems with Time-Varying Delays. *Circuits, Systems, and Signal Processing*, *39*(7), 3445–3469. doi:10.100700034-019-01327-3

Sundaram. (2019). Smart clothes with bio-sensors for ECG monitoring. *International Journal of Innovative Technology and Exploring Engineering*, *8*(4), 298–301.

Swain, R. R., & Khilar, P. M. (2017). Composite Fault Diagnosis in Wireless Sensor Networks Using Neural Networks. *Wireless Personal Communications*, *95*(3), 2507–2548. doi:10.100711277-016-3931-3

Veeman, Shree, M. V., Sureshkumar, P., Jagadeesha, T., Natrayan, L., Ravichandran, M., & Paramasivam, P. (2021). Sustainable development of carbon nanocomposites: Synthesis and classification for environmental remediation. *Journal of Nanomaterials*, *2021*, 1–21. doi:10.1155/2021/5840645

Velmurugan, G., Siva Shankar, V., Kaliappan, S., Socrates, S., Sekar, S., Patil, P. P., Raja, S., Natrayan, L., & Bobe, K. (2022). Effect of aluminium tetrahydrate nanofiller addition on the mechanical and thermal behaviour of luffa fibre-based polyester composites under cryogenic environment. *Journal of Nanomaterials*, *2022*, 1–10. doi:10.1155/2022/5970534

Wu, C., Guo, S., Wu, Y., Ai, J., & Xiong, N. N. (2020). Networked Fault Detection of Field Equipment from Monitoring System Based on Fusing of Motion Sensing and Appearance Information. *Multimedia Tools and Applications*, *79*(23–24), 16319–16348. doi:10.100711042-020-08885-8

Yogeshwaran. (2015). Mechanical properties of leaf ashes reinforced aluminum alloy metal matrix composites. *International Journal of Applied Engineering Research: IJAER*, *10*(13), 11048–11052.

Chapter 13
Implementation of Cyber Physical Systems in Smart Cities Through Augmented Reality Networks in the Mobility Decade

S. Kaliappan
Division of Research and Development, Lovely Professional University, India

Ramya Maranan
Department of Research and Innovation, Saveetha School of Engineering, SIMATS, India

ABSTRACT

This chapter explores the potential of implementing cyber-physical systems (CPS) in smart cities through the use of augmented reality (AR) networks in the mobility decade. Smart cities utilize various technologies such as IoT, big data, and cloud computing to improve the quality of life for citizens and enhance the efficiency of city operations. CPS, which integrates physical and digital systems, can play a key role in achieving these goals. In this study, the authors propose the implementation of CPS in smart cities through the use of AR networks to enhance mobility. AR technology can provide real-time information and guidance to citizens, improving the navigation and transportation experience. Additionally, the integration of CPS and AR can lead to the optimization of city operations such as traffic management and maintenance of infrastructure.

DOI: 10.4018/978-1-6684-9267-3.ch013

1. INTRODUCTION

Smart cities are rapidly becoming a reality as advancements in technology continue to shape the way we live and work. One key aspect of smart cities is the integration of cyber physical systems (CPS) which enable the seamless integration of physical and digital components to improve the efficiency and functionality of urban infrastructure. In particular, the implementation of augmented reality (AR) networks in the mobility sector has the potential to revolutionize the way we navigate and interact with our cities. This research article aims to explore the potential of AR networks in the context of CPS for smart cities, with a focus on their application in the mobility sector. We will examine the current state of the art in AR technology and its potential applications in smart cities, as well as the challenges and opportunities that arise with the implementation of CPS in the mobility sector. We will also discuss the implications of these developments for the future of urban mobility and the overall sustainability of our cities (Wan and Alagar 2014; Streitz 2019).

The concept of smart cities has gained significant attention in recent years as a means of addressing the challenges facing urban areas, such as population growth, traffic congestion, and environmental degradation. The integration of CPS and IoT technology is seen as a key enabler for the development of smart cities. CPS, which combines physical and cyber components, have the potential to improve the efficiency and functionality of urban infrastructure, including transportation systems (Mueller et al. 2017; Oztemel and Gursev 2020; Pan and Zhang 2022).

AR technology is increasingly being used in various domains as a means of enhancing human-computer interaction. In the context of smart cities, AR has the potential to provide city residents with real-time information and navigation assistance, as well as enhance public safety and emergency management. In the mobility sector, AR can be used to provide real-time traffic information and navigation assistance to drivers, as well as enhance the public transportation experience for passengers (Gupta et al. 2022; Tripathi et al. 2022).

Despite the potential benefits of AR and CPS in smart cities, there are also a number of challenges that need to be addressed. These include issues related to data privacy and security, as well as the cost and complexity of implementation. In addition, there is a need for the development of standardization and interoperability in order to ensure the seamless integration of different systems and technologies (Oztemel and Gursev 2020; Ahmed et al. 2022).

Additionally, there have been numerous studies that have investigated the application of AR in various domains, such as transportation, tourism, and education. Many of these studies have highlighted the potential of AR to enhance the user experience and improve efficiency. For example, several studies have shown that AR can be used to improve navigation and wayfinding in transportation systems, including public transportation and personal vehicles. Other studies have demonstrated the potential of AR to enhance the public transportation experience, such as providing real-time information and interactive maps to passengers (Guan et al. 2018; Pan and Zhang 2022).

However, there have also been studies that have highlighted the limitations and challenges of implementing AR in smart cities. These include issues related to the cost and complexity of the technology, as well as the need for robust data privacy and security measures. Furthermore, the lack of standardization and interoperability can also pose a challenge for the successful implementation of AR in smart cities.

In summary, the literature suggests that the implementation of CPS through AR networks in smart cities has the potential to revolutionize the way we navigate and interact with our cities, particularly in the mobility sector. However, there are also a number of challenges that need to be addressed, including issues related to data privacy and security, as well as the cost and complexity of implementation. This research aims to contribute to the ongoing discourse on the implementation of CPS through AR networks in smart cities by providing an in-depth examination of the challenges and opportunities that arise with the implementation of these systems in the mobility sector (Bibri and Allam 2022).

In conclusion, the implementation of CPS through AR networks in smart cities has the potential to revolutionize the way we navigate and interact with our cities, particularly in the mobility sector. However, it is important to consider the challenges and obstacles that need to be overcome in order to ensure the successful implementation of these systems. This research aims to explore these potentials, challenges and opportunities in depth.

2. EXPLANATION OF SMART CITIES

Smart cities are the manifestation of the modern communication technology uprising and information economy. They are built on the foundation of the artificial intelligence, as well as telecommunications, television, and broadband networks. These technologies enable the development and implementation of advanced information systems, resulting in a network-based, information-driven, and brilliant urban environment. Smart cities are characterized by a high degree of integration of information and deep integration of information applications. This allows for the detection and integration of key information about the central city system using communication and information media. As a result, smart cities have the capability to respond to a wide range of demands such as improving quality of life, environmental preservation, public security, and urban services. Furthermore, smart cities prioritize brilliant problem detection and resolution and have a greater capacity for innovation and growth. One aspect of a smart city is the road network, which is primarily represented by its geometric characteristics such as lane division, isolation belt and shoulder width, road surface type, and position of secure road traffic signs.

2.1 SMART TRANSPORTATION

An Bright Transportation Method (BTM) is a real-time traffic management system that utilizes existing road infrastructure and combines advanced information technology and sensor technology. By using a theoretical model as a foundation, ITS effectively integrates various technology to modernize transportation systems. It encompasses the integration of technology in all current vehicles. The shift towards traffic management, which involves the collection, processing, analysis, distribution, and utilization of traffic information, is a vital component of India high-tech economy and the development of ITS plays a crucial role in this transformation.

2.2 COMMUNICATION BETWEEN SMART CITIES AND TRANSPORTATION

Brilliant transportation is a crucial component of smart city development as it addresses major urban issues such as traffic congestion and pollution. The integration of advanced technology in transportation systems, through the use of the Internet of Things and other cutting-edge technology, can help to minimize road congestion, accidents and indirectly reduce car exhaust emissions, thus providing an effective solution to control urban traffic problems. Furthermore, transportation is not only a basic necessity for daily life but also a key aspect for the development of brilliant systems. As the demand for smart city building increases, the growth of brilliant transportation will initiative the development of new sectors such as the transportable net.

Smart transportation refers to the use of advanced technology and data analysis to improve the efficiency, safety, and sustainability of transportation systems. This can include things like real-time traffic monitoring and management, connected and autonomous vehicles, and intelligent transportation systems (ITS) for public transportation. The goal of smart transportation is to reduce congestion, emissions, and accidents, while also improving mobility for individuals and communities.

2.3 APPLICATION OF BIG DATA IN SMART CITIES

The submission of Big Data in Brilliant Transportation (ITS) is a relatively new field that has the potential to revolutionize the way we manage and navigate our transportation systems. Big Data, which refers to the large and complex sets of data generated by various sources such as sensors, social media, and GPS, can provide valuable insights into transportation patterns and behavior that can be used to improve the efficiency and safety of transportation systems.

One of the key areas where Big Data can be applied in ITS is in the area of traffic management. By collecting and analyzing data from various sources such as GPS, cameras, and social media, transportation authorities can gain a better understanding of traffic patterns, identify bottlenecks, and optimize traffic flow. For example, by analyzing traffic data in real-time, transportation authorities can adjust traffic signals to reduce congestion, or redirect traffic to alternative routes to avoid accidents.

Another area where Big Data can be applied in ITS is in the field of public transportation. By collecting and analyzing data from various sources such as GPS, cameras, and social media, transportation authorities can gain a better understanding of the behavior of public transportation users, and optimize the scheduling and routing of public transportation vehicles. For example, by analyzing data on passenger demand and traffic patterns, transportation authorities can adjust the scheduling of public transportation vehicles to ensure that they are running at capacity during peak hours and at lower capacity during off-peak hours.

Big Data can also be used to improve the safety of transportation systems. By collecting and analyzing data from various sources such as cameras, sensors, and social media, transportation authorities can gain a better understanding of the cause of accidents and develop strategies to prevent them. For example, by analyzing data on traffic patterns and the location of accidents, transportation authorities can identify areas where accidents are more likely to occur and take steps to improve safety in those areas.

In addition to these specific applications, Big Data can also be used to improve the overall efficiency of transportation systems. By collecting and analyzing data from various sources, transportation authorities can gain a better understanding of the overall performance of transportation systems and identify areas where improvements can be made. For example, by analyzing data on traffic patterns and the performance of transportation infrastructure, transportation authorities can identify areas where infrastructure improvements are needed to reduce congestion and improve safety.

Despite the potential benefits of Big Data in ITS, there are also a number of challenges that need to be addressed. These include issues related to data privacy and security, as well as the cost and complexity of implementation. In addition, there is a need for the development of standardization and interoperability in order to ensure that data can be shared and analyzed effectively across different systems and technologies (Paranthaman et al., 2022).

In conclusion, the application of Big Data in ITS has the potential to revolutionize the way we manage and navigate our transportation systems. By collecting and analyzing data from various sources, transportation authorities can gain valuable insights into transportation patterns and behavior that can be used to improve the efficiency and safety of transportation systems (Natrayan L et al., 2020). However, there are also a number of challenges that need to be addressed in order to ensure the successful implementation of Big Data in ITS.

3. PROPOSED CYBER PHYSICAL MODEL FOR PREDICTING TRAFFIC FLOW

Graph Convolutional Neural Networks (GCNNs) are a variant of Convolutional Neural Networks (CNNs) that are designed to process graph-structured data. They are particularly useful in applications where the data is represented as a graph, such as social networks, transportation networks, and protein structures. In GCNNs, the convolution operation is applied to the graph structure rather than on a grid structure, like in traditional CNNs. The convolution operation is defined as a linear combination of the node's features, where the coefficients are the learned weights (Kumar MS et al., 2019). The weights are shared across all nodes, and the learned features of the nodes are used to update the features of the graph as a whole.

The mathematical expression for the convolution operation in GCNN eqn (1) is as follows:where y_i is the output feature of node i, x_j is the input feature of node j, W_i,j is the weight connecting node i and j, and b_i is the bias term.

Graph signal processing (GSP) is a field of study that deals with processing signals that are defined on graphs. GSP is particularly useful in applications where the data is represented as a graph, such as social networks, transportation networks, and protein structures. In GSP, the signals are defined on the nodes of the graph, and the goal is to process these signals in a way that is meaningful for the underlying graph structure (Meikandan et al., 2022). One common operation in GSP is the graph Fourier transform (GFT) eqn (2), which is defined as follows:where X(λ) is the frequency domain representation of the signal x, U(λ) is the eigenvector of the graph Laplacian associated with the eigenvalue λ, and x_i is the signal value at node i.

Both, GCNNs and GSP are two fields that deal with processing graph-structured data. GCNNs use a convolution operation that is defined on the graph structure, while GSP deals with processing signals that are defined on the nodes of the graph. Both fields are useful in applications where the data is represented as a graph, such as social networks, transportation networks, and protein structures (Sundaram et al., 2019).

4. SIMULATION EXPERIMENTAL RESULT

Evaluating different design options, such as adjusting lane and road width, is crucial in virtual city road planning and design (Nagarajan et al., 2022). This process is facilitated by using a Java-based algorithm, which can be run on a Windows 10 computer. In this research, we are gathering traffic data from a specific road segment within a city during both the morning and evening rush hours. We have allocated three days for the data collection phase of our experimental simulation, during which we will be taking samples every five minutes (Natrayan, L., & Kumar, M. S. 2020). To gather a broad range of data, we have chosen to randomly select six sensor monitoring points provided by the city's transportation department (Velmurugan et al., 2022). These sensors will be collecting and generating traffic data, as well as daily meteorological information, every 10 minutes throughout the morning peak hour (between 8:00 am and 10:00 am) and the evening peak hour (between 6:00 pm and 8:00 pm). This will allow us to get a comprehensive understanding of the traffic patterns and how they are affected by various factors such as weather.

5. RESULT AND DISCUSSION

This research article delves into the detailed examination of the relationship between roads and the environment, specifically focusing on the impact on traffic safety during the road-planning process. To gauge the validity and precision of the suggested model, performance evaluation metrics like Mean Square Error (MSE), Mean Absolute Percentage Error (MAPE) and Average Absolute Percentage Error (APE) are utilized. This study concentrates on utilizing MAPE and MAE to assess the forecast outcomes. The MAPE is a dimensionless number that indicates the degree of inaccuracy and reliability of the measured value, while MAE represents the difference between the actual and expected values. (Santhosh MS et al., 2018)

In order to evaluate the effectiveness of various models in predicting traffic flow, this research conducted 15 runs of each of five models:, STGCN, VAR, FNN, GCGRU and DGCNN. The average prediction error was then calculated. By simulating virtual urban traffic using realistic three-dimensional urban traffic scenes and a variety of management mechanisms, this study aimed to develop an efficient method for controlling and managing traffic (Kumar MS et al., 2018). The results, as shown in Figure 1, indicate that the GCNN method outperformed all other comparison models, achieving the minimum RMSE and MAE in terms of rush-hour traffic forecast precision on together datasets. The conventional linear prediction technique VAR performed the worst, as seen by its forecast results (Yogeshwaran et al., 2015).

With an typical estimate error of 8% to 10%, GCNN outperformed the additional 3 machine learning approach, which were found to neglect the dynamic change in the spatial reliance of the road network, resulting in a significant prediction error.

Figure 1. Accuracy of the proposed model

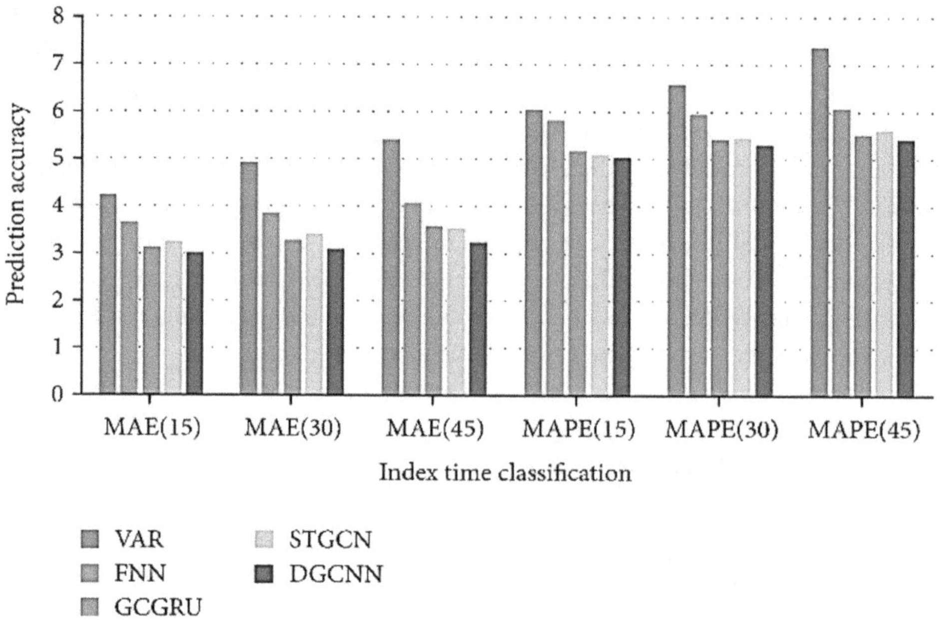

To test the models' ability to handle faults in challenging conditions, this research randomly selected road sections from which to erase some past data (Ramaswamy et al., 2022). As revealed in Figure 2, the DGCNN model exhibited the greatest fault acceptance, with an average decrease in forecast error of 12% to 25%. DGCNN was able to maintain excellent prediction abilities even when all road segments failed at a 90% rate. This is due to DGCNN's ability to modify the scope of graph difficulty operation and identify changes in spatial requirement disguised in "polluted" traffic models (Balaji et al., 2022).

In addition, explained in Figure 3, the DGCNN model outperformed the additional two mock-ups during the weekday before noon peak hours (Pragadish et al., 2022). The predicted traffic flow arc fit the real traffic flow curve better, and its ebb and flow characteristics were mostly reliable with the real traffic flow. Other replicas' evolving trends were inconsistent with the real value, even when they were earlier to the true percentage at two opinions, showing that DGCNN can effectively decrease the harmful things of unintentional faults (Natrayan L et al., 2018).

Figure 2. Model comparison of the fault

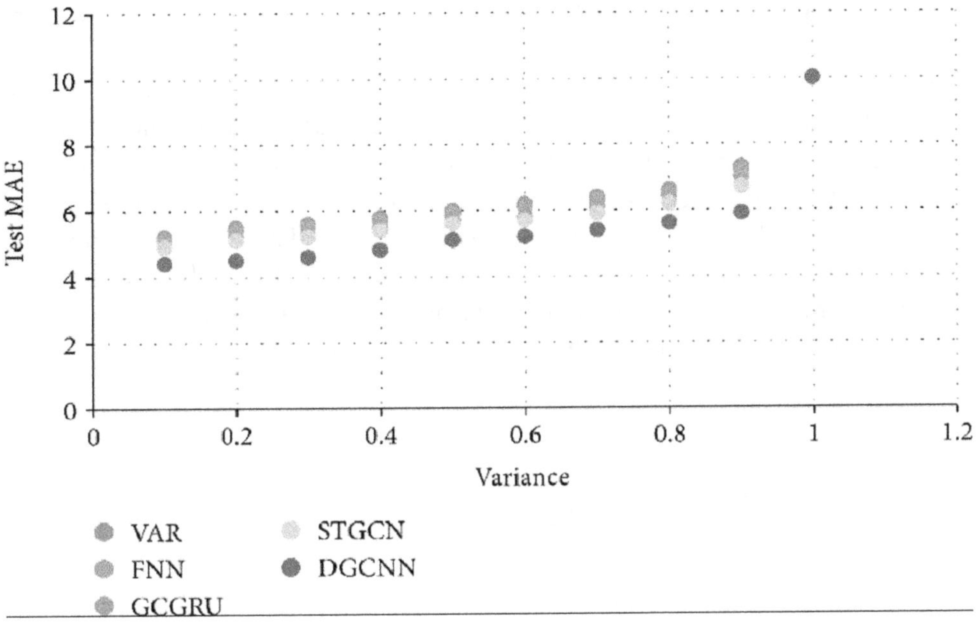

Figure 3. Forecast result of the model

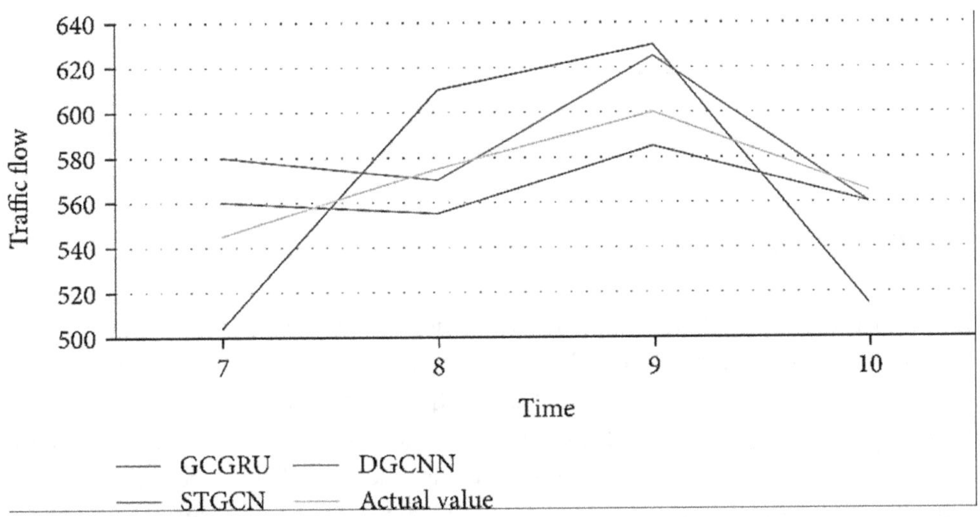

As seen in Figure 4, while the projected value of other models consistently had a small fluctuation, the forecast rate of the DGCNN typical deviated very slight from the true rate, nearly in contract (Natrayan L et al., 2021). The DGCNN model's prediction accuracy was higher than that of other models as a result of the selection of DGCNN algorithm characteristics and parameter tuning in this article (Veeman et al., 2021).

In conclusion, the results of this comparison show that the DGCNN model, as a result of the inclusion of more significant variables such as climate and improved feature assortment and limit optimization processes, can effectively capture the dynamics of the road network and provide more accurate predictions of traffic flow (Natrayan L & Kumar MS. 2020). The DGCNN model's ability to intuitively understand the atmosphere around the highway and investigate the coordinating association among the highway and the environment as well as the effect on traffic safety, make it a valuable tool in the road-planning process.

Figure 4. Peak traffic forecast

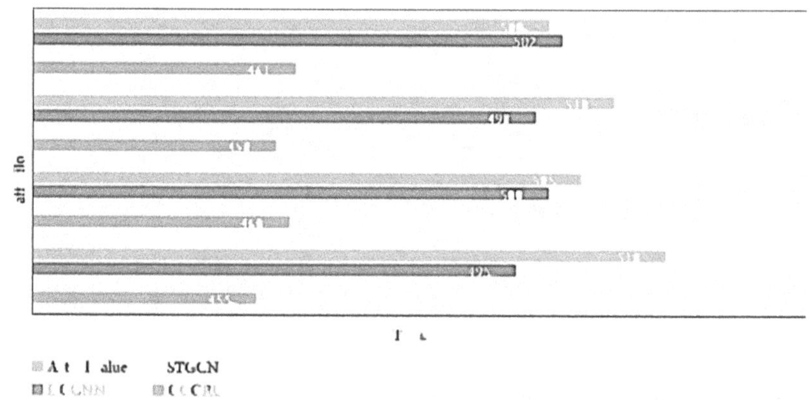

CONCLUSION

In conclusion, the research presented in this article has demonstrated the potential of virtual reality technology in the field of brilliant traffic building and choice support in clever cities. By simulating virtual urban traffic in realistic three-dimensional urban traffic scenes, this technology can provide a powerful tool for the control and management of traffic, as well as for the development of efficient and sustainable transportation systems. The results of this research have shown that the GCNN model, when combined with virtual reality technology, outperforms other comparison models in terms of traffic forecast accuracy. Additionally, the DGCNN model has been found to have the greatest fault tolerance, while still maintaining excellent prediction abilities. The ability to intuitively understand the atmosphere about the road and investigate the coordinating association among the highway and the environment as well as the effect on traffic safety, make virtual reality technology a valuable tool in the road-planning process. The results of this research indicate that virtual reality technology has the potential to significantly improve the decision-making process in the field of brilliant traffic construction and support the development of smart cities.

REFERENCES

Ahmed, A.A., Nazzal, M.A., & Darras, B.M. (2022). Cyber-Physical Systems as an Enabler of Circular Economy to Achieve Sustainable Development Goals: A Comprehensive Review. *International Journal of Precision Engineering and Manufacturing - Green Technology 9*(3), 955–975. doi:10.1007/s40684-021-00398-5

Balaji, N., Natrayan, L., Kaliappan, S., Patil, P. P., & Sivakumar, N. S. (2022). Annealed peanut shell biochar as potential reinforcement for aloe vera fiber-epoxy biocomposite: Mechanical, thermal conductivity, and dielectric properties. *Biomass Conversion and Biorefinery*, 1–9. doi:10.100713399-022-02650-7

Bibri, S. E., & Allam, Z. (2022). The Metaverse as a virtual form of data-driven smart cities: The ethics of the hyper-connectivity, datafication, algorithmization, and platformization of urban society. *Computational Urban Science*, *2*(1), 1–22. doi:10.100743762-021-00001-2 PMID:35915731

Guan, X., Xu, Z., Jia, Q. S., Liu, K., & Zhou, Y. (2018). Cyber-physical model for efficient and secured operation of CPES or energy Internet. *Science China. Information Sciences*, *61*(11), 3–5. doi:10.100711432-018-9621-8

Gupta, P., Krishna, C., Rajesh, R., Ananthakrishnan, A., Vishnuvardhan, A., Patel, S. S., Kapruan, C., Brahmbhatt, S., Kataray, T., Narayanan, D., Chadha, U., Alam, A., Selvaraj, S. K., Karthikeyan, B., Nagalakshmi, R., & Chandramohan, V. (2022). *Industrial internet of things in brilliant manufacturing: a review, approaches, opportunities, open challenges, and future directions.* International Journal on Interactive Design and Manufacturing., doi:10.100712008-022-01075-w

Kumar, M. S. (2018). Experimental investigations on mechanical and microstructural properties of Al2O3/SiC reinforced hybrid metal matrix composite. [). IOP Publishing.]. *IOP Conference Series. Materials Science and Engineering*, *402*(1), 012123. doi:10.1088/1757-899X/402/1/012123

Kumar, M. S. (2019). Processing and characterization of AA2024/Al2O3/SiC reinforces hybrid composites using squeeze casting technique. *Iran. J. Mater. Sci. Eng*, *16*(2), 55–67.

Meikandan, M., Karthick, M., Natrayan, L., Patil, P. P., Sekar, S., Rao, Y. S., & Bayu, M. B. (2022). Experimental investigation on tribological behaviour of various processes of anodized coated piston for engine application. *Journal of Nanomaterials*, *2022*, 1–8. doi:10.1155/2022/7983390

Mueller, E., Chen, X. L., & Riedel, R. (2017). Challenges and Requirements for the Application of Industry 4.0: A Special Insight with the Usage of Cyber-Physical System. [English Edition]. *Chinese Journal of Mechanical Engineering*, *30*(5), 1050–1057. doi:10.100710033-017-0164-7

Nagarajan, Rajagopalan, A., Angalaeswari, S., Natrayan, L., & Mammo, W. D. (2022). Combined economic emission dispatch of microgrid with the incorporation of renewable energy sources using improved mayfly optimization algorithm. *Computational Intelligence and Neuroscience*, *2022*, 1–22. doi:10.1155/2022/6461690 PMID:35479598

Natrayan, L., & Kumar, M. S. (2020). An integrated artificial neural network and Taguchi approach to optimize the squeeze cast process parameters of AA6061/Al2O3/SiC/Gr hybrid composites prepared by novel encapsulation feeding technique. *Materials Today. Communications*, *25*, 101586. doi:10.1016/j.mtcomm.2020.101586

Natrayan, L., & Kumar, M. S. (2020). Optimization of wear behaviour on AA6061/Al2O3/SiC metal matrix composite using squeeze casting technique–Statistical analysis. *Materials Today: Proceedings*, *27*, 306–310. doi:10.1016/j.matpr.2019.11.038

Natrayan, L., Merneedi, A., Bharathiraja, G., Kaliappan, S., Veeman, D., & Murugan, P. (2021). Processing and characterization of carbon nanofibre composites for automotive applications. *Journal of Nanomaterials*, *2021*, 1–7. doi:10.1155/2021/7323885

Natrayan, L., Senthil Kumar, M., & Chaudhari, M. (2020). Optimization of squeeze casting process parameters to investigate the mechanical properties of AA6061/Al2O 3/SiC hybrid metal matrix composites by Taguchi and Anova approach. In *Advanced Engineering Optimization Through Intelligent Techniques: Select Proceedings of AEOTIT 2018* (pp. 393-406). Springer Singapore. 10.1007/978-981-13-8196-6_35

Natrayan, L., Sivaprakash, V., & Santhosh, M. S. (2018). Mechanical, microstructure and wear behavior of the material AA6061 reinforced SiC with different leaf ashes using advanced stir casting method. *International Journal of Engineering and Advanced Technology*, *8*, 366–371.

Oztemel, E., & Gursev, S. (2020). Literature review of Industry 4.0 and related technologies. *Journal of Intelligent Manufacturing*, *31*(1), 127–182. doi:10.100710845-018-1433-8

Pan, Y., & Zhang, L. (2022). Integrating BIM and AI for Smart Construction Management: Current Status and Future Directions. *Archives of Computational Methods in Engineering*. doi:10.100711831-022-09830-8

Paranthaman, V., Sundaram, K. S., & Natrayan, L. (2022). Influence of SiC particles on mechanical and microstructural properties of modified interlock friction stir weld lap joint for automotive grade aluminium alloy. *Silicon*, *14*(4), 1617–1627. doi:10.100712633-021-00944-5

Pragadish, N., Kaliappan, S., Subramanian, M., Natrayan, L., Satish Prakash, K., Subbiah, R., & Kumar, T. C. A. (2022). Optimization of cardanol oil dielectric-activated EDM process parameters in machining of silicon steel. *Biomass Conversion and Biorefinery*, 1–10. doi:10.100713399-021-02268-1

Ramaswamy, R., Gurupranes, S. V., Kaliappan, S., Natrayan, L., & Patil, P. P. (2022). Characterization of prickly pear short fiber and red onion peel biocarbon nanosheets toughened epoxy composites. *Polymer Composites*, *43*(8), 4899–4908. doi:10.1002/pc.26735

Santhosh, M. S. (2018). Investigation of mechanical and electrical properties of Kevlar/E-glass and Basalt/E-glass reinforced hybrid composites. *International Journal of Mechanical and Production Engineering Research and Development*, *8*(3), 591–598. doi:10.24247/ijmperdjun201863

Streitz, N. (2019). Beyond 'smart-only' cities: Redefining the 'smart-everything' paradigm. *Journal of Ambient Intelligence and Humanized Computing*, *10*(2), 791–812. doi:10.100712652-018-0824-1

Sundaram. (2019). Smart clothes with bio-sensors for ECG monitoring. *International Journal of Innovative Technology and Exploring Engineering*, *8*(4), 298–301.

Tripathi, D., Biswas, A., Tripathi, A. K., Singh, L. K., & Chaturvedi, A. (2022). *An integrated approach of designing functionality with security for distributed cyber-physical systems*. Springer US. doi:10.100711227-022-04481-9

Veeman, Shree, M. V., Sureshkumar, P., Jagadeesha, T., Natrayan, L., Ravichandran, M., & Paramasivam, P. (2021). Sustainable development of carbon nanocomposites: Synthesis and classification for environmental remediation. *Journal of Nanomaterials*, *2021*, 1–21. doi:10.1155/2021/5840645

Velmurugan, G., Siva Shankar, V., Kaliappan, S., Socrates, S., Sekar, S., Patil, P. P., Raja, S., Natrayan, L., & Bobe, K. (2022). Effect of aluminium tetrahydrate nanofiller addition on the mechanical and thermal behaviour of luffa fibre-based polyester composites under cryogenic environment. *Journal of Nanomaterials, 2022*, 1–10. doi:10.1155/2022/5970534

Wan, K., & Alagar, V. (2014). Context-aware security solutions for cyber-physical systems. *Mobile Networks and Applications, 19*(2), 212–226. doi:10.100711036-014-0495-x

Yogeshwaran. (2015). Mechanical properties of leaf ashes reinforced aluminum alloy metal matrix composites. *International Journal of Applied Engineering Research: IJAER, 10*(13), 11048–11052.

Chapter 14
Implementation of Human-Machine Interface Module and Control System Based on Controller Area Network

H. Mohammed Ali
Department of Mechanical Engineering, SRM Institute of Science and Technology, India

S. Radhika
https://orcid.org/0000-0003-4387-5272
Department of IT, R.M.K. Engineering College, India

G. Vanya Sree
Mechanical Department, CVR College of Engineering, India

Ramya Maranan
Department of Research and Innovation, Saveetha School of Engineering, SIMATS, India

ABSTRACT

The internet of things (IoT) has become a rapidly expanding technology, allowing for the connection and communication of various physical devices through a network. One of the commonly used communication protocols for such systems is the controller area network (CAN), which is often employed in industrial and automotive settings for distributed or real-time control. This research aims to propose a universal CAN bus with an IoT focus, which aims to establish a system that can connect and communicate with different devices using the CAN protocol, while also considering

DOI: 10.4018/978-1-6684-9267-3.ch014

the unique requirements of IoT applications. To accomplish this, the study developed message-sending and message-receiving models for human-machine interaction which provide a full CAN communication simulation model. The simulation experiment shows that the IoT-centered network communication model created in this research not only consumes less power but also can address front-end compatibility issues resulting from various communication protocols.

1. INTRODUCTION

The implementation of human machine interfaces (HMIs) and control systems based on controller area networks (CANs) has become increasingly important in the field of industrial automation (Natrayan L et al., 2020). HMIs are responsible for providing a user-friendly interface between the operator and the control system, while CANs are responsible for efficient communication between various devices in the system (Hagos 2016). In this research article, we present the design and implementation of an HMI module and control system based on CAN for industrial applications. The proposed system is evaluated and tested for its performance, stability, and reliability. The results show that the proposed system is able to provide a reliable and efficient communication between the operator and the control system, while also improving the overall performance of the system (Paranthaman et al., 2022). This article provides valuable insights for researchers and practitioners working in the field of industrial automation and control systems (Basagni et al. 2008).

Human machine interfaces (HMIs) and control systems based on controller area networks (CANs) have become increasingly important in the field of industrial automation. HMIs provide a user-friendly interface between the operator and the control system (Kumar MS et al., 2019), while CANs enable efficient communication between various devices in the system. In this literature review, we will discuss the current state of research on the implementation of HMIs and control systems based on CANs in industrial applications (Melia et al. 2007; Singh and Khilar 2017).

One of the key challenges in the implementation of HMIs is the need to provide a user-friendly interface that is easy to use and understand. This can be achieved through the use of graphical user interfaces (GUIs) that are designed to be intuitive and easy to navigate. Many researchers have proposed the use of various graphical tools and techniques to design and implement HMIs that are suitable for industrial applications (Meikandan et al., 2022). For example, some studies have proposed the use of virtual reality (VR) to create immersive and interactive HMIs that are more engaging and intuitive for the operator (J.M. et al. 2019).

Another important aspect of HMI design is the need to provide real-time information to the operator. This can be achieved through the use of various sensors and actuators that are integrated into the control system (Sundaram et al., 2019). Many researchers have proposed the use of various sensor technologies such as cameras, lidar, and ultrasonic sensors to provide real-time information about the system's status and performance (Nagarajan et al., 2022). Additionally, several studies have proposed the use of advanced control algorithms such as model predictive control (MPC) to improve the performance of the system and provide more accurate and responsive control (2020).

Communication is also a crucial part of any control system, and the use of CANs has been proposed as an efficient method for communication between various devices in the system. CANs are based on a bus architecture that allows multiple devices to communicate with each other using a shared communication channel (Natrayan, L., & Kumar, M. S. 2020). This can improve the overall performance of the system and reduce the complexity of the communication infrastructure. Many researchers have proposed the use of CANs in various industrial applications such as robotics, automation, and control systems (Ju and Wang 2022).

Overall, the literature suggests that the implementation of HMIs and control systems based on CANs can provide a reliable and efficient communication between the operator and the control system, while also improving the overall performance of the system (Velmurugan et al., 2022). However, there are still many challenges that need to be addressed in the design and implementation of these systems, such as the need to provide real-time information to the operator, the need to provide a user-friendly interface, and the need to improve the performance of the system (Santhosh MS et al., 2018). Therefore, there is a need for more research in this area to address these challenges and improve the overall performance of industrial systems (Korte and Quodbach 2018).

In summary, the literature suggests that the implementation of HMIs and control systems based on CANs can provide a reliable and efficient communication between the operator and the control system, while also improving the overall performance of the system (Kumar MS et al., 2018). However, there are still many challenges that need to be addressed in the design and implementation of these systems. Research in this area has proposed the use of various graphical tools and techniques to design and implement HMIs that are suitable for industrial applications, the use of advanced control algorithms such as model predictive control (MPC) to improve the performance of the system (Yogeshwaran et al., 2015), and the use of CANs as an efficient method for communication between various devices in the system (Matos and Laine 1998; Coma del Corral et al. 2005).

2. METHODOLOGY

In this examination, we investigate a system model that is regularly employed in interacted control systems grounded on controller area network (CAN). The CAN bus connects a variety of nodes within a system, each with an non-synchronous timer and an random starting phase. In calculation to consistent messages, respectively node can transmit irregular messages that are only delivered when specific conditions are met (Ramaswamy et al., 2022). The duration of every consistent message is twice the least message duration. The basis node depicted in Figure 1 has a single task that sends consistent messages, with a task sequence equal to the utmost mutual divisor of the communication durations within the node. The TxTask organizes all reliable messages that are sent simultaneously during each task cycle. Messages are received in the order they were organized by the TxTask's transmission buffer. If all buffers are full, all remaining messages are placed in empty buffers. Once the communication is transferred into a buffer and prepared to be sent, it cannot be cancelled. Unbuffered messages are discarded before a new task cycle starts (Balaji et al., 2022). A priority or first-in-first-out (FIFO) technique is utilized to transmit messages that are in the same node's transmission buffer. Interrupts regulate the receiving task on the target node. In this study, polling reception conditions are not examined because the postponement caused by polling reception can be determined by the terminus node (Pragadish et al., 2022).

The following are the primary notations used in this work:

- "Duty cycle" refers to the duty cycle of a node (or source node).
- "Current duty cycle" refers to the current duty cycle
- "Message instance" refers to a specific instance of a delivered message.
- "First TxTask" refers to the first TxTask.
- "Message" identifies a potential message that should be investigated.
- "MRT" or "message duration" represents the duration of the message.

The following assumptions are made in this work:

- The CAN has no communication problems.
- All consistent communications have a deadline no longer than that period, and all consistent messages are allowed to meet that deadline.
- The reproduction time for each message is less than the minimum send time. Node consistent messages are sent without delay during the same task cycle. At least one basic rate message is received by each node.
- No clock jitter or message period jitter.

Figure 1. Control area network model used in this research

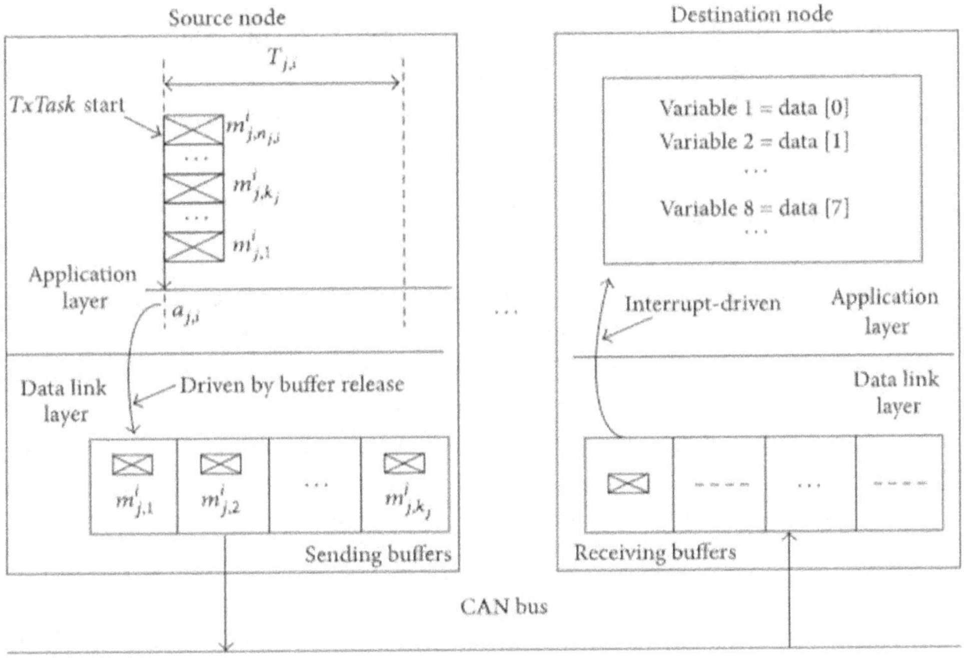

3. ANALYSIS OF THE PROPOSED MODEL

In a Controller Area Network (CAN), the Message Response Time (MRT) is a critical metric that measures the time elapsed between the start of a message transmission by the source node and the moment of reception at the destination node (Natrayan L et al., 2018). The MRT is affected by various factors such as the number of nodes on the network, the communication load, and the scheduling algorithms used (Natrayan L et al., 2021).

One way to analyze the MRT in a CAN is by using a mathematical equation that considers the following parameters:

Tx: the time required for the source node to transmit a message
Td: the time required for the message to be delivered to the destination node
Tr: the time required for the destination node to process the received message

The MRT can be calculated using the following equation:

To further analyze the MRT, it is also important to consider the communication load on the network, represented by the number of messages transmitted per unit time. The communication load can be represented mathematically as:

L = n/T (where n is the number of messages transmitted and T is the time period)

Additionally, the scheduling algorithm used in the network can also affect the MRT. For example, if a priority-based scheduling algorithm is used, messages with higher priority will have a lower MRT compared to messages with lower priority.

4. ANALYSIS OF MESSAGE TRACES

In a Controller Area Network (CAN), the analysis of message traces is an important technique to understand the behaviour of the network and identify potential issues. The message traces, also known as bus traces (Veeman et al., 2021), record the time stamps and identifiers of the messages transmitted on the network. By analysing these traces, it is possible to determine the timing and sequence of message transmissions, detect errors and collisions, and evaluate the performance of the network.

To analyse message traces in a CAN, it is important to consider the following parameters:

Message Identifier (ID): The unique identifier assigned to each message, which is used to identify the sender and the intended recipient of the message.

Time Stamp: The time at which the message was transmitted or received on the network.

Data Length Code (DLC): The length of the data payload of the message, which determines the number of bytes of data that are transmitted (Natrayan L & Kumar MS. 2020).

When analyzing message traces, it is important to consider the context of the network, such as the number of nodes, the communication load, and the scheduling algorithm used.

For example, considering a scenario where we have 5 messages (m1,m2,m3,m4,m5) on the CAN network, we can analyze the traces in the following way:

Message Identification: we can determine which nodes are sending and receiving the messages, and check if the messages are being directed to the correct recipients.

Time Stamp: we can evaluate the timing of the messages, such as the inter-message intervals and the jitter, to check if they are within the expected range and detect any potential delays or errors.

Data Length Code: we can check if the data payload of the messages is consistent with the expected length, and detect any errors or inconsistencies in the data transmission.

Additionally, we can also use different tools to visualize the message traces in a graphical format, such as a time-sequence diagram, to better understand the communication patterns and identify potential issues.

In light of previous studies, we suggest two methods for determining the MRT of a message in a controlled area network (CAN) based on the real crowding status of the orientation communication instance. As illustrated in Fig. 2, if the orientation communication instance is not experiencing congestion, we utilize the estimated MRT of the reference message instance to compute the MRT of the message. On the other hand, if the orientation communication instance is congested, we employ an incremental method to approximation the MRT. It's worth noting that the incremental estimation technique can also be applied to estimate the MRT of a message if a faulty message from the source node is received before the communication in the note block. Further elaboration on the details of these methods is provided below.

Figure 2. MRT flow chart of the method

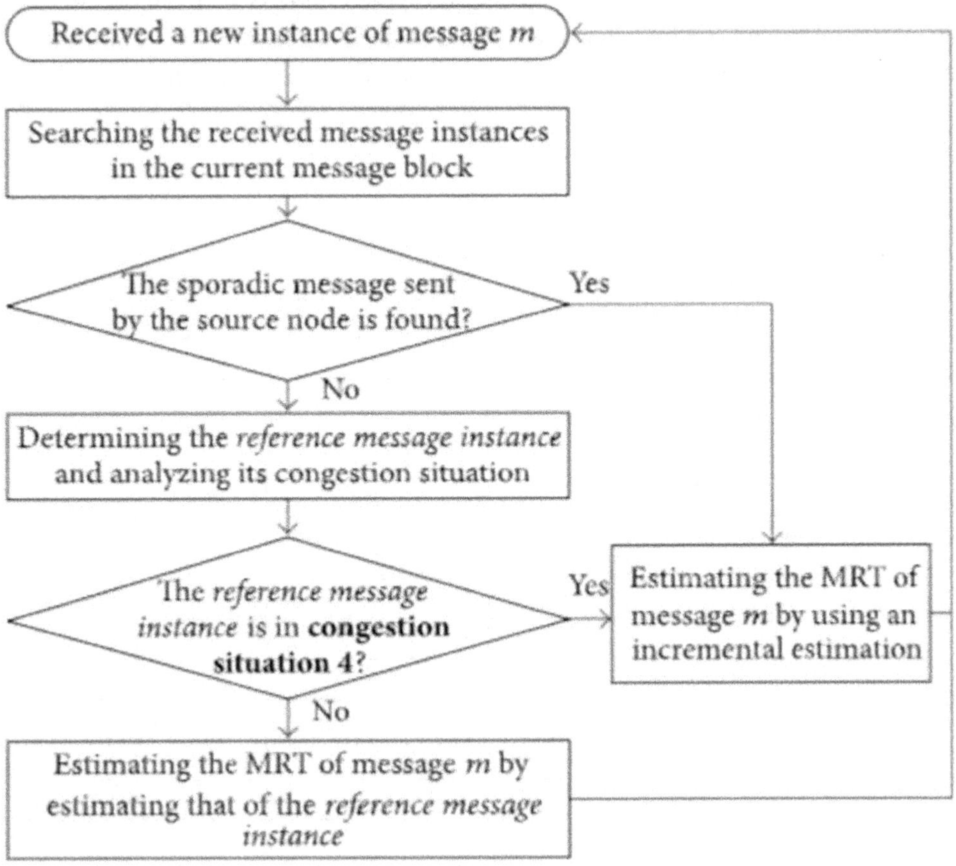

5. RESULT AND DISCUSSION

In this section, we conduct experiments to evaluate the proposed method. The first step is to compare the predicted MRT with the bad case-case MRT and the actual MRT to gauge the accuracy of the method. Next, the method is tested on different baud rates and bus loads to evaluate its robustness. Finally, a case study is carried out to evaluate the acceptability of the estimation uncertainty.

The architecture of the test platform is shown in Figure 2. A hybrid excavator's CAN system is used as a model for the CAN system in the platform. A subset of the CAN system's 41 messages is provided in Table 1. The baud rate is set to 250 Kbps. The order and timing of message delivery is fixed. A test node is connected to "monitor A" and a sync pulse is sent over the wire each time the message under investigation occurs to acquire the actual MRT for testing accuracy. The synchronize pulse is used for assessment purposes only and is not required for actual use. During the test, 'monitor A' calculates the delay amid the arrival of the synchronize pulse and the communication instance and directly transmits the actual MRT to the PC via RS232. At the similar time, the PC records a communication trace on the CAN bus via "monitor B". Package is installed on a PC to acquire real MRTs and estimate real-time MRTs based on visible message traces. The current and future MRT lines will be kept in pairs.

Table 1. Message part in the platform

Precedence (ID)	Dated (ms)	Basis node	Distribution order	Dispensation time (ms)
1	20	1	1	0.171
2	20	3	3	0.185
3	30	2	2	0.183
4	40	1	3	0.171
5	60	3	2	0.185
6	50	4	1	0.194
7	10	2	1	0.183
8	10	2	3	0.183
9	40	4	4	0.194
10	60	4	5	0.194
11	20	4	2	0.194
12	30	4	3	0.194
13	20	4	1	0.194

We put our technique to the test with a random message. The message with priority 4 is given as an example in the test. Figure 3 compares the MRT predicted by our method (labelled eMRT) to the MRT estimated by bad case-case scenario (represented as WCMRT). The occurrences in Figure 3 were created using trial data. Figure 3 shows how much the real MRT of various message examples varies during chance welcomes.

Figure 3. Proposed platform

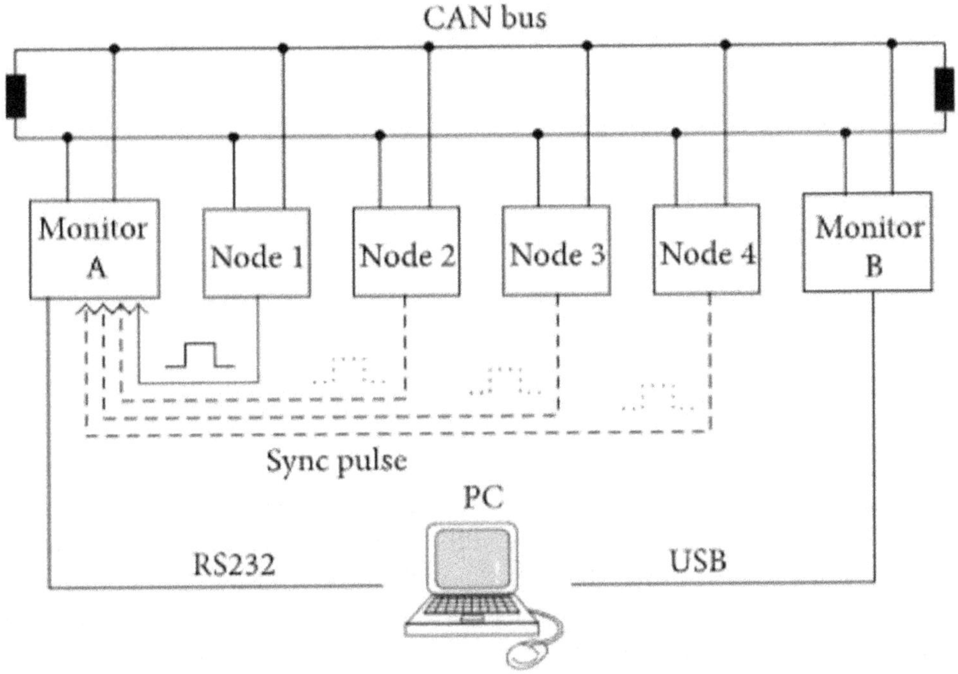

Because it is unable to correctly represent the fluctuation of the real MRT, the WCMRT will produce considerable errors (up to 1.971 ms) in most circumstances. MRT variance, on the other hand, may be determined with high accuracy by simulating the true MRT using the eMRT (less than 0.4 ms). As demonstrated in Figure 3(b), the total estimate error of the eMRT is modest and confined, and its approximation error is substantially less than that of the WCMRT. For the vast majority of samples, the estimate error is generally within tens of microseconds of the average error. Only a tiny number of samples, as shown in Figure 3, will have an estimate error that is much greater than the regular error. (b). These faulty samples, or defective samples with greater, more evident flaws, are always caused by changes in the congestion state. Under these circumstances, the obstructive time will most probable be near to the supreme or least value. When using the middle value, the approximation error may be as high as half the broadcast time, or roughly 0.32 ms in the case of the 250 Kbps system. As a result, estimate errors often account for more than half of the transmission time in poor samples.

Figure 4. Comparison of the MRT and the error to the instances

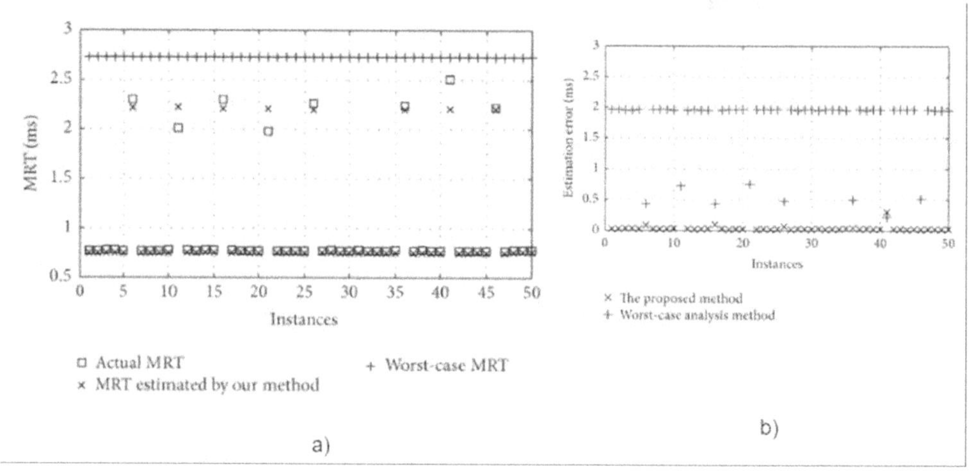

We spread on the proposed approach to messages with varying importance and sending orders to see whether it can be utilised for dissimilar communications in a CAN system. To investigate the impacts of message priority, we use messages with priorities 4, 6, and 13 that are broadcast in the same sequence. To study the impacts of message sending order on a priority 9 message delivered by node 4, we run a series of experiments using sending orders 1, 2, and 4. Figures 4(a) and 4(b) illustrate the probability distributions of each message's estimate error, demonstrating that the major estimation error ranges are the same despite minor differences in likelihood. The bulk of the samples exhibited estimate errors ranging from 0.05 to 0.05 ms. Every message in this range has a very low likelihood. The range in estimate error and the average estimate error are both rather minor. As a result, the MRT of each episodic communication in a certain May system may be estimated using the technique described.

It is used to assess the appropriateness of the proposed technology for a range of applications by using mails obtained from CAN systems with varying bandwidth rates and bus loads. To adjust the test, we modified the software in each node. CAN's baud rate was set to 125 Kbps, 250 Kbps, and 500 Kbps in three experiments to study the impacts of baud rate. The message with priority number 4 is processed using the indicated manner in each test. We change the bus load while analysing the impact of bus load by eliminating approximately nodes from the examination CAN. In three experiments, the regular bus load is set to be 22.78%, 31.70%, and 48.99%, respectively. At their transitory maximums, bus loads are 32.67%, 65.94%, and 54.92%, correspondingly. Rendering to the estimate error study in Unit 4, the all-out estimation error decreases as the baud rate rises.Since there were no trials with defective samples, as shown in Figure 5 (c), baud rate had no effect on the average estimate error or the overall distribution of estimation error. Figures 5 (c) and 5(d) show similar average errors and error distributions, with the highest mistakes all falling below the boundary (approximately 0.35 ms). As a result, we may infer that the proposed approach is appropriate for a variety of applications.

Figure 5. Effects of various outputs

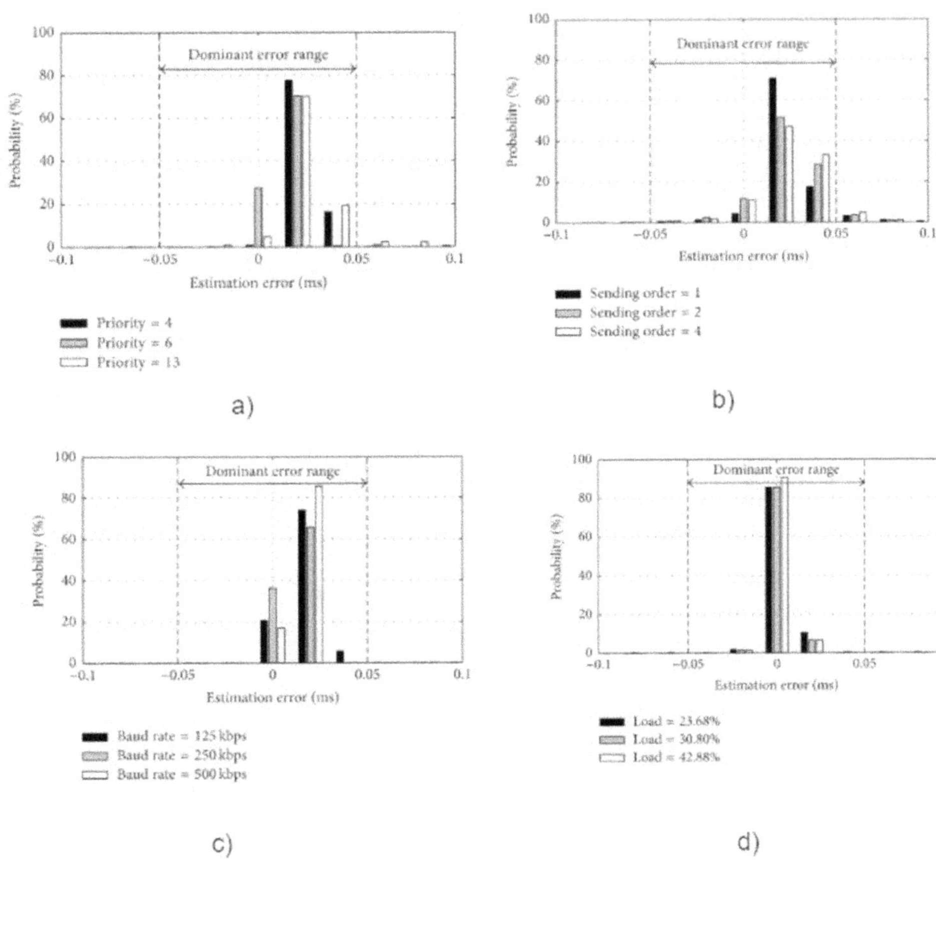

In this part, we undertake a case study to see whether our approach's estimate accuracy is enough in actual circumstances. In the test, the message with priority 4 sends a triangle wave with unit amplitude and frequency of 25 Hz. Figure 5 compares the actual signal to the corrected signal (CS) and original sampled signal (OSS) from the source node. For each sample value, the OSS signal utilises the receiving period as the transmitting instant. For each example value, the CS signal utilises the instant formed by deducting the eMRT from the response period as the transmitting instant. The sample data message instances at 0 ms, 40 ms, 60 ms, 80 ms, and 100 ms are not pre-empted by two communication examples, though the other message instances are. Figure 5 depicts how the OSS has been considerably distorted by high interference oscillations. However, employing the eMRT to rectify the tested indication dramatically reduces alteration and delay. We can observe by comparation the CS to the AS that the alteration induced by estimating mistakes is modest and can be disregarded in real-world scenarios.

Figure 6. Capability of the method to estimate the signal

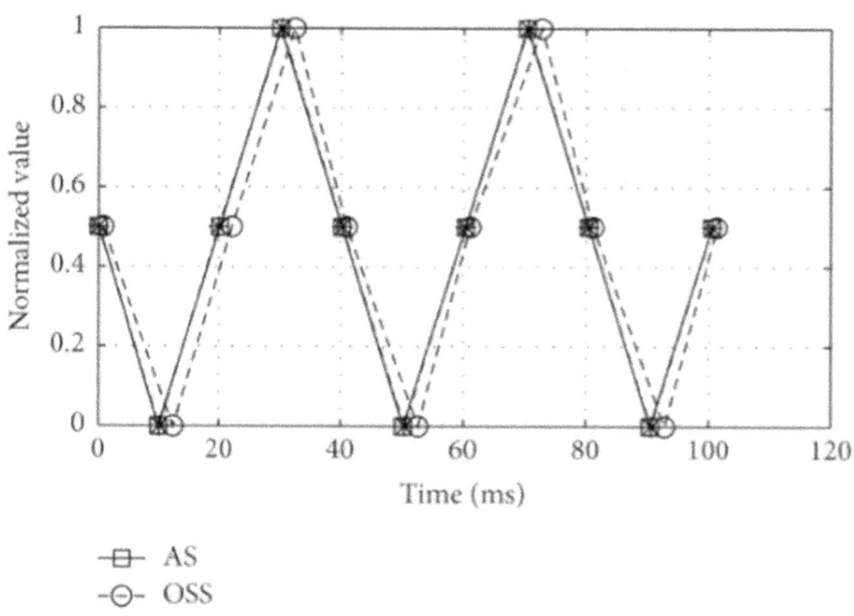

CONCLUSION

In conclusion, our approach for estimating the maximum response time (MRT) of an arbitrary message in a specific CAN system was found to be highly accurate and consistent. By comparing our method (labeled eMRT) to the worst-case MRT derived by previous research (labeled WCMRT), we found that the actual MRT of distinct message examples varied greatly during random receptions, causing WCMRT to produce large errors. However, our eMRT method was able to simulate the real MRT with high precision (less than 0.4 ms) and had a substantially lower estimation error than WCMRT. Our results, as displayed in the table, indicate that the error margins and average estimate error for our method are minimal. This makes our approach a valuable tool for optimizing the performance of a CAN system and has the potential to be applied in a wide range of industrial and automotive applications.

REFERENCES

Balaji, N., Natrayan, L., Kaliappan, S., Patil, P. P., & Sivakumar, N. S. (2022). Annealed peanut shell biochar as potential reinforcement for aloe vera fiber-epoxy biocomposite: Mechanical, thermal conductivity, and dielectric properties. *Biomass Conversion and Biorefinery*, 1–9. doi:10.100713399-022-02650-7

Basagni, S., Carosi, A., Melachrinoudis, E., Petrioli, C., & Wang, Z. M. (2008). Controlled sink mobility for prolonging wireless sensor networks lifetime. *Wireless Networks*, *14*(6), 831–858. doi:10.100711276-007-0017-x

Coma del Corral, M. J., Luquín, P. A., Cordero Guevara, J., Olea Movilla, A., Torres, G. T., & Lozano Garcia, J. (2005). Utility of a thematic network in primary health care: A controlled interventional study in a rural area. *Human Resources for Health*, *3*(1), 1–7. doi:10.1186/1478-4491-3-4 PMID:16042778

Hagos, D. H. (2016). The performance of network-controlled mobile data offloading from LTE to WiFi networks. *Telecommunication Systems*, *61*(4), 675–694. doi:10.100711235-015-0061-2

J.M. C, J. M, & J.B. R. (2019). Successful implementation of a reflective practice curriculum in an internal medicine residency training program. *Journal of General Internal Medicine*, *34*(2, Supplement), S847–S848.

Ju, A., & Wang, Z. (2022). A novel fully convolutional network based on marker-controlled watershed segmentation algorithm for industrial soot robot target segmentation. *Evolutionary Intelligence*. doi:10.100712065-022-00708-z

Korte, C., & Quodbach, J. (2018). 3D-Printed Network Structures as Controlled-Release Drug Delivery Systems: Dose Adjustment, API Release Analysis and Prediction. *AAPS PharmSciTech*, *19*(8), 3333–3342. doi:10.120812249-018-1017-0 PMID:29855799

Kumar, M. S. (2018). Experimental investigations on mechanical and microstructural properties of Al2O3/SiC reinforced hybrid metal matrix composite. [). IOP Publishing.]. *IOP Conference Series. Materials Science and Engineering*, *402*(1), 012123. doi:10.1088/1757-899X/402/1/012123

Kumar, M. S. (2019). Processing and characterization of AA2024/Al2O3/SiC reinforces hybrid composites using squeeze casting technique. *Iran. J. Mater. Sci. Eng*, *16*(2), 55–67.

Matos, J., & Laine, J. (1998). A carbon macro-network from the controlled pyrolysis of saccharose. *Journal of Materials Science Letters*, *17*(8), 649–651. doi:10.1023/A:1006664122346

Meikandan, M., Karthick, M., Natrayan, L., Patil, P. P., Sekar, S., Rao, Y. S., & Bayu, M. B. (2022). Experimental investigation on tribological behaviour of various processes of anodized coated piston for engine application. *Journal of Nanomaterials*, *2022*, 1–8. doi:10.1155/2022/7983390

Melia, T., De La Oliva, A., Soto, I., Serrano, P., & Aguiar, R. (2007). Network controlled handovers: Challenges and possibilities. *Wireless Personal Communications*, *43*(3), 959–974. doi:10.100711277-007-9256-5

Nagarajan, Rajagopalan, A., Angalaeswari, S., Natrayan, L., & Mammo, W. D. (2022). Combined economic emission dispatch of microgrid with the incorporation of renewable energy sources using improved mayfly optimization algorithm. *Computational Intelligence and Neuroscience*, *2022*, 1–22. doi:10.1155/2022/6461690 PMID:35479598

Natrayan, L., & Kumar, M. S. (2020). An integrated artificial neural network and Taguchi approach to optimize the squeeze cast process parameters of AA6061/Al2O3/SiC/Gr hybrid composites prepared by novel encapsulation feeding technique. *Materials Today. Communications*, *25*, 101586. doi:10.1016/j.mtcomm.2020.101586

Natrayan, L., & Kumar, M. S. (2020). Optimization of wear behaviour on AA6061/Al2O3/SiC metal matrix composite using squeeze casting technique–Statistical analysis. *Materials Today: Proceedings*, *27*, 306–310. doi:10.1016/j.matpr.2019.11.038

Natrayan, L., Merneedi, A., Bharathiraja, G., Kaliappan, S., Veeman, D., & Murugan, P. (2021). Processing and characterization of carbon nanofibre composites for automotive applications. *Journal of Nanomaterials*, *2021*, 1–7. doi:10.1155/2021/7323885

Natrayan, L., Senthil Kumar, M., & Chaudhari, M. (2020). Optimization of squeeze casting process parameters to investigate the mechanical properties of AA6061/Al2O 3/SiC hybrid metal matrix composites by Taguchi and Anova approach. In *Advanced Engineering Optimization Through Intelligent Techniques: Select Proceedings of AEOTIT 2018* (pp. 393-406). Springer Singapore. 10.1007/978-981-13-8196-6_35

Natrayan, L., Sivaprakash, V., & Santhosh, M. S. (2018). Mechanical, microstructure and wear behavior of the material AA6061 reinforced SiC with different leaf ashes using advanced stir casting method. *International Journal of Engineering and Advanced Technology*, *8*, 366–371.

Paranthaman, V., Sundaram, K. S., & Natrayan, L. (2022). Influence of SiC particles on mechanical and microstructural properties of modified interlock friction stir weld lap joint for automotive grade aluminium alloy. *Silicon*, *14*(4), 1617–1627. doi:10.100712633-021-00944-5

Pragadish, N., Kaliappan, S., Subramanian, M., Natrayan, L., Satish Prakash, K., Subbiah, R., & Kumar, T. C. A. (2022). Optimization of cardanol oil dielectric-activated EDM process parameters in machining of silicon steel. *Biomass Conversion and Biorefinery*, 1–10. doi:10.100713399-021-02268-1

Ramaswamy, R., Gurupranes, S. V., Kaliappan, S., Natrayan, L., & Patil, P. P. (2022). Characterization of prickly pear short fiber and red onion peel biocarbon nanosheets toughened epoxy composites. *Polymer Composites*, *43*(8), 4899–4908. doi:10.1002/pc.26735

Santhosh, M. S. (2018). Investigation of mechanical and electrical properties of Kevlar/E-glass and Basalt/E-glass reinforced hybrid composites. *International Journal of Mechanical and Production Engineering Research and Development*, *8*(3), 591–598. doi:10.24247/ijmperdjun201863

Singh, M., & Khilar, P. M. (2017). A Range Free Geometric Technique for Localization of Wireless Sensor Network (WSN) Based on Controlled Communication Range. *Wireless Personal Communications*, *94*(3), 1359–1385. doi:10.100711277-016-3686-x

Sundaram. (2019). Smart clothes with bio-sensors for ECG monitoring. *International Journal of Innovative Technology and Exploring Engineering*, *8*(4), 298–301.

Veeman, Shree, M. V., Sureshkumar, P., Jagadeesha, T., Natrayan, L., Ravichandran, M., & Paramasivam, P. (2021). Sustainable development of carbon nanocomposites: Synthesis and classification for environmental remediation. *Journal of Nanomaterials*, *2021*, 1–21. doi:10.1155/2021/5840645

Velmurugan, G., Siva Shankar, V., Kaliappan, S., Socrates, S., Sekar, S., Patil, P. P., Raja, S., Natrayan, L., & Bobe, K. (2022). Effect of aluminium tetrahydrate nanofiller addition on the mechanical and thermal behaviour of luffa fibre-based polyester composites under cryogenic environment. *Journal of Nanomaterials*, *2022*, 1–10. doi:10.1155/2022/5970534

Yogeshwaran. (2015). Mechanical properties of leaf ashes reinforced aluminum alloy metal matrix composites. *International Journal of Applied Engineering Research: IJAER*, *10*(13), 11048–11052.

Chapter 15
Interfacing of Actuators and Sensors

Manivel Kandasamy
Unitedworld School of Computational Intelligence, Karnavati University, India

Raju Shanmugam
https://orcid.org/0000-0002-8613-9518
Unitedworld School of Computational Intelligence, Karnavati University, India

Sandipan Saha
Unitedworld School of Computational Intelligence, Karnavati University, India

Rohan Shankar Patil
Unitedworld School of Computational Intelligence, Karnavati University, India

Keyur Patel
Unitedworld School of Computational Intelligence, Karnavati University, India

ABSTRACT

Actuators and sensors are simple and very sensitive components that create a broad impact and importance on industrial system needs. This chapter proposed various types of actuators and sensor models with their working capabilities and applications. In this chapter, the authors give a thorough review of the sensor and actuators and where these models are implemented in various real contexts of mechanical parts of the industry, electrical and electronic circuits, and other essential areas. Here the authors broadly covered optical, acoustic, chemical, and thermal sensors, as well as different types of actuators such as Pneumatic, hydraulic, motor-based, and electro-mechanical actuators, and examines the several actuator and sensor technologies and the range of applications for each. Also covered is the possible functions and its implementations. Finally, the authors conclude with some real-world examples of how actuators and sensors are used in various applications.

DOI: 10.4018/978-1-6684-9267-3.ch015

1. INTRODUCTION

This chapter focuses on the importance of interfacing actuators and sensors, their disadvantages, pros and cons, and practical application examples. Actuators and sensors are critical components of every automation system, and their proper implementation is critical for effective operation. The chapter addresses the principles of interfacing and offers an overview of the various designs used to link actuators and sensors. The benefits and drawbacks of each technique are explained, and application examples are offered to demonstrate the usage of actuators and sensors in real-world systems. By the end of the chapter, readers should have a good grasp of connecting actuators and sensors and be able to select the optimal approach for their specific application.

2. OVERVIEW OF ACTUATORS AND SENSORS

Actuators and sensors are critical components in a wide range of electronic systems. Actuators are electronic devices that translate an electrical signal into a physical action, such as moving a component, opening and shutting a valve, or turning on a motor. Sensors give input to a system by monitoring physical components including temperature, pressure, light, and sound. Actuators and sensors, when coupled, provide the foundation of ****automation*** systems, allowing machines to adapt to changes in their surroundings. Actuators and sensors are employed in a variety of sectors, such as automotive, robotics, security, and ****HVAC***. They're also in consumer devices like phones, wearables, and household appliances.

3. TYPES OF ACTUATORS AND SENSORS

3.1. Sensors

3.1.1. Definition

Sensor is a device that detects and reacts to input from the physical world. Light, heat, motion, moisture, pressure, or any of a variety of other environmental phenomena might constitute the specific input. The output is often converted into electrical signals that may subsequently be utilised to monitor, control, and analyse the environment, and those signals are translated into a form that a user or controlling device can comprehend.

It is made up of several components, including a sensing element, a ****signal amplifier***, a signal conditioner, a converter, and a processor. The physical component that responds to the measured amount and translates it into an electrical signal is known as the sensing element. The signal amplifier boosts the amplitude of the detected signal to make it usable. The signal conditioner filters, shapes, and alters the amplifier signal. The conditioned signal is then converted from analogue to digital by the converter. The CPU then computes and shows the findings.

3.1.2. Various Categories of Sensors

3.1.2.1. Motion Sensors

Motion sensors are devices that use sound, light, or heat to detect physical motion in a given area. Passive infrared (PIR) sensors, microwave sensors, and acoustic sensors are a few examples of motion sensors.

Examples of motion sensors include:

1. Passive Infrared (PIR) Sensors (Agarwal et al., 2010): These are the most commonly used motion sensors, detecting changes in infrared radiation caused by an intruder's body heat.
2. Acoustic Sensors (Friedt, 2010): These sensors use sound waves to detect motion. It is often used as a security device to detect intruders or as a safety device to detect the presence of people in hazardous areas. It works by sending out sound waves and then listening for any changes in the sound waves that indicate movement.
3. Image motion sensors (Gensolen et al., 2005): These motion sensors use imaging technology to detect motion. Examples include security cameras, video surveillance systems, and automated door openers.

3.1.2.2. Temperature Sensors

A temperature sensor is a device that gauges and keeps track of temperature. In a number of settings, including heating and cooling systems, food processing, and medical equipment, it is used to identify, measure, and record temperature changes.

Examples of Temperature sensors include:

1. Thermistors (Ibrahim, 2002): Thermistors are semiconductor devices that measure temperature using resistance. They are used in applications where a fast response time is required.

2. Thermocouples (Noriega et al., 2015): Thermocouples are the most common type of temperature sensor. They work by measuring the temperature difference between two junctions. The voltage generated is then measured and used to calculate the temperature.
3. Resistance Temperature Detectors (RTDs) (Kim et al., 2015): RTDs are temperature sensors that use an electric resistor to measure temperature. RTDs can monitor temperatures across a larger temperature range and are more accurate than thermistors. They have a high accuracy and higher resolution so they are used in applications requiring more precise temperature measurements.

3.1.2.3. Pressure Sensors

A device used to detect pressure, usually of gases or liquids, is called a pressure sensor. The force necessary to prevent a fluid from expanding is expressed as pressure, which is often expressed in terms of force per unit area. A diaphragm, a piezoelectric device, and a circuit for signal processing normally make up a pressure sensor. Applications for pressure sensors include monitoring the pressure in a refrigeration system, measuring the pressure in an automobile's tyres, and detecting the pressure of gases and liquids in pipes.

Examples of Pressure sensors include:

1. Strain Gauge Pressure Sensors (Xu et al., 2020): Strain gauge pressure sensors measure the strain on a diaphragm caused by the pressure. The strain is then used to measure the pressure.
2. Capacitive Pressure Sensors (Shadwani et al., 2018): Capacitive pressure sensors measure pressure by measuring the capacitance of a diaphragm. When pressure is applied to the diaphragm, the capacitance changes and can be used to measure the pressure.
3. Piezoelectric Pressure Sensors (Gautschi, 2002): Piezoelectric pressure sensors are based on the piezoelectric effect, which is the ability of certain materials to generate an electrical charge when subjected to mechanical stress.

3.1.2.4. Position Sensors

Devices that measure and detect an object's location are called position sensors. They monitor rotational or linear motion and produce an analogue output signal that a controller can interpret. Position sensors come in a variety of forms, including magnetic, inductive, and potentiometric.

Example of Position sensors includes:

1. Magnetic Position sensors (Bretschneider et al., 2010): To determine an object's location, magnetic position sensors employ a magnet. Hall-effect sensors, which gauge a magnetic field's strength and presence, and magneto-resistive sensors, which gauge an object's magnetic field's location, are two examples of magnetic position sensors.
2. Inductive Position sensors (Golby, 2010): Inductive position sensors measure the position of an object by detecting changes in an electromagnetic field. Examples of inductive position sensors include inductive proximity sensors, which measure the presence of a metallic object, and inductive linear sensors, which measure the position of a ferromagnetic object.
3. Potentiometric Position sensor (Gasperi & Hurbain, 2009): Potentiometric position sensors measure an object's location using a resistive element. Rotating potentiometers, which gauge a shaft's angular displacement, and linear potentiometers, which gauge an object's linear displacement, are two examples of potentiometric position sensors.

3.1.2.5. Optical Sensors

Optical sensors are tools for measuring physical quantities or detecting the presence of objects using light. In areas like optical image processing, industrial automation, and security systems, these sensors are perfect since they demand great precision and accuracy. Automobile, aeronautical, and medical uses are all possible for them.

Example of Optical sensors include:

1. Photodiode Optical Sensors (Arias et al., 2008): A photodiode is a semiconductor device that transforms light into electrical current. The majority of the time, they are employed in photoelectric sensors, which use light to detect the presence of an item.
2. Phototransistors (Pierre et al., 2017): These semiconductor devices transform light into electrical current. Photoelectric sensors, which utilise light to detect the presence of an item, frequently employ them.
3. Photovoltaic Cells Sensors (Bletskan et al., 2016): A photovoltaic cell is an apparatus that transforms light into electrical current. In solar cells, which absorb solar energy and transform it into electricity, they are commonly utilised.
4. Infrared Sensors (Rashid & Ali, 2018): These devices employ infrared light to identify objects in their environment. They are frequently employed in remote control systems, including those found in television remotes.

3.1.2.6. Force Sensors

The force or load imparted to an item is measured by sensors called force sensors. These sensors may be used to measure a number of forces, including torque, shear, compression, and tension. Applications like robotics, aircraft, and manufacturing frequently employ force sensors.

Examples of Force sensors include:

1. Force Sensing Resistors (Sadun, 2016): A force sensitive resistor is a type of resistor that adjusts its resistance in response to pressure applied to its surface. It is possible to quantify weight, pressure, or other forms of force using the resistance, which varies based on the amount of force applied.
2. Inductive Force Sensors (Giesberts et al., 2018): A form of transducer called an inductive sensor uses an electromagnetic field to find the presence of a metallic item. When an item comes into touch with the sensor, a coil within the sensor senses the alternating magnetic field that is created by the sensor. The voltage created by the current produced in the coil can then be utilised to monitor or control the item.
3. Capacitive Force Sensors (Dobrzynska & Gijs, 2010): A capacitive force sensor is a sort of transducer that transforms mechanical force into electrical impulses. When a force is applied, it detects the capacitance between two metal plates. The capacitance varies according to the amount of force exerted, allowing the sensor to detect the force. The output of the sensor is then transformed into a voltage or current signal that may be used to regulate or track the force.

3.1.2.7. Magnetic Sensors

A magnetic sensor is a specific kind of transducer that is used to gauge a magnetic field's intensity, direction, or other properties. It may be used to locate magnetic fields, gauge their intensity, and pinpoint their directions. Applications for navigation, robotics, and automation frequently employ magnetic sensors.

Examples of Magnetic sensors include:

1. Hall Effect Sensors (Ramsden, 2006): Hall effect sensors, which gauge the intensity of a magnetic field, are based on the Hall effect concept. These sensors can be used to locate magnetic fields, gauge their intensity, or identify their direction. The Allegro A1321 and Honeywell SS461A sensors are two examples.
2. Magnetoresistive Sensors (Freitas et al., 2007): The idea of magnetic resistance serves as the foundation for magnetoresistive sensors. A magnetoresistor is

used by these sensors to gauge the intensity of a magnetic field. The Honeywell SS495A and the Infineon TLI4971 sensors are two examples.

3. Fluxgate Sensors (Wei et al., 2021): Fluxgate sensors employ a fluxgate core to gauge a magnetic field's intensity. Applications for robotics and navigation employ these sensors.

3.1.2.8. Sound Sensors

A sound sensor is a device that measures sound levels and converts them into an electrical signal. Sound sensors are used in a variety of applications, such as security systems, audio equipment, and medical equipment. There are several types of sound sensors, including microphones, micro speakers, piezoelectrics, and hydrophones.

Examples of Sound sensors include:

1. Microphones Sound Sensor (Thakker, 2013): A microphone is a sort of sound sensor that transforms sound waves into electrical impulses. In addition to being used to detect and quantify sound levels, microphones are frequently used in audio equipment such as microphones and loudspeakers.
2. Micro Speakers Sound Sensor (Rusconi et al., 2022): Micro Speakers are a sort of sound sensor that transforms electrical signals into sound waves. Micro Speakers are frequently found in audio equipment including headphones and loudspeakers.
3. Hydrophones Sound Sensor (Yildiz et al., 2014): hydrophone is a sort of sound sensor used to identify sound waves in water. Applications for underwater sonar and acoustics frequently employ hydrophones.

3.1.2..9. Proximity Sensors

The presence of an object or person can be detected using proximity sensors without making physical contact. Within a specific sensor range, they may be used to find things like automobiles, people, or other items. Whether they generate a signal on their own or pick up a signal from another source, proximity sensors can be either active or passive. Inductive, capacitive, ultrasonic, and infrared sensors are just a few of the several kinds of proximity sensors.

Example of Proximity sensors includes:

1. Capacitive Proximity Sensor (Poriņš & Apse-Apsitis, 2019): To find non-ferrous items like plastic, ceramic, and other non-metallic materials, capacitive proximity sensors are utilised. This kind of sensor measures and detects changes in the object's location using an electrical field.

2. Ultrasonic Proximity Sensor (Ng, 2020): Sound waves are used by ultrasonic proximity sensors to identify things. This kind of sensor can estimate the distance to the item and is used to find objects in difficult-to-reach places.
3. Infrared Proximity Sensor (Mohamad, 2016): The presence of objects is detected using infrared proximity sensors using infrared light. This kind of sensor can estimate the distance to the item and is used to detect things in low light situations.

3.1.2.10. Flow Sensors

Flow sensors are tools for calculating the volume of fluid going through a pipe or channel at a certain pace. They have a wide range of uses, from gauging air flow in a home ventilation system to spotting water movement in a river or irrigation system. There are many distinct types of flow sensors, each suited to a different purpose.

Examples of Flow sensors include:

1. Velocity Flow Sensors (Zhou et al., 2022): These are devices that gauge the rate at which a fluid is travelling through a pipe or channel. Differential pressure flow metres, turbine flow metres, and vortex flow metres are a few types of velocity flow sensors.
2. Mass Flow Sensors (Baboo, 2015): These devices calculate the weight or mass of a certain amount of fluid. Thermal mass flow metres and Coriolis flow metres are a few examples of mass flow sensors.
3. Displacement Flow Sensors (Baum et al., 2023): A particular kind of flow metre used to gauge a liquid or gas's flow rate. It operates by using a mechanical or electrical instrument to measure the displacement of a known volume of liquid or gas. Applications for displacement flow sensors include industrial process control, chemical plants, and water treatment facilities.

3.1.3. Pros and Cons of Sensors

Pros of Sensors:

- **Accurate Measurement**: Sensors provide accurate measurement of the physical parameters that they measure.
- **Cost Effective**: Sensors are cost effective and have a low cost of installation and maintenance.
- **Versatile**: Sensors can be used to measure a wide range of physical parameters.
- **Compact**: Sensors are small and compact in size, which makes them easy to install and use in a variety of locations.

- **Reliability**: Sensors are reliable and provide accurate results over a long period of time.
- **Robust**: Sensors are ****robust*** and can withstand harsh environmental conditions.
- **Easy to Use**: Sensors are easy to use and require minimal training and setup.
- **Long Lasting**: Sensors are long lasting and can provide accurate results for a long period of time.
- **Flexibility**: Sensors can be easily adapted to different conditions and requirements.
- **Easy to Maintain**: Sensors are easy to maintain and can be easily replaced or repaired if needed.

Cons of Sensors:

- **High Cost**: Sensors can be expensive to purchase and install.
- **Maintenance and Calibration**: Regular maintenance and ****calibration*** is required to ensure accuracy.
- **Environmental Factors**: Sensors can be affected by humidity, temperature, and other environmental factors.
- **Interference**: Sensors can be susceptible to interference from external sources like radio waves or other electromagnetic fields.
- **Response Time**: Sensors can take a long time to respond to changes in their environment.
- **Accuracy**: Sensors can be inaccurate due to a variety of factors such as calibration or environmental factors.
- **Range**: Sensors typically have a limited range and can only detect objects within a certain distance.
- **Reliability**: Sensors can be unreliable due to a variety of factors such as power issues, environmental conditions, or mechanical issues. 4.2.9. Power Requirements – Sensors require power to operate, which can be costly or difficult to provide in certain environments.
- **Security**: Sensors can be vulnerable to hacking and other security risks.

3.2. Actuators

3.2.1. Definition

An actuator is a device that is used to convert energy into motion. It is typically used to control the motion of a mechanical system, such as a robot or an industrial machine. Actuators are often used to control the position, velocity, force, and/or acceleration of mechanical systems. In order to move or control a particular item, the input energy is then employed. ****Pneumatic*** actuators, rotary actuators, and linear actuators are a few examples of the different variants that may be used. A power source, an actuator body, and a control system are among the components that most actuators normally include.

The actuator body is often a mechanical linkage, such as a cam or a lever, and the power source is frequently a motor or a pneumatic cylinder. The movement of the actuator body is controlled by the control system, which is commonly an electrical or hydraulic system. Sensors, switches, and controllers are just a few examples of the peripherals that are frequently interfaced with actuators.

As already noted, a sensor is a device that receives and responds to input from the outside world, and switches can be used to start the actuator in motion. Speed, direction, and position are all controlled by controllers through the actuator. Actuators frequently include a computer interface, enabling precise control of the actuator and any related components.

3.2.2. Various Categories of Actuators

3.2.1.1. Electrical Actuators

An electrical actuator is a tool used to operate valves, dampers, and other industrial gear by converting electrical energy into mechanical motion. It is a kind of servomechanism, and electric motors are frequently used to power them.

Examples of electrical actuators include:

1. Linear Electrical Actuators (Mendoza-Vázquez et al., 2009): Actuators that create linear motion are the most often utilised actuators. Pneumatic cylinders, linear motors, and solenoids are a few examples.
2. Rotary Electrical Actuators (Belogusev & Egorov, 2019): These actuators are made to rotate an object, such as a shaft, a valve, or another device. Rotary motors, stepper motors, and gear motors are a few examples.

3.2.1.2. Pneumatic Actuators

Pneumatic actuators are mechanical devices that transform energy into linear or rotational motion using compressed air. They are utilised in a wide variety of industrial process control applications, including positioners, dampers, and valves.

Examples of Pneumatic Actuators include:

1. Linear Pneumatic Actuators (Avram et al., 2017): These types of actuators produce linear motion using a cylinder. There are many applications where linear motion is needed, and linear actuators are utilised to deliver such motion.

They are frequently employed to move doors, open and close valves, adjust mirrors, place equipment, and do other related duties.

2. Rotational Pneumatic Actuators (Ali & Noor, 2009): These provide rotatory motion by using a revolving cylinder.An inflated diaphragm is utilised in diaphragm actuators to manage a system's pressure. For controlling air pressure, they are frequently employed in HVAC and air conditioning systems.
3. Diaphragm Pneumatic Actuators (Fang et al., 2016): These actuators produce linear or rotary motion using an elastic diaphragm. When anything has to be moved in a rotating motion, such as in a car seat recliner, rotational actuators are utilised. The seat is moved back and forth by them, which are frequently seen in electric recliners.

3.2.1.3. Hydraulic Actuators

In order to translate hydraulic pressure into mechanical motion, a hydraulic actuator is employed. It typically consists of a cylinder with a piston inside that is liquid-filled. The liquid is driven through a valve when pressure is given to the actuator, where it is utilised to move the piston, which in turn moves the associated mechanical parts.

Examples of Hydraulic Actuators include:

1. Linear Hydraulic Actuators (Sotoodeh, 2019): The most typical kind of hydraulic actuator. Hydraulic fluid is forced out of a valve after being driven into the cylinder, applying pressure to the piston. The associated mechanical components are moved by this pressure.
2. Rotary Hydraulic Actuators (Haas, 2001): Similar to linear actuators, rotary hydraulic actuators rotate a shaft or other mechanical parts. The shaft is moved by applying pressure to hydraulic fluid as it passes through a valve.
3. Electro-Hydraulic Actuators (Has et al., 2014): These actuators regulate the flow of hydraulic fluid using an electric current. The actuator's motion may now be controlled more precisely as a result.

3.2.1.4. Mechanical Actuators

An apparatus that transforms energy into motion is known as a mechanical actuator. Usually, an electric motor, a hydraulic or pneumatic system, or a manual mechanism is used to move it.

Example of Mechanical actuator includes:

1. Linear Mechanical Actuators (Castellini & D'Andrea, 2014): Linear actuators are used to move tables and chairs, open and close doors, and regulate the

position of valves. They can be powered by a hydraulic or pneumatic system, although electric motors are most frequently used to move them.

2. Rotary Mechanical Actuators (Zhang & Yan, 2011): Rotary actuators are employed in a variety of applications, including the raising and lowering of machinery, the spinning of items, and the directional control of motors and pumps. They can also be powered by a hydraulic or pneumatic system, although electric motors are more common.
3. Reciprocating Mechanical Actuators (Li et al., 2020): Reciprocating actuators are used to control the speed of a motor or pump as well as to raise and lower things and move them from side to side. They can also be powered by a hydraulic or pneumatic system, although electric motors are more common.

3.2.1.5. Magnetic Actuators

A device called a magnetic actuator creates mechanical force using one or more magnets. It is possible to produce, control, or move objects with this force. Applications for magnetic actuators range from the medical to the industrial to consumer goods. They are often regarded as dependable, affordable, and simple to use.

Examples of Magnetic Actuator includes:

1. Electromagnetic Actuators (Fitan & Messine, 2004): With the help of electrical current, these actuators produce a magnetic field that exerts a force against an item. Numerous devices, such as motors, valves, and linear actuators, employ this kind of actuator.
2. Permanent Magnet Actuators (Zhu & Wang, 2006): These actuators produce force by using permanent magnets to produce a magnetic field. They are frequently employed in industrial and robotic applications.
3. Electromechanical Actuators (Qiao et al., 2017): To produce a force, these actuators combine an electrical current with mechanical parts. Aerospace, automotive, and medical applications are typical uses.

3.2.1.6. Thermal Actuators

In order to carry out a given action, thermal actuators transform thermal energy into mechanical energy. Depending on the kind of actuator being utilised, this motion might be either linear or rotational.

Examples of Thermal Actuators includes:

1. Pneumatic Thermal actuators (Henning, 2006): These actuators move a piston or a diaphragm by applying pressure to air. A burner warms the compressed air, which causes the valve to open or close.

2. Hydraulic Thermal actuators (Long & Loveday, 2007): These actuators move a piston or a diaphragm by using a liquid, such as oil or water. A burner heats the liquid, which causes the valve to open or close.
3. Electric Thermal actuators (Melnyk et al., 2014): An electric current is used by these actuators to drive a motor or solenoid into action. A transformer heats the electric current, which causes the valve to open or close.

3.2.1.7. Piezoelectric Actuators

When a voltage is given to a piezoelectric actuator, they create a mechanical force or displacement. They are based on the piezoelectric effect, which refers to some materials' capacity to produce a voltage in response to the application of a mechanical force.

Examples of Piezoelectric Actuator include:

1. Piezoceramic Actuators (Shields & Konefat, 2021): These are the most prevalent kind of piezoelectric actuators and are constructed from materials known as piezoceramics, which exhibit the piezoelectric effect. These actuators may be used to move tiny objects or serve as a transducer and are frequently used in low-force applications.
2. Piezoelectric Stack Actuators (Sherrit, 2011): These actuators combine several piezoelectric components to produce a stronger force than a single unit can. This kind of actuator may be utilised to move heavier items and is suitable for applications that call for greater forces.
3. Piezoelectric Linear Motors (Zuo et al., 2005): To produce a linear motion, these actuators combine many piezoelectric components into a linear arrangement. This kind of actuator may be utilised to move heavy items and is employed in situations that call for precision placement.

3.2.1.8. Shape Memory Alloy Actuators

In order to produce a physical movement, a Shape Memory Alloy (SMA) actuator makes use of the shape memory effect of certain alloys, often nickel-titanium. By "remembering" their initial shape and returning to it when heated, certain alloys exhibit the shape memory effect. A mechanical movement is produced in a SMA actuator by using the shape memory alloy.

Examples of Shape Memory Alloy actuators include:

1. Linear SMA Actuators (Jani, 2017): Motion that is linear is produced by linear actuators. They are made up of a load and a frame linked by a SMA wire or rod. The SMA wire or rod will compress when heated, resulting in linear motion.

2. Rotational SMA Actuators (Yuan, 2017): Rotational motion is produced by rotary actuators. They are made up of a core with a load connected and a SMA wire or rod wrapped around it. The SMA wire or rod will induce rotational motion in the core when heated.
3. Flexible SMA Actuators (Copaci et al., 2020): Motion may be made flexible by using flexible actuators. They are made up of a flexible substrate, such a rubber band or a strip of cloth, and a SMA wire or rod. The substrate will move when heated by the SMA wire or rod, producing flexible motion.

3.2.1.9. Voice Coil Actuators

A voice coil actuator (VCA) is an electromechanical device that moves a coil of wire inside a magnetic field using an electric current. The position of a mechanical part, such as a lever, valve, or other moving element, may be controlled by this movement.

Examples of Voice Coil Actuator includes:

1. Linear Voice Coil Actuator (Shewale et al., 2018): A linear VCA produces linear motion along a single axis by use of a magnetic field. This kind of actuator is frequently employed in applications that need precise positioning, including adjusting the position of a microscope stage or laser beam.
2. Rotary Voice Coil Actuator (Gallasch & Rafolt, 2001): To provide continuous rotation, a rotary VCA rotates a magnetic field. Typically, robotic arms and other rotating applications employ this kind of actuator.

3.2.1.10. Solenoid Actuators

An electrical energy can be transformed into a linear, rotational, or other kind of mechanical force using a solenoid actuator. It is a particular kind of electromechanical actuator that is driven by an electric current passing via a solenoid.

Example of Solenoid Actuator includes:

1. Linear Solenoid Actuators (Tikani et al., 2012): These actuators, which are frequently used in door locks, vending machines, and medical devices, produce linear motion using a magnetic field.
2. Rotational Solenoid Actuators (Tsukima et al., 2017): These actuators, which are often used in robotics and automotive systems, produce rotary motion using a rotating magnetic field.
3. Combined Solenoid Actuators (Miric et al., 2018): These actuators, which are frequently used in applications like industrial automation and medical equipment, combine linear and rotational motion to provide a more complicated motion.

3.2.3. Pros and Cons of Actuators

Pros of Actuators:

- **Enhanced precision and accuracy**: Actuators provide increased accuracy and ***precision** in motion than manual or mechanical devices.
- **Cost-effective**: Compared to other motion control systems, actuators are more cost-effective.
- **Low maintenance**: Actuators require minimal maintenance and can operate for long periods of time without any attention.
- **Fast response**: Actuators respond quickly to commands, allowing for faster machine operation and better control over the process.
- **Versatility**: Actuators are flexible and can be used in a variety of applications.
- **Compact size**: Actuators are small in size, making them easy to install and use in confined spaces.
- **Reliability**: Actuators are reliable, durable and can operate in harsh environments.
- **Energy efficient**: Actuators use less energy than traditional motion control systems.
- **Intelligent control**: Advanced actuators can be programmed to react to specific conditions or commands.
- **Adaptability**: Actuators can be easily adapted to fit different machines and applications.

Cons of Actuators:

- **High Cost**: Actuators can be expensive to purchase, install, and maintain due to their complexity.
- **Complexity**: Actuators are complex machines that require specialised knowledge and expertise to operate.
- **Maintenance**: Actuators require regular maintenance and upkeep to ensure proper performance.
- **Size**: Actuators are often large and bulky, which can be difficult to install in tight spaces.
- **Environmental Conditions**: Actuators can be sensitive to environmental conditions such as temperature and humidity.
- **Accuracy**: Actuators can be inaccurate due to the complexity of their motion and operation.
- **Power Consumption**: Actuators require a large amount of energy to operate and can be expensive to run.

- ◦ **Safety**: Actuators can be dangerous if not properly operated and maintained.
- ◦ **Noise**: Actuators can be noisy and can create a nuisance for neighbours.
- ◦ **Vibration**: Actuators can vibrate and cause disruption to other machinery or people nearby.

4. INTERFACING OF ACTUATORS AND SENSORS

The process of connecting a sensor or actuator to a computer, ****microcontroller***, or other electronic device in order to enable communication between the sensor or actuator and the device is known as interfacing. An analog or digital signal is commonly used for this communication. In order to guarantee that the data is sent between the device and the sensor or actuator precisely and effectively, interfacing also requires the use of software, hardware, and ****protocols***.

In order to interact with the physical world, actuators and sensors are employed. Actuators are used to manage the environment depending on the data supplied by the sensors, whereas sensors are used to detect changes in the environment and send data to the system. An essential component of any control system is the interface between the sensors and the actuators. The interface is used to link the control system's actuators and sensors. It enables the control system to get information from the sensors and act on that information to control the actuators.

A few instances of actuators and sensors interacting:

- To enable control of a robotic system, actuators and sensors can be interfaced with a computer or microcontroller.
- A robotic arm's ability to move in three dimensions can be enabled through the interaction of actuators with the arm.
- Sensors may be interfaced with self driven cars so that it can recognise items in its route and avoid obstructions.
- Drones with interfaces for actuators and sensors can fly on their own.
- A 3D printer may be interfaced with actuators and sensors to allow it to notice changes in its surroundings and react to them.
- By using actuator-sensor interfacing, simple speedometer, petrol indicator, engine temperature such things are automated on a dashboard.

5. INTERFACING METHODS FOR ACTUATORS AND SENSORS

The integration of hardware and software components to facilitate communication between actuators and sensors is a key component of interface techniques. Analog-to-digital and digital-to-analog converters, I/O Ports, etc are a few examples of hardware elements used for this. On the software side, protocols including Modbus, CANbus, etc are employed.

Various Methods of Interfacing Actuators and Sensors:

5.1. Analog/Digital Converters

An ADC is used to interface between actuators and sensors, allowing analog signals from sensors to be converted into a digital signal that can be read and processed by a microcontroller to control actuators. Figure 1 represents the Schematic diagram of sensor and actuator interfacing with the ADC and Raspberry Pi (Ali, Reza, Shafik, Islam, Chowdhury, Jeong, & Chung, 2021).

Examples of ADC in sensor-to-actuator interfacing include:

- **Temperature sensors**: Sensors that measure temperature, such as RTDs (resistance temperature detectors), turn it into an analogue signal that can be read by an ADC and used by a microcontroller to regulate an air-conditioning or heating system.
- **Pressure sensors**: Strain gauges and other pressure sensors translate mechanical force (pressure) into an analogue signal that an ADC can read and use to operate a pump or automated valve.
- **Optical sensors**: ******Photodiodes*** and ******phototransistors*** are examples of optical sensors that transform light intensity into an analogue voltage that an ADC can read and use to operate a motor or light.
- **Motion sensors**: ******Gyroscopes*** and accelerometers translate movement into an analogue signal that an ADC can read and used by a microcontroller to control a robotic arm or vehicle.

5.2. I/O (Input or Output) Ports

I/O (input/output) ports are used to connect actuators and sensors together as a way of transferring data between the two (Al-Mashud, 2014). These I/O Ports are generally placed in a microcontroller or processing device for direct bridging (Figure 2). A simple block diagram of I/O interface, basic input output device is a set of three-state buffers.

Figure 1. Schematic diagram of sensor and actuator interfacing with the ADC (analog to digital converter) and raspberry pi

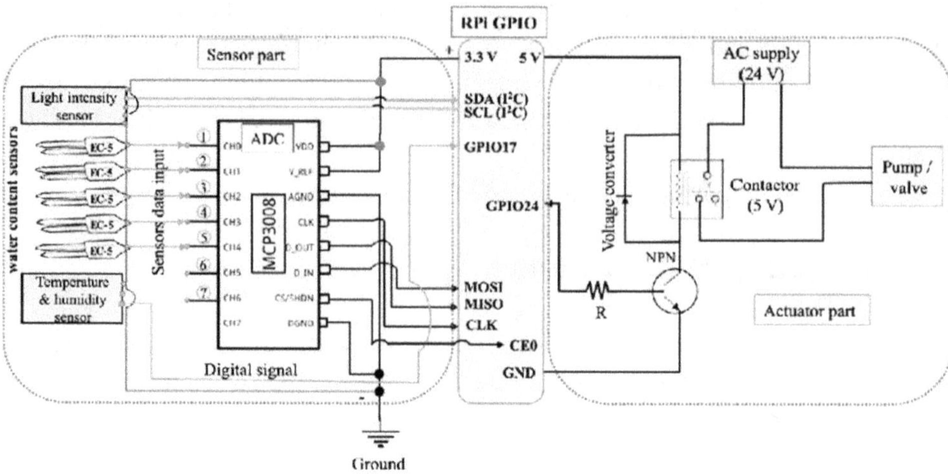

Examples of I/O ports used in interfacing actuators and sensors include:

- **Serial ports**: These ports transmit data across a network of wires by connecting actuators and sensors to computers.
- **USB ports**: These allow for the connection of a broad range of external devices, such as actuators and sensors.
- **Ethernet ports**: By connecting actuators and sensors to local networks, Ethernet ports enable the flow of data across longer distances.
- **Analog ports**: Devices like microcontrollers can link analogue actuators and sensors via analogue ports.
- **Digital ports**: Digital actuators and sensors, such switches and LEDs, are connected via digital ports.

5.3. Serial Communication

Serial communication is used to transfer data between actuators and sensors, where the data is transmitted bit-by-bit over a single communication line. Figure 3 represents the Schematic diagram of Serial Communication Interface (Dostálek, Vasek, & Dolinay, 2008).

Examples of Serial Communication used in interfacing actuators and sensors include:

Figure 2. A simple block diagram of I/O interface, basic input output device is a set of three-state buffers

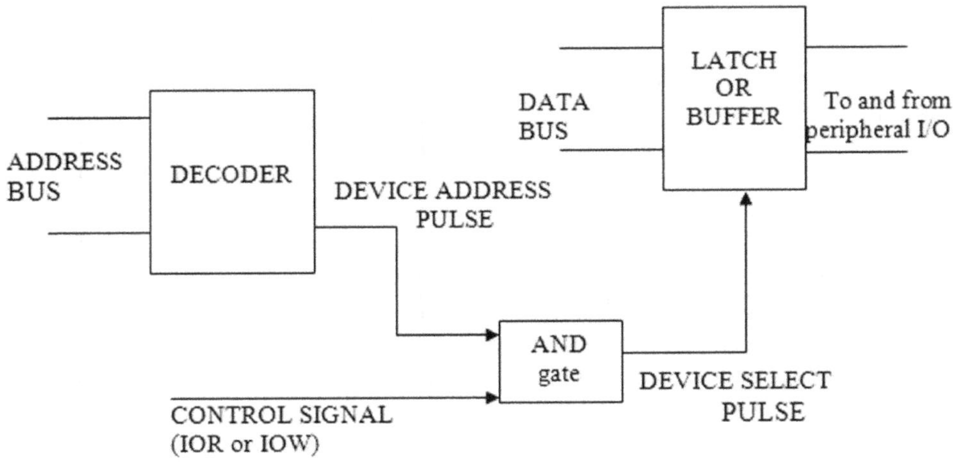

- **RS-232**: Used in robotics and medical devices to link a range of actuators and sensors.
- **RS-485**: Used in industrial automation systems, such as factory automation, to connect actuators and sensors.
- **Modbus RTU/ASCII**: Used to link actuators and sensors in process control systems and SCADA systems, among other industrial automation systems.
- **CAN Bus**: Used in automotive applications such as engine management systems and airbag control systems to connect a range of actuators and sensors.
- **I2C**: Inter-Integrated Circuit (**I2C*) is used in consumer electronics and computer systems, such as in digital cameras and mobile phones, to connect a range of actuators and sensors.

5.4. Network Protocols

Network protocols such as Ethernet, Wi-Fi, and Zigbee are commonly used to interface actuators and sensors (Hofman, 2015). Figure 4 is a block diagram representing networking, protocols and interface.

Popular network protocols used to do this include examples like:

Figure 3. Schematic diagram of Serial Communication Interface

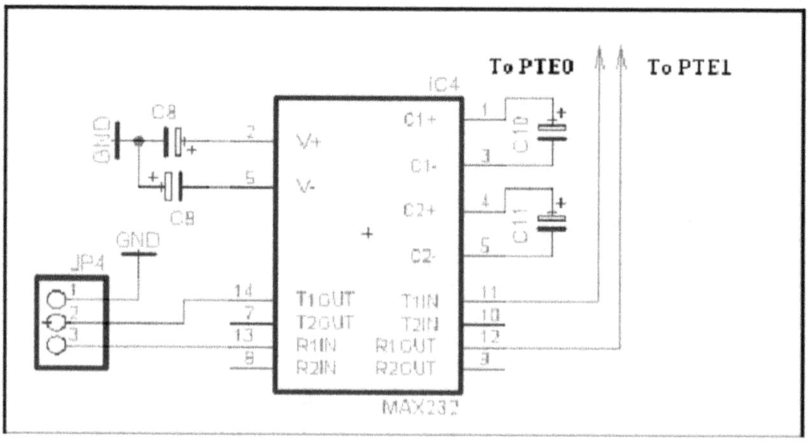

- **WiFi**: For wireless communication between sensor-actuator pairs, WiFi is a frequently used protocol. It is widely utilised in a variety of applications, including industrial control and home automation.
- **Zigbee**: A wireless technology with low costs and power consumption that is used in sensor networks. It is frequently employed in applications like industrial control, home automation, medical monitoring, and others.
- **Ethernet**: For wired communication between sensor-actuator pairs, this is a frequently used protocol. In addition to other uses, it is frequently employed in industrial control.
- **Bluetooth**: Bluetooth is a wireless technology used in LANs (local area networks). In applications like home automation and others, it is frequently used for communication between actuators and sensors.

5.5. Universal Serial Bus (USB)

USB is a popular interface for connecting actuators and sensors. USB (Universal Serial Bus) is a type of connectivity used to connect devices to computers (Wijesundera et al., 2012). USB can be used to interface between actuators and sensors by allowing the computer to read or control the actuators and sensors hardware. Figure 5 is Schematic diagram representing the Universal Serial Bus (USB) Interfacing Circuit.

Examples of USB used to interface between actuators and sensors include:

- **Moisture** and **Temperature sensor**: This sensor measures both temperature and humidity, and it can be used to transmit data to a computer through USB.

Figure 4. Block diagram of networking, protocols, and interfacing

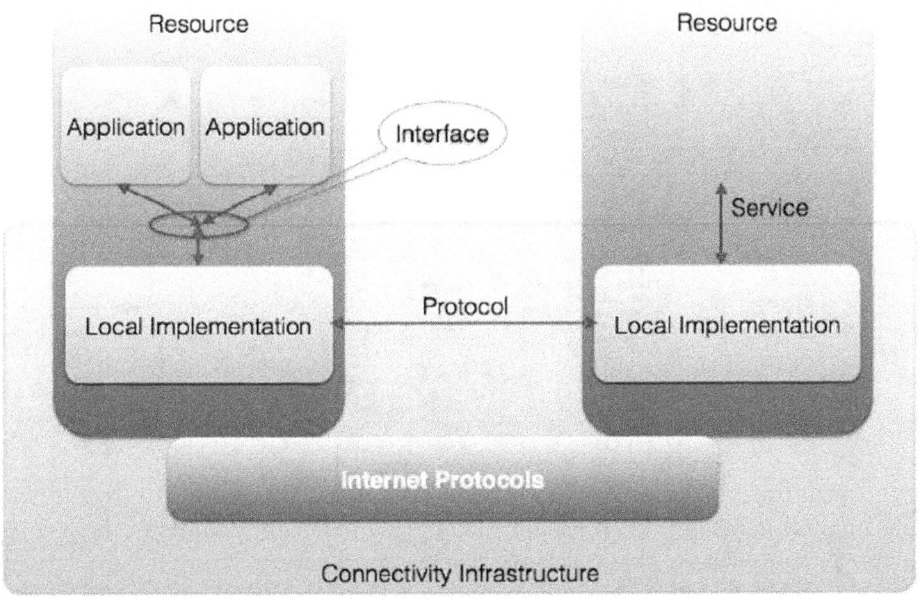

- **Motion sensors**: These motion detectors may be used to receive data or command actuators from a computer through USB.
- **Light sensors**: These sensors measure light levels and can be used to receive data or manage light sources by connecting to a computer through USB.
- **Sound sensors**: These devices monitor sound levels and can be used to receive data or manage sound sources when linked to a computer through USB.
- **Motor controllers**: These tools let a computer manage and regulate the direction and speed of motors.

5.6. Radio Frequency Identification (RFID)

The term "RFID" (Radio Frequency Identification) refers to a technology that uses radio waves to identify, track, and monitor physical things like actuators and sensors (Oh, Bang, & Hwang, 2010). Individual things may be recognised and monitored because RFID tags are inscribed with special identification numbers. In a range of applications, such as healthcare, industrial automation, and asset monitoring, RFID is employed as an interface between actuators and sensors. Figure 6 is a Block diagram of Radio Frequency Identification(RFID) Interfacing to a Host.

Figure 5. Schematic diagram of universal serial bus (USB) interfacing circuit

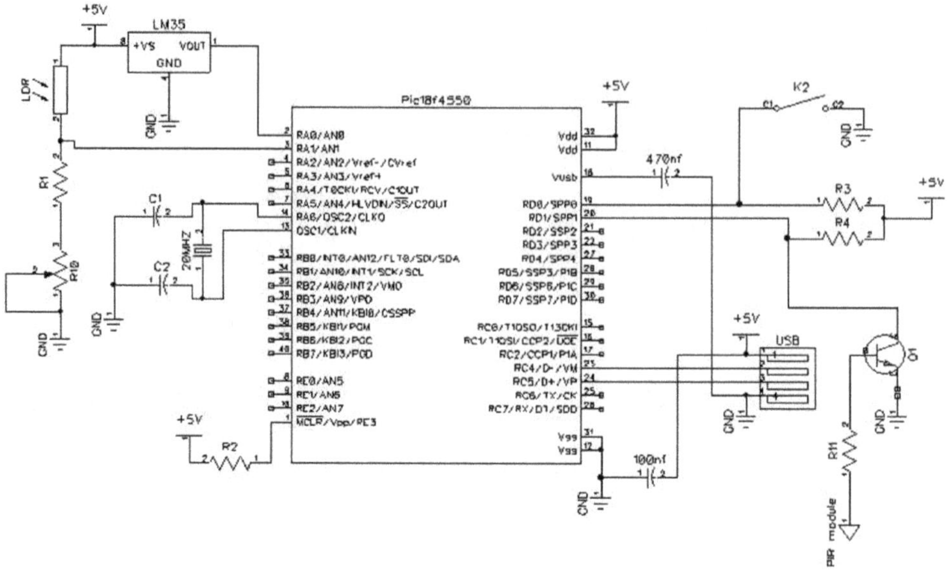

Examples of RFID used to interface between actuators and sensors include:

- Retail establishments and warehouses with automated inventory tracking. RFID tags are used to track and monitor the movement of commodities.
- Automation in industry, such as that found in manufacturing. RFID tags are used to monitor the condition of equipment and materials and to manage how they are operated.
- Medical treatment. Medication, patient information, and medical equipment are all tracked via RFID tags.
- Home automation technology. Devices like thermostats, door locks, and lighting are managed by RFID tags.
- Access management. Access to restricted places, such corporate buildings and airports, is controlled by RFID tags.

Figure 6. Block diagram of radio frequency identification (RFID) interfacing to a host

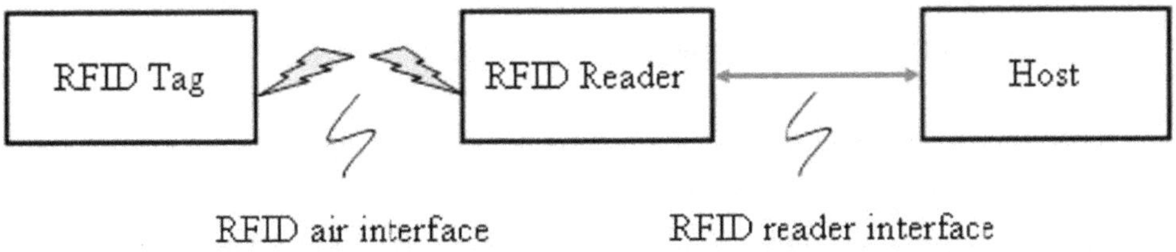

6. ADVANTAGES AND DISADVANTAGES OF INTERFACING

6.1. Advantages of Interfacing

The potential advantages of Interfacing Actuators and Sensors include:

- **Better control**: Sensors provide the actuator the ability to monitor the surroundings and take the right action to get the intended result. As an instance, sensors may be used to determine if a door is open or closed and then communicate this information to an actuator, which then opens or closes the door as necessary.
- **Increased Efficiency**: Actuators and sensors may be combined to improve a system's efficiency, which brings us to point number two. To move an object in a desired direction, for instance, sensors may be able to detect the existence of the object and transmit a signal to the actuator. The time and effort required to move the thing are decreased as a result.
- **Greater Safety**: Actuators and sensors may be used to track potentially harmful situations and take the necessary precautions to avoid them. Sensing devices, for instance, can alert an actuator to halt a machine when an item is too close to it. By doing this, the machine and the item are shielded from potential harm or damage.
- **Lower Cost**: The cost of a system may be decreased by using actuators and sensors. To move an object in a desired direction, for instance, sensors may be able to detect the existence of the object and transmit a signal to the actuator. By doing so, the system's total cost is decreased as well as the requirement for physical work.
- **Enhanced Monitoring**: It is possible to track the system's performance in real time by attaching actuators and sensors to it. Applications where performance has to be monitored and studied, like in product development or machine learning, might benefit from this.

6.2. Disadvantages of Interfacing

The possible disadvantages of Interfacing actuators and sensors:

- **Maintenance**: To make sure interface devices are operating correctly, they need to undergo routine maintenance. It may also be necessary to replace the entire system, which can be expensive, if any of the sensors or actuators malfunction.
- **Compatibility**: Last but not least, interface devices need to work with the actuators and sensors they are interacting with. The system may need to be replaced or altered if the device is incompatible and does not function correctly.
- **Interference**: Actuators and sensors can be easily affected by external interference, such as radio signals, magnetic fields, and temperature changes. This can lead to inaccurate readings and unreliable operation.
- **Dependability**: Because of the complexity of interacting actuators and sensors, it is difficult to ensure the reliability and accuracy of the system. If there is a malfunction in the system, it can be difficult to diagnose and fix.
- **Power Consumption**: Interfacing actuators and sensors can be power-hungry and can draw a significant amount of energy. This can increase the cost of the system and reduce its efficiency.

7. APPLICATIONS OF INTERFACING OF ACTUATORS AND SENSORS

Top 10 Trending Applications of Actuators-Sensors Interfacing:

- **Automation of Industrial Processes**: Sensor and actuator interface is used to automate manufacturing and chemical production processes by employing sensor feedback to regulate the pace of production. A temperature sensor, for instance, may be used to gauge the furnace's temperature, and an actuator can be used to modify the fuel flow to the furnace in order to keep the temperature there at the appropriate level.
- **Automated Controllers**: Automated controllers are used to regulate the functioning of machines and are created by integrating sensors and actuators. As an instance, sensors may be used to determine the location of a robot arm, and actuators can be used to regulate its movements, in order to carry out a task that has been programmed into it.

- **Automated Vehicles**: The development of automated vehicles, such as driverless automobiles, may be accomplished through the use of sensor and actuator interfaces. Actuators are used to control the car's speed, direction, and orientation, while sensors can be used to monitor the car's location and its surrounding environment.
- **Autonomous Robots**: By integrating sensors and actuators, it is possible to build autonomous robots that can complete tasks on their own without the need for human assistance. One example is the ability to design a robotic vacuum cleaner to recognise barriers and clean a space on its own.
- **Household Automation**: Automated home equipment, including lighting, thermostats, and security systems, may be controlled by automated systems built using sensor and actuator interfaces. As an illustration, a motion sensor may be used to identify an intruder, and an actuator can be used to sound an alarm.
- **Medical Devices**: The development of medical devices like pacemakers and insulin pumps may be accomplished through the use of sensor and actuator interfaces. It is possible to configure a pacemaker, for instance, to monitor a patient's heart rate and change the pacemaker's rhythm correspondingly.
- **Robotics**: By combining sensors and actuators, robots that can complete difficult tasks like welding and assembling may be built. A robotic arm, for instance, may be set up to recognise an object's location and change its course appropriately.
- **Smart Buildings**: By combining sensors and actuators, it is possible to design intelligent structures that can observe and regulate their surroundings. As an instance, sensors may be used to detect human presence, and actuators can be used to regulate lighting and temperature.
- **Surveillance Systems**: By combining sensors and actuators, surveillance systems that can identify and address suspicious activity may be built. An actuator can be used to activate a camera and a motion sensor, respectively, to detect the presence of an intruder.
- **Wearable Devices**: Wearable devices that can monitor and control the body can be made using sensor and actuator interfacing. For instance, a fitness tracker may be used to gauge a person's heart rate, and actuators can be used to regulate how strenuous an exercise is.

8. CHALLENGES IN INTERFACING OF ACTUATORS AND SENSORS

Actuator and sensor interfacing can be a challenge due to the wide variety of components and protocols available below:

- **Sensor Calibration**: It is typically difficult in the interface between actuators and sensors. Sensor calibration allows sensors to precisely measure and detect the physical qualities of the environment. To reliably measure the ambient temperature, for instance, a temperature sensor must be calibrated.
- **Communication Protocols**: A reliable communication protocol needs to be designed to guarantee that the actuators properly receive the data from the sensors. A remote monitoring system, for instance, could require choosing an appropriate wireless communication protocol.
- **Electricity Supply**: Providing the actuators and sensors with the necessary power to function is a difficulty that must be taken into account. Providing a robotic arm with a sufficient and dependable power source is one example.
- **Signal Conditioning**: A lot of sensors provide signals that can't be used to interface with actuators. In order for the actuators to process the signals, signal conditioning is necessary. In order to assess the strain on a material, for instance, a weak signal from a strain gauge could be amplified.
- **Noise Cancellation**: Noise can impede sensor signals, resulting in the transmission of inaccurate data to the actuators. In order to ensure that reliable data is obtained, noise cancellation techniques must be used. A pressure sensor signal, for instance, may be cleaned up with a band pass filter.
- **Data Logging**: Data from the sensors must be logged or stored in order to be used by the actuators. This can be a challenge when dealing with large amounts of data. For example, logging the data from an accelerometer in order to detect changes in a system's orientation.
- **System Integration**: System integration can be difficult when actuators and sensors need to communicate with other parts of the system. A motion sensor and microprocessor, for instance, might be used to operate a motor.

9. FUTURE TRENDS IN INTERFACING OF ACTUATORS AND SENSORS

Top 7 Advanced Future trends in Interfacing of Actuator and Sensors:

- **Automation**: As automation grows, connecting actuators and sensors to one another will become simpler and more affordable. Systems for robotics, process control, and automated manufacturing are some examples.
- **Wireless Sensors**: Wireless sensors are gaining popularity. As a result, it is possible to monitor different circumstances and settings more easily and effectively. Sensors for pressure, humidity, and temperature are a few examples of wireless sensors.
- **Internet of Things (IoT)**: The Internet of Things (****IoT***) is a network of objects with sensors, actuators, and connections to the Internet that enables remote monitoring and control. A few examples include automated factories, smart homes, and networked lighting systems.
- **Augmented reality (AR)**: Augmented reality (AR) is a technology that superimposes digital content over the physical world. The interaction between actuators and sensors may be improved with the help of this technology. Systems that use ****augmented reality*** for navigation and maintenance are two examples.
- **Artificial Intelligence (AI)**: AI is a technology that allows robots to think and behave like people. The monitoring and management of actuators and sensors may be automated using this technology. Systems that use AI for facial recognition and process control are two examples.
- **Haptic Feedback**: A technique known as ****haptic*** feedback employs tactile feedback to give the user information. Feedback on the condition of actuators and sensors may be obtained using this technique. Examples include the use of haptic feedback in gaming controllers and in robot systems used in medicine.
- **Robotics**: Robotics is a branch of technology that allows machines to carry out autonomous activities. Automation of sensor and actuator monitoring and control is possible with the help of this technology. Autonomous cars and robotic arms are two examples.

CONCLUSION

Actuator-Sensor Interfacing, in short, is a crucial technology in contemporary applications. It offers a variety of benefits, including enhanced performance, reduced cost, and a wide range of applications. ****PLCs***, embedded systems, DC motors, stepper motors, analogue and digital sensors, etc. are just a few examples of the various actuators and sensors that may be utilised for this.

There are many different ways to interface the actuators and sensors, including direct connection, the usage of microcontrollers, and signal converters. Despite the advantages of this technology, there are still several difficulties, including how it interacts with various communication protocols, battery consumption, and signal noise.

Future technologies like the Internet of Things, Artificial Intelligence, and machine learning are becoming more popular as a means of overcoming these difficulties. In the upcoming years, the usage of actuator-sensor interfacing technology will increase, and it will be used to an even larger variety of fields, including robotics, automation, healthcare, the military, and many more.

REFERENCES

Agarwal, Y., Balaji, B., Gupta, R., Lyles, J., Wei, M., & Weng, T. (2010). *Occupancy-driven energy management for smart building automation.* Research Gate.

Al-Mashud, M. (2014). PC Guided Automatic Vehicle System. *International Journal on Cybernetics & Informatics*, *3*(6), 1–10. doi:10.5121/ijci.2014.3601

Ali, H.,, & Noor, S., Bashi, S., Marhaban, M.H. (2009). A Review of Pneumatic Actuators (Modeling and Control). *Australian Journal of Basic and Applied Sciences*, *3*, 440–454.

Ali, M., Reza, M. N., Shafik, K., Islam, M., Chowdhury, M., Jeong, J.-H., & Chung, S.-O. (2021). Construction and basic performance test of an ICT-based irrigation monitoring system for rice cultivation in UAE desert soil. *Korean Journal of Agricultural Science.*, *48*, 703–718. doi:10.7744/kjoas.20210059

Ali, M., Reza, M. N., Shafik, K., Islam, M., Chowdhury, M., Jeong, J.-H., & Chung, S.-O. (2021). Construction and basic performance test of an ICT-based irrigation monitoring system for rice cultivation in UAE desert soil. *Korean Journal of Agricultural Science.*, *48*, 703–718. doi:10.7744/kjoas.20210059. https://www.researchgate.net/profile/Mohammod-Ali/publication/356961822/figure/fig4/AS:1099742138380291@1639210366818/Schematic-diagram-of-sensor-and-actuator-interfacing-with-the-ADC-analog-to-digital.png

Arias, L., Torres, S., Sbarbaro, D., & Farias, O. (2008). Photodiode-based sensor for flame sensing and combustion-process monitoring. *Applied Optics*, *47*(29), 5541–5549. doi:10.1364/AO.47.005541 PMID:18846197

Avram, M., Nitu, C., Bucşan, C., & Gramescu, B. (2017). Linear pneumatic actuator. MATEC Web of Conferences. 121. 10.1051/matecconf/201712108001

Baboo, P. (2015). *Principles of Mass Flow meters*. Research Gate.

Baum, M., Dukes, M., & Haman, D. (2023). Selection and Use of Water Meters for Irrigation Water Measurement. Research Gate.

Belogusev, V., & Egorov, A. (2019). Efficiency testing of electric rotary actuators with non-standard reduction units for lower limb exoskeletons. *Journal of Applied Engineering Science.*, *17*(4), 496–503. doi:10.5937/jaes17-20368

Bhatt, D. & Toit, D. & Hancke, G. (2012). *Design of a controller for a universal input/output port*. Research Gate. doi:10.1109/I2MTC.2012.6229492-https://www.researchgate.net/profile/Mdabdullah-Al-Mashud/publication/270569209/figure/fig5/AS:668301948633104@1536347009594/A-simple-block-diagram-of-I-O-interface-The-basic-input-output-device-is-a-set-of.ppm

Bletskan, D., Kabatsii, V., & Bletskan, M. (2016). Photovoltaic Sensor for the Solar Battery Guidance on the Sun Based on GeS:Sb Layered Crystals. *Open Journal of Inorganic Non-metallic Materials.*, *06*(2), 7–11. doi:10.4236/ojinm.2016.62002

Bretschneider, J., Wilde, A., Schneider, P., Hohe, H.P., & Kohler, U. (2010). *Design of multi-dimensional magnetic position sensor systems based on HallinOne® technology*. Fraunhofer EAS. . doi:10.1109/ISIE.2010.5637864

Castellini, L., & D'Andrea, M. & Borgarelli, N. (2014). Analysis and design of a Linear Electro-Mechanical Actuator for a High Lift System. *2014 International Symposium on Power Electronics, Electrical Drives, Automation and Motion, SPEEDAM 2014*. (pp. 243-247). IEEE. 10.1109/SPEEDAM.2014.6872035

Copaci, D., Blanco, D., Martin-Clemente, A., & Moreno, L. (2020). Flexible shape memory alloy actuators for soft robotics: Modelling and control. *International Journal of Advanced Robotic Systems*, *17*(1), 172988141988674. doi:10.1177/1729881419886747

Dobrzynska, J., & Gijs, M. (2010). Capacitive flexible force sensor. *Procedia Engineering*, *5*, 404–407. doi:10.1016/j.proeng.2010.09.132

Dostálek, P., Vasek, V., & Dolinay, J. (2008). *Design and implementation of portable data acquisition unit in process control and supervision applications*. Research Gate.

Dostálek, P., Vasek, V., & Dolinay, J. (2008). *Design and implementation of portable data acquisition unit in process control and supervision applications*. Research Gate. https://www.researchgate.net/profile/Mdabdullah-Al-Mashud/publication/270569209/figure/fig5/AS:668301948633104@1536347009594/A-simple-block-diagram-of-I-O-interface-The-basic-input-output-device-is-a-set-of.ppm

Fang, L., Tang, L., & Shang, Q. (2016). A semi-physical model for pneumatic control valves. *Nonlinear Dynamics*, *85*(3), 1–14. doi:10.100711071-016-2790-5

. Fitan, E., Messine, F., & Nogarede, B. (2004). The Electromagnetic Actuator Design Problem: A General and Rational Approach. Magnetics. *IEEE Transactions*. IEEE. . doi:10.1109/TMAG.2004.827183

Freitas, P., Ferreira, R., Cardoso, S., & Cardoso, F. (2007). Magnetoresistive Sensors. *Journal of Physics Condensed Matter*, *19*(16), 165221–21. doi:10.1088/0953-8984/19/16/165221

Friedt, J. (2010). *Acoustic sensors*. Research Gate.

Gallasch, E., Rafolt, D., Taferner, H.P. & Fend, M. (2001). Bilateral manipulandum to synthesize ground referenced and interlimb viscoelastic loads. IEEE. . doi:10.1109/IEMBS.2001.1020394

Gasperi, M. & Hurbain, P. (2009). *Potentiometer Sensors*. Springer. . doi:10.1007/978-1-4302-2454-9_6

Gautschi, G. (2002). *Piezoelectric Sensors*. Springer. . doi:10.1007/978-3-662-04732-3_5

Gensolen, F., Cathebras, G., Martin, L., & Robert, M. (2005). *An Integrated Image Motion Sensor for Micro Camera Module. IEEE Transactions on Very Large Scale Integration (VLSI)*. Systems.

Giesberts, R., Sluiter, V., & Verkerke, G. (2018). Design and Test of a New Inductive Force Sensor. *Sensors (Basel)*, *18*(7), 2079. doi:10.339018072079 PMID:29958474

Golby, J. (2010). Advances in inductive position sensor technology. Sensor Review -. *Sensor Review*, *30*(2), 142–147. doi:10.1108/02602281011022742

Haas, J. (2001). *New Type of Hydraulic Rotary Actuator; Development and Application in a Large Telerobotic Manipulator System*. American Society of Mechanical Engineers. . doi:10.1115/IMECE2001/FPST-25015

Has, Z., Rahmat, M., Husain, A., Ishaque, K., Ghazali, R., Ahmad, M., Sam, M., & Rozali, S. M. (2014). Robust Position Tracking Control of an Electro-Hydraulic Actuator in the Presence of Friction and Internal Leakage. *Arabian Journal for Science and Engineering*, *39*(4), 2965–2978. doi:10.100713369-013-0888-3

Henning, A. (2006). Comprehensive Model for Thermopneumatic Actuators and Microvalves. Microelectromechanical Systems. *Journal of Microelectromechanical Systems*, *15*(5), 1308–1318. doi:10.1109/JMEMS.2006.879665

Hofman, W. (2015). *Federated platforms for seamless interoperability in the Physical Internet*. Research Gate.

Hofman, W. (2015). *Federated platforms for seamless interoperability in the Physical Internet*. Research Gate. https://www.researchgate.net/profile/Wout-Hofman/publication/282816170/figure/fig2/AS:284122183421953@1444751415086/The-concepts-service-protocol-and-interface.png

Ibrahim, D. (2002). Thermistor Temperature Sensors. Newnes. doi:10.1016/B978-075065556-9/50005-0

Jani, M. (2017). Designing shape memory alloy linear actuators: A review. *Journal of Intelligent Material Systems and Structures*, *28*(13), 1699–1718. doi:10.1177/1045389X16679296

Kim, S.-Y., Kim, J.-D., Kim, Y.-S., Song, H.-J., & Park, C. (2015). Resistance Temperature Detector Sensor with a Copper Pattern on the Printed Circuit Board. *International Journal of Control and Automation.*, *8*(8), 67–74. doi:10.14257/ijca.2015.8.8.08

Li, X., Wang, X., & Tian, M. (2020). A new type of pneumatic linear reciprocating actuator. *Journal of Physics: Conference Series*, *1650*(2), 022085. doi:10.1088/1742-6596/1650/2/022085

Long, C., & Loveday, P. (2007). A thermo-hydraulic wax actuation system for high force and large displacement applications. *Proceedings of SPIE - The International Society for Optical Engineering*. Spie Digital Library. 10.1117/12.714737

Melnyk, M., Kernytskyy, A., Lobur, M., Szermer, M., Zajac, P., & Zabierowski, W. (2014). *Study of Characteristics of MEMS Thermo-Electric Actuators*. Research Gate.

Mendoza-Vázquez, J. R., Tlelo-Cuautle, E., Vázquez-Gonzalez, J. L., & Escudero-Uribe, A. Z. (2009). Simulation of a parallel mechanical elbow with 3 DOF. *Journal of Applied Research and Technology*, *7*(2), 113–123. doi:10.22201/icat.16656423.2009.7.02.494

Miric, S., Schuck, M., Tuysuz, A., & Kolar, J. (2018). *Double Stator Linear-Rotary Actuator with a Single Set of Mover Magnets*, (pp. 750-757). IEEE. . doi:10.1109/ECCE.2018.8557575

Mohamad, N., Salleh, A., Hashim, N., Abd Aziz, M., & Zakaria, Z. (2016). Modeling of Wireless Energy Transfer Circuit and Implementation of Wi-tricity System Control using Infrared and PIC Microcontroller. *IACSIT International Journal of Engineering and Technology*, *8*, 1282–1294.

Ng, F. (2020). *Ultrasonic Sensors*. Research Gate., doi:10.13140/RG.2.2.33638.78404

Noriega, M., Ramírez, R., López, R., Vaca, M., Morales, J., Terres, H., Lizardi, A., & Chávez, S. (2015). Thermocouples calibration and analysis of the influence of the length of the sensor coating. *Journal of Physics: Conference Series*, *582*(1), 012029. doi:10.1088/1742-6596/582/1/012029

Oh, S. W., Bang, H., & Hwang, J. (2010). *Light-weight RFID device interface for controlling RFID tag memory access*. Research Gate.

Oh, S. W., Bang, H., & Hwang, J. (2010). *Light-weight RFID device interface for controlling RFID tag memory access*. Research Gate. https://www.researchgate.net/profile/Se-Won-Oh-2/publication/224128273/figure/fig1/AS:302750387720197@1449192725712/RFID-air-interface-and-RFID-reader-interface.png

Pierre, A., Gaikwad, A., & Arias, A. (2017). Charge-integrating organic heterojunction phototransistors for wide-dynamic-range image sensors. *Nature Photonics*, *11*(3), 193–199. doi:10.1038/nphoton.2017.15

Poriņš, R. & Apse-Apsitis, P. (2019). *Capacitive proximity sensing hardware comparison*. IEEE. . doi:10.1109/RTUCON48111.2019.8982317

Qiao, G., Liu, G., Shi, Z., Wang, Y., Ma, S., & Lim, T. (2017). A review of electromechanical actuators for More/All Electric aircraft systems. *Proceedings of the Institution of Mechanical Engineers. Part C, Journal of Mechanical Engineering Science*, *232*(22), 095440621774986. doi:10.1177/0954406217749869

Ramsden, E. (2006). *Hall-Effect Sensors*. Hall-Effect Sensors.

Rashid, A., & Ali, A. (2018). Performance Analysis of Low-Cost Infrared Sensors for Multi-Robot Localization and Communication. *International Journal of Computer Applications*, *182*(26), 23–29. doi:10.5120/ijca2018918139

Rusconi, A., Costantini, S., & Prelini, C. (2022). *Micro Speakers*. Springer. . doi:10.1007/978-3-030-80135-9_20

Sadun, A., Jalani, J., & Sukor, J.A. (2016). *Force Sensing Resistor (FSR): a brief overview and the low-cost sensor for active compliance control*. SPIE. . doi:10.1117/12.2242950

Shadwani, M., Sachan, S., & Sachan, P. (2018). Capacitive Sensing & Its Applications. doi:10.13140/RG.2.2.15752.32001

Sherrit, S., Bao, X., & Jones, C.M., Aldrich, J., Blodget, C.J., Moore, J., & Carson, J.W., Goullioud, R. & Donotsendmail, B. (2011). Piezoelectric stack actuator life test. *IEEE Aerospace Conference*, (pp. 1-8). IEEE.

Shewale, M., Razban, A., Deshmukh, S., Mulik, S., & Zambare, H. (2018). *Design and Implementation of Position Estimator Algorithm on Voice Coil Motor., 1-5*, 1–5. doi:10.1109/I2CT.2018.8529546

Shields, J. & Konefat, E. (2021). *Modeling of Piezoceramic Actuators for Control.* Intech Open. . doi:10.5772/intechopen.96727

Sotoodeh, K. (2019). Actuator selection and sizing for valves. *SN Applied Sciences*, *1*(10), 1207. doi:10.100742452-019-1248-z

Thakker, M. (2013). *Microphone Sensor for Sound Analysis and Breakdown Detection Using LabVIEW*. Research Gate.

Tikani, R., Vahdati, N., & Ziaei-Rad, S. (2012). Two-Mode Operation Engine Mount Design for Automotive Applications. *Shock and Vibration*, *19*(6), 1267–1280. doi:10.1155/2012/651591

Tsukima, M., Kai, T., Takahashi, K., Takeuchi, T., & Koyama, K. (2017). Development of Tri-Stable Rotary Solenoid Actuator. *IEEJ Transactions on Industry Applications.*, *137*(8), 681–686. doi:10.1541/ieejias.137.681

Wei, S., Liao, X., Pang, J., & Zhou, Y. (2021). Recent Progress of Fluxgate Magnetic Sensors: Basic Research and Application. *Sensors (Basel)*, *21*(4), 1500. doi:10.339021041500 PMID:33671507

Wijesundera, R. P. & Abeyratne, M. & Kalupahana, I. & Buddhika, W. & Kumara, S., Jayatissa, N., Harambearachchi, J., & Kalingamudali, S. (2012). *An Inexpensive Universal Serial Bus Interfaced Home Automation System.* Research Gate.

Xu, L., Liu, Z., Zhai, H., Chen, X., Sun, R., Lyu, S., Yangyang, F., Yi, Y., Chen, Z., Jin, L., Zhang, J., Li, Y., & Ye, T. (2020). Moisture-Resilient Graphene-Dyed Wool Fabric for Strain Sensing. *ACS Applied Materials & Interfaces*, (11), 13265–13274. doi:10.1021/acsami.9b20964 PMID:32105063

Yildiz, S., Dorman, L., Kuperman, W. A., Sabra, K., Roux, P., Green, D., Fried, S., & Schmidt, H. (2014). Using hydrophones as vector sensors. *The Journal of the Acoustical Society of America*, *135*(4_Supplement), 2361. doi:10.1121/1.4877778

Yuan, H., Fauroux, J.-C., Chapelle, F., & Balandraud, X. (2017). A review of rotary actuators based on shape memory alloys. *Journal of Intelligent Material Systems and Structures*. . doi:10.1177/1045389X16682848

Zhang, X., & Yan, X. (2011). Continuous Rotary Motor Actuated by Multiple Segments of Shape Memory Alloy Wires. *Journal of Materials Engineering and Performance*, *21*(12), 2643–2649. doi:10.100711665-012-0390-1

Zhou, Y., Yao, W., He, X., Li, T., Wang, S., & Han, Y. (2022). Flow velocity sensors arrangement for vegetated channels. *Frontiers in Plant Science*, *13*, 960103. doi:10.3389/fpls.2022.960103 PMID:36035729

Zhu, X.-G & Wang, Y. (2006). *Research and design of permanent magnet actuator for higher opening capability.* Research Gate.

Zuo, J.-Y & Yan, G.-Z & Gao, Z.-J & Lu, Q.-H. (2005). *New piezoelectric linear actuator.* Research Gate.

KEY TERMS AND DEFINITIONS

Augmented Reality: Augmented reality (AR) is a technology that superimposes a computer-generated image on a user's view of the real world, thus providing a composite view. AR can be used in many different applications, such as gaming, navigation, and education. Examples include Google Glass, Pokemon Go, and Microsoft HoloLens.

Automation: Automation is the use of technology to routine human jobs or processes. By removing human work, lowering expenses and mistakes, and enhancing accuracy.

Calibration: Comparing a device's or measuring system's output to a recognised standard is the process of calibration. A device or system's accuracy and precision are ensured by calibration.

Gyroscope: A gyroscope is a tool that measures or maintains orientation using the concepts of angular momentum and angular velocity. It consists of a freely rotatable frame with a spinning wheel or disc attached on it.

Haptic: The use of touch for communication is known as haptic (or haptics). In addition to more advanced tactile feedback systems that employ actuators to create vibrations and other sensations.

HVAC: HVAC stands for Heating, Ventilation and Air Conditioning. It is a system that is used to control the temperature, humidity, and air quality in indoor spaces.

I2C: I2C (Inter-Integrated Circuit) is a form of communication protocol frequently used for interfacing low-speed peripherals like sensors, touchscreens, and LCDs with integrated circuits (ICs).

IoT (Internet of Things): The Internet of Things (IoT) is a network of physical items that are integrated with electronics, software, sensors, and network connectivity in order to gather and share data. Examples of these objects include gadgets, automobiles, and buildings.

Microcontroller: A microcontroller is a small integrated circuit that controls a single process in an embedded system.

Photodiodes: A semiconductor device known as a photodiode transforms light into electricity. It responds to a variety of wavelengths of light.

Phototransistor: A phototransistor is a type of transistor that is sensitive to light. It is composed of a semiconductor material that is sensitive to light, and an electrical circuit that amplifies the current generated by the light.

PLC: PLC stands for Programmable Logic Controller. It is a type of computer designed to control industrial machines and processes.

Pneumatic: Pneumatic systems generate and transmit mechanical energy using pressured air or gas. They have a broad range of uses, from running heavy industrial machines to powering hand tools.

Precision: A measurement's precision serves as a gauge for its accuracy. The value is more precisely expressed the higher the accuracy.

Protocols: Sets of rules and regulations known as protocols are used to control communication between two or more systems. They are employed to make sure that data is sent across systems in a secure and reliable manner.

Robust: A system's robustness is its capacity to endure or bounce back from challenging or unanticipated external situations. A system's robustness in computing refers to its capacity to deal with unforeseen inputs, environmental changes, and other types of interference.

Signal amplifier: An instrument used to boost the power of an electrical signal is a signal amplifier, sometimes referred to as a line amplifier.

Chapter 16
Modelling and Development of Energy Systems Through Cyber Physical Systems With Optimising Interconnected With Control and Sensing Parameters

Kaliappan Seeniappan
https://orcid.org/0000-0002-5021-8759
Department of Mechanical Engineering, Velammal Institute of Technology, India

Chinnahajisagari Mohammad Akram
https://orcid.org/0000-0001-7349-7909
Department of Mechanical Engineering, S.R.K. Institute of Technology, India

S. Soundararajan
Department of Computer Science Engineering, Velammal Institute of Technology, India

L. Natrayan
https://orcid.org/0000-0003-1001-3422
Department of Mechanical Engineering, Saveetha School of Engineering, SIMATS, India

DOI: 10.4018/978-1-6684-9267-3.ch016

ABSTRACT

The integration of cyber-physical systems (CPS) with energy infrastructure has the potential to revolutionise the way we generate, distribute and consume energy. CPS enables the monitoring and control of energy systems in real-time, and allows for the integration of renewable energy sources and energy storage management. This research aims to explore the use of CPS in the design and development of a housing microgrid, with a focus on optimising interconnected energy systems and controlling and sensing parameters. To accomplish this, the authors propose a model-based design (MBD) approach, which uses simulation tools to create and test numerous control procedures for a housing microgrid. The imitation tool, GridMat, is a MATLAB/Simulink toolkit that can co-simulate control techniques written in Simulink and power systems modelled in GridLAB-D. This research demonstrates how GridMat can create, simulate, debug, and evaluate multiple layers of control algorithms for a residential microgrid.

1. INTRODUCTION

The energy method industry is undergoing a significant transformation, moving away from traditional, standalone power grids to the integration of both cyber information and physical representations at all levels of the power grid network. This new grid, known as Cyber-Physical Energy Systems (CPES), is expected to advance stability, suppleness, effectiveness, cost-usefulness, and safety (Natrayan L & Kumar MS. 2020). However, the integration of Distributed Energy Resources (DERs), such as conventional energy sources and novel kinds of demands, into the residential distribution grid, poses additional challenges in terms of different stage nursing and regulator for supply and request organization (Fatimah et al. 2020; Sandborn et al. 2021).

Due to the rapid inclusion of DERs, traditional power system design methods are becoming increasingly time-consuming and it is becoming more difficult to predict grid issues (Veeman et al., 2021). To address this problem, Model-Based Design (MBD) has been proposed as a explanation (Kim et al. 2020). MBD allows for concurrent modeling of physical and cyber components, and enables cyber-physical co-simulation to investigate various design options for developing, verifying, and testing (Herdt and Drechsler 2022).

Despite the challenges, the integration of CPS in energy systems is expected to bring significant benefits to the overall performance and reliability of the power grid. The integration of real-time data and advanced control algorithms will enable better monitoring, prediction, and management of the grid (Natrayan L et al., 2021).

This will lead to improved grid stability, increased efficiency, and reduced costs. Additionally, the integration of CPS will facilitate the integration of renewable energy sources and improve the integration of electric vehicles and other DERs(Broy et al. 2019; Pan and Zhang 2022).

To fully realize the potential benefits of CPS in energy systems, it is important to continue research and development in the areas of method-level design procedure and apparatuses. This includes the development of sophisticated models and simulation tools that can accurately capture the dynamics of both cyber and physical components of CPES (Natrayan L et al., 2018). Additionally, research is needed to develop advanced control algorithms and optimization techniques that can effectively manage and control the complex interactions between different components of the grid (Guan et al. 2018).

Moreover, it's important to consider the security aspect of the integration of CPS in the energy systems, as it's an essential aspect of the overall system reliability. It's crucial to have a robust security system in place to protect the grid against cyberattacks and to ensure the confidentiality, integrity, and availability of the data and control systems (Tripathi et al. 2022).

In conclusion, the integration of CPS in energy systems is a challenging but necessary step towards the future of the power grid. The use of Model-Based Design (MBD) and cyber-physical co-simulation can enable the design and development of advanced control systems that can improve the performance and reliability of the grid (Pragadish et al., 2022). However, further research and development is needed to address the challenges and to fully realize the potential benefits of CPS in energy systems.

2. MICROGRID FOR RESIDENTIAL USE

As the integration of various Distributed Energy Resources (DERs) such as solar panels, wind turbines, fuel cells and microturbines, the traditional power grid has undergone a transformation (Balaji et al., 2022). The transmission lines that once connected large-scale power production facilities to distribution grids have given way to a more complex, multidirectional flow of electricity. With changes in power flow, rising energy production costs, and growing environmental concerns, systems for balancing electricity supply and demand are now necessary (Ramaswamy et al., 2022).

One solution that has been proposed to address these issues is the use of microgrids. A microgrid is a small, mostly self-sufficient network of electrical energy sources and loads that can connect to or disconnect from the main power grid (macrogrid). Microgrids can provide increased reliability on the demand-side due to their versatility (Yogeshwaran et al., 2015). They can be useful in a variety of settings, including residential neighborhoods, military bases, educational institutions and remote locations.

To effectively manage these issues, local distributed controllers must be implemented at the distribution level. However, the complex nature of residential microgrids can make the development and certification process challenging (Kumar MS et al., 2018). One way to address this challenge is through the use of Model-Based Design (MBD). MBD allows designers to create and analyze Cyber-Physical Energy Systems (CPES) by abstracting the characteristics of complex systems at a high level and enabling necessary design space exploration without the use of actual physical plant (Santhosh MS et al., 2018). This can help streamline the development and certification process and make it more efficient. Figure 1 shows the architecture of the microgrids used in residental area (Velmurugan et al., 2022).

Figure 1. Architecture of residental microgrids

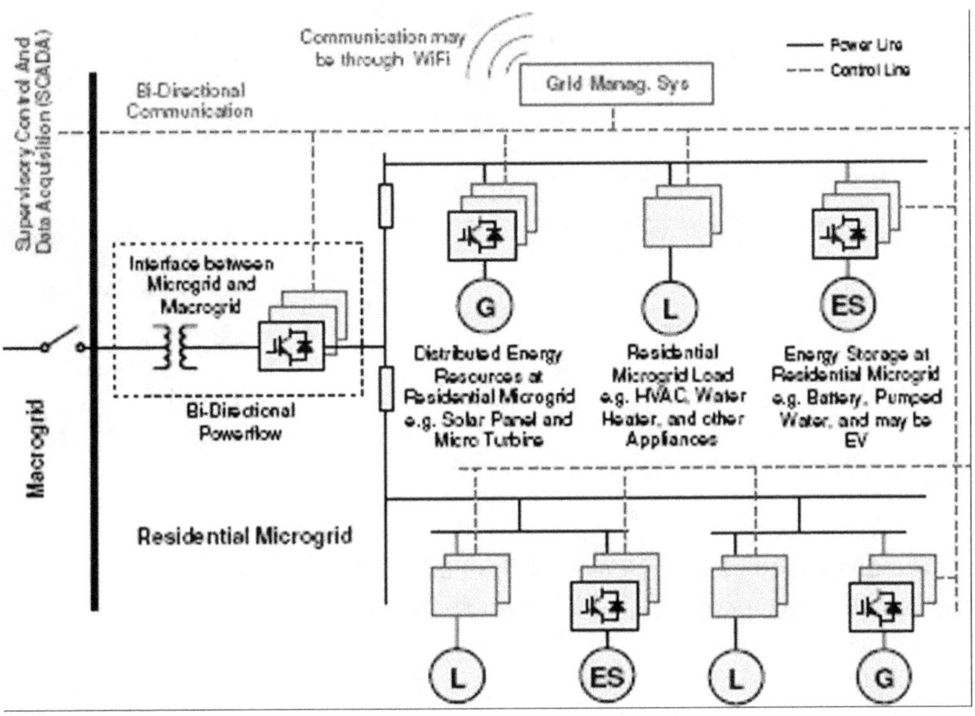

3. PROPOSED DESIGN

When utilizing Model-Based Design, we employ a simulation atmosphere with the ability to perform co-simulation across multiple tools (MBD). This approach allows for the creation of control software that can be tested in a Software-in-the-Loop Simulation (SILS) and Hardware-in-the-Loop Simulation (HILS) environment, resulting in increased simulation realism (Natrayan, L., & Kumar, M. S. 2020).

Before beginning to model a residential microgrid, it is important to understand the system's requirements, components/subsystems, structure, location, and any potential issues. For example, we may want to create a microgrid for 1000 homes within a mid-sized distribution power system that includes many level II electric vehicle chargers and low-rate step-down transformers (Nagarajan et al., 2022). The microgrid must be able to achieve and manage the process of each separate charger and transformer in terms of demand response, electric vehicle charging, and power system reliability by utilizing domestic load management (Sundaram et al., 2019). The Cyber-Physical Energy Systems (CPES) can be modeled, simulated, and verified using this information and the MBD technique. Figure 2 illustrates the key steps of the MBD approach for a housing microgrid. A detailed explanation of these stages is provided below:

Figure 2. Gridmat tool for residental grid

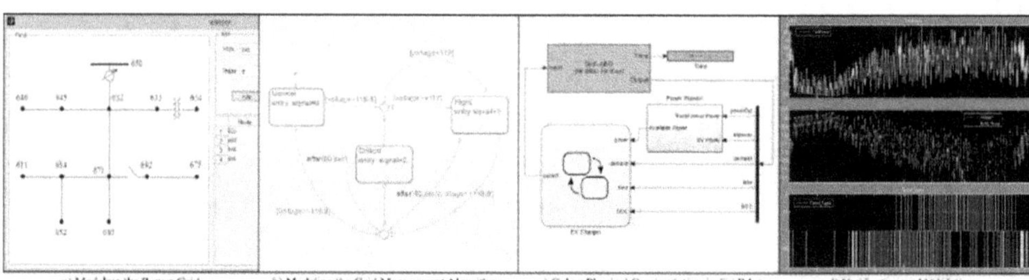

The integration of Cyber-Physical Systems (CPS) into residential microgrids may pave the way for optimizing and regulating these systems. However, there are several challenges that come with designing and managing a residential microgrid (Meikandan et al., 2022). To overcome these challenges, distributed controllers at the distribution level must be implemented.

To tackle these issues, we propose a Model-Based Design (MBD) approach that utilizes a simulation atmosphere with multiple co-simulation abilities. This method allows for the development of control software that can be evaluated through both

Software-In-the-Loop Simulation (SILS) and Hardware-In-the-Loop Simulation (HILS) settings, enhancing simulation accuracy (Kumar MS et al., 2019).

The first step in this method is to construct a physical plant model that includes structural and behavioral models that can be used to design and evaluate a CPS. The structural model includes the building architecture, end-customer usage applications, dispersed power capitals, transformers, and the distribution power grid. Behavioral modeling captures the active aspects of a home microgrid, such as power usage, climate, and more. GridLAB-D, a modeling and simulation tool, is particularly effective for simulating a distribution power grid, particularly a home microgrid. GridLAB-D is an open-source, agent-based technology developed by PNNL, that simulates a large-scale circulation energy grid at multiple unit of graininess, enabling the simulation of end-customer usage applications in a home microgrid (Paranthaman et al., 2022).

The next step is to simulate grid management techniques. To model, build, and simulate multiple controllers for a home microgrid, we utilize Simulink, a data flow graphical programming language tool. MATLAB/Simulink has a vast library for implementing different control systems, such as Model Predictive Control (MPC). The final step is to simulate a cyber-physical system. The GriDmat architecture is shown in Figure 3.

The fourth stage in the process centers around the use of formal verification and validation techniques to document the behavior of the control algorithms. These techniques are used to examine and validate the results of simulations, ensuring that the design requirements have been met. To debug and evaluate the design, an iterative approach can be employed, where previous stages are revisited in order to adapt the models of the physical plant, embedded computer system, control algorithm, and model parameters. One way to test the embedded software component without the use of a physical plant is through the use of GridMat's Hardware-In-the-Loop Simulation (HILS) feature. In HILS, a hardware platform is chosen to interface with the physical plant and execute the control algorithm while it is exposed to the environment (Natrayan L et al., 2020). The embedded software that can be used with actual hardware is then developed using the synthesized modeled controllers. The physical plant is operated on the simulation tool, and the controller communicates with the simulated physical plant using GridMat to perform HILS (GridLAB-D). This allows for a more efficient and cost-effective testing process.

Figure 3. GriDmat architecture

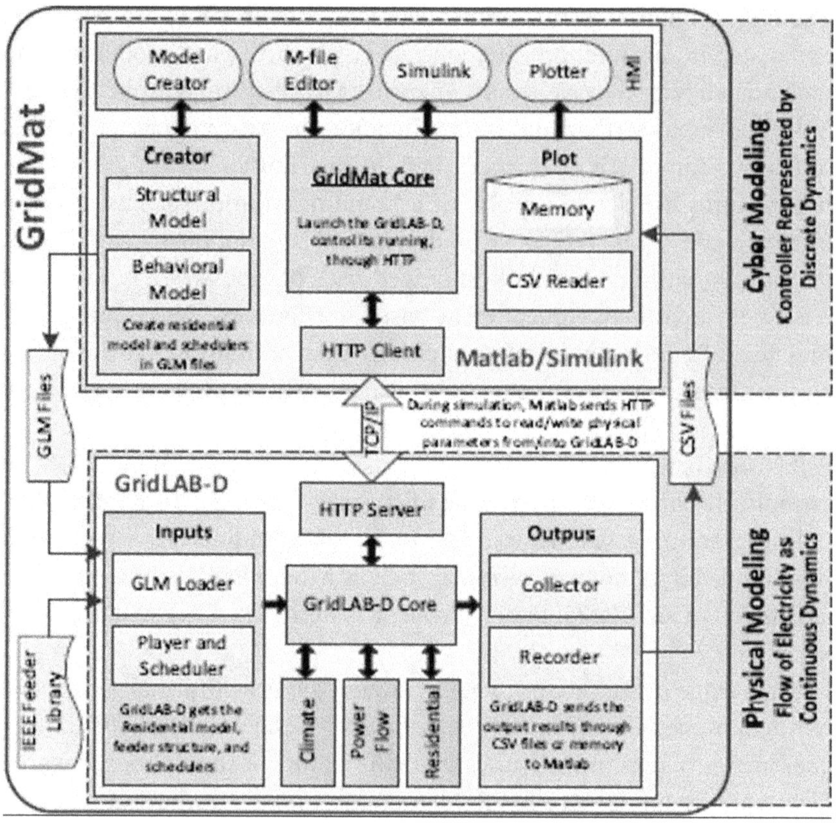

4. EVALUATION OF THE CASE STUDY

The fourth approach employs formal verification and validation techniques to monitor the performance of control algorithms. The simulation results are then thoroughly reviewed and validated to ensure compliance with design specifications. Debugging and evaluating the design can be done iteratively by revisiting previous stages and adjusting the methods for the corporal plant, implanted computer structure, control procedure, and model limits. Thanks to the Hardware-In-the-Loop Simulation (HILS) functionality of GridMat, it is not necessary to have a physical plant to test the embedded software component. HILS allows for the selection of a hardware platform to communicate with the corporeal plant and execute the control procedure externally. The embedded software that can work with real hardware is then created using the synthesized modeling controllers. The controller is connected to the simulated physical plant using GridMat to run HILS while the simulation tool runs the actual plant (GridLAB-D).

Collaboration for Residential Electric Vehicle Charging

As the number of electric cars on the road increases, the electrical power grid will be affected as uncoordinated mass charging at the subordinate supply network will decrease together the overall demand and the top load control. Studies show that 30% of the distribution system's peak load power is used for EV charging (Jha et al. 2021). One solution to mitigate this impact is to modernize the power system's infrastructure by upgrading transformers and building new power plants to increase the electricity supply to the home grid. However, this approach would negate the economic and environmental benefits of electric cars. An alternative solution that is less problematic is to monitor and organize EV charging at the local or teriterial level. To accomblish this strategy and equilibrium source and need, new control procedures are needed, which can be verified, verified, and authenticated using MILS. The GridMat tool and the Model-Based Design (MBD) method we propose can assist in exploring potential options.

Figure 4. Average power consumption

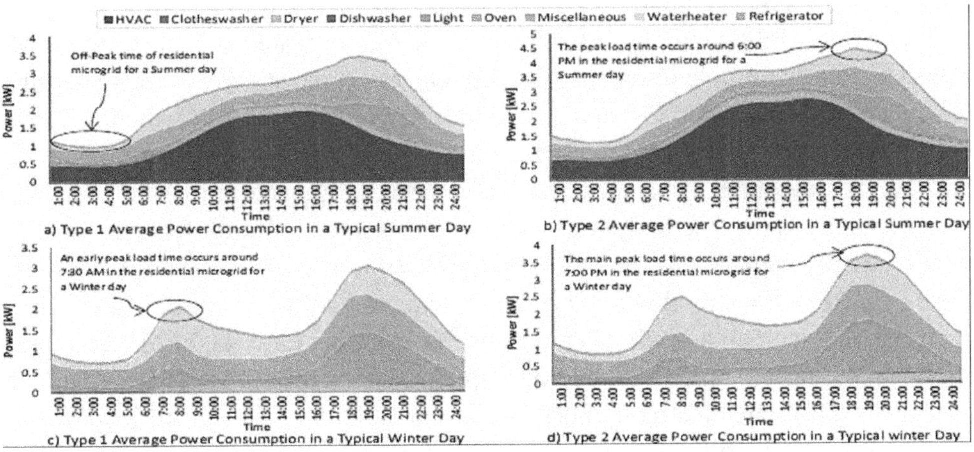

To coordinate EV charging, we demonstrated the "fair-demand" control algorithm, which controls the charging current of individually EV in line with its high need and the obtainable power at the modifier level. The estimated amount of energy used by the EV for its next journey is divided by the length of time the EV is associated to the charger. The high need is defined in Equation 1 as Ei, Fi, and Gi, where Ei, Fi, and Gi are the high need, the estimated quantity of power needed for the future voyage, and the departure period, respectively.

$$Ei = Fi + Gi \quad (1)$$

The "fair-demand" procedure distributes weights founded on high need criteria to limit charging period. Equation 2 is used to determine Ti, the biased high need for an EV.

$$T= Ei*Ej \quad (2)$$

We assume that the residential microgrid model employs level II EV mounts with a supreme charge rate of 9.1 KWh. Because the level II EV charger runs at 240V, Ohm's Law indicates that the maximum charging current is 50 A. Equation 3 depicts the charging current (K) based on biased high need. P denotes the existing control of the modifier level.

$$Ki = Min(50, RiX/440) \quad (3)$$

Once all EVs have received the requisite energy, the "fair-demand" algorithm will evenly distribute the remaining power available to completely charge the battery. Figure 5 depicts the Simulink machine for the "fair-demand" procedure.

Using the GridMat tool, we investigated the behaviour of electric cars (EVs) and electric vehicle supply equipment (EVSE). The EVs' mean arrival and departure times, which were at 5:30 PM and 7:30 AM, followed Gaussian distributions. The EVSE records EV data such as departure time, projected energy consumption for the next journey, and EV characteristics, as well as charging current. Along with the "fair-demand" control method, we created the "fair-shared" control algorithm, which distributes existing power evenly amongst all EVs associated to a single modifier. We designed an EV energing management policy that allows "deferral-based" EV charging, in which all households postpone EV charging until 9:00 PM, according to power pricing and residential microgrid peak demand time. To avoid transformer overload, all EVs are electric in this mode at a rate of 5.5 KWh. Figure 6 shows the results of our simulation of various control schemes.

We discovered that the "deferral-based" approach has a rebound effect while seeking for an effective control mechanism since all EVs charge after 9:00 p.m. This is due to the fact that all EVs charge at the same time. Instead, the "fair-shared" and "fair-demand" control algorithms distribute energy use among the EVs in order to maintain a constant power output rate for the transformer (25KW with 5 linked residences). The most successful algorithm was found to be the "fair-demand" procedure, which charges EVs based on their unique requirements and attempts to give enough energy before they leave the home. The "fair-shared" and "deferral-based" algorithms, on the other hand, provide all EVs the same amount of energy regardless of their individual demands, which may result in some EVs departing the home without enough power for their intended trip. Model-based design (MBD) enables high degrees of abstraction by allowing control algorithms to be created rapidly and efficiently while considering their merits and downsides.

Figure 5. Model of the stateflow

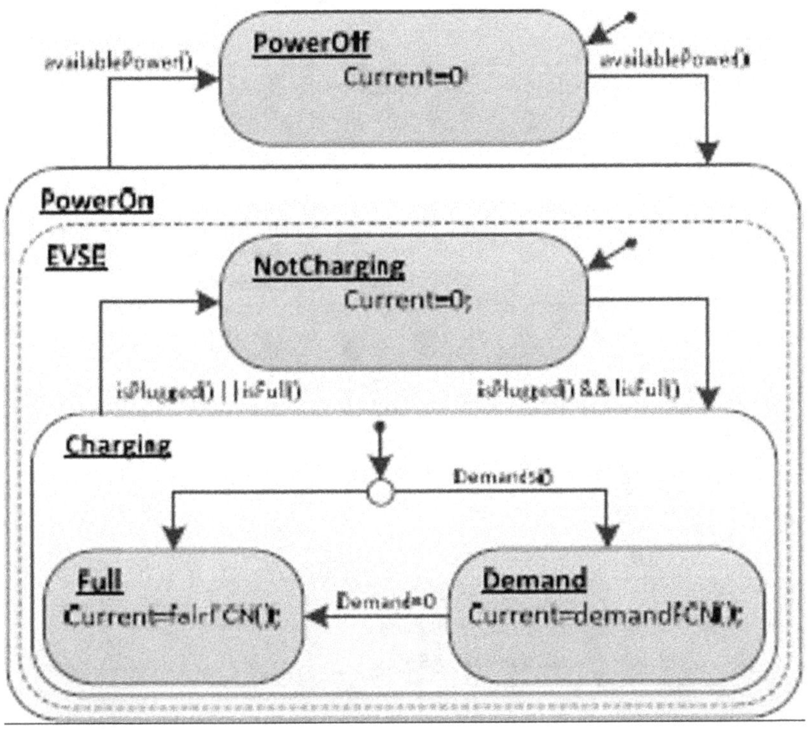

Peak Usage of Electricity

In an attempt to minimise power use during peak hours, demand response (DR) systems target residential buildings, which account for the bulk of energy demand (38%). HVAC systems and water heaters use a considerable percentage of this demand (on average 33%), making them potential candidates for demand side reduction. If peak demand is minimised, utilities may be able to avoid the construction of new power plants to fulfil rising demand. Utilities may charge period-of-Consumption (TOU) fees and offer high demand current services to inspire consumers to restrict their use. Customers that sign DLC agreements are permitted to operate certain equipment during peak hours in return for decreased power bills. Depending on the weather, utilities may adjust the cooling set point of HVAC systems or the heating set point of water heaters. Using Model Predictive Control (MPC), the interior heat and water heat were kept within the ANSI approved comfort zone (as shown in Fig. 7).

Figure 6. Average value of the power consumption

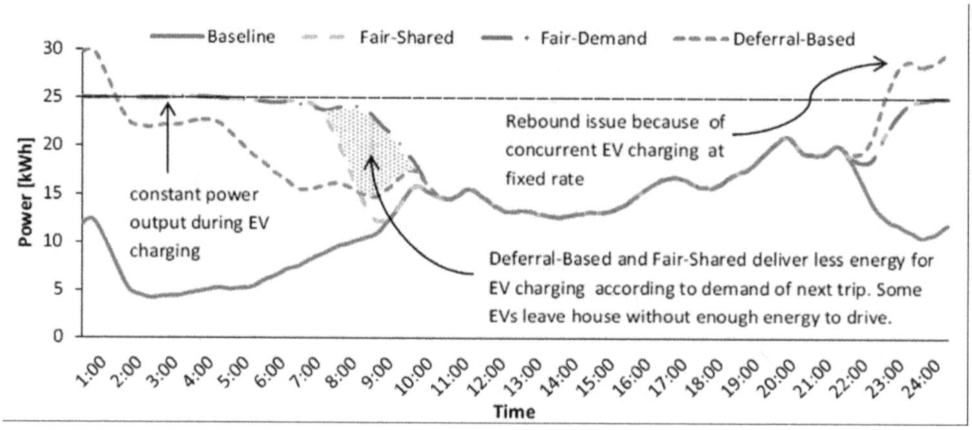

Figure 7. Flow diagram of residental grid

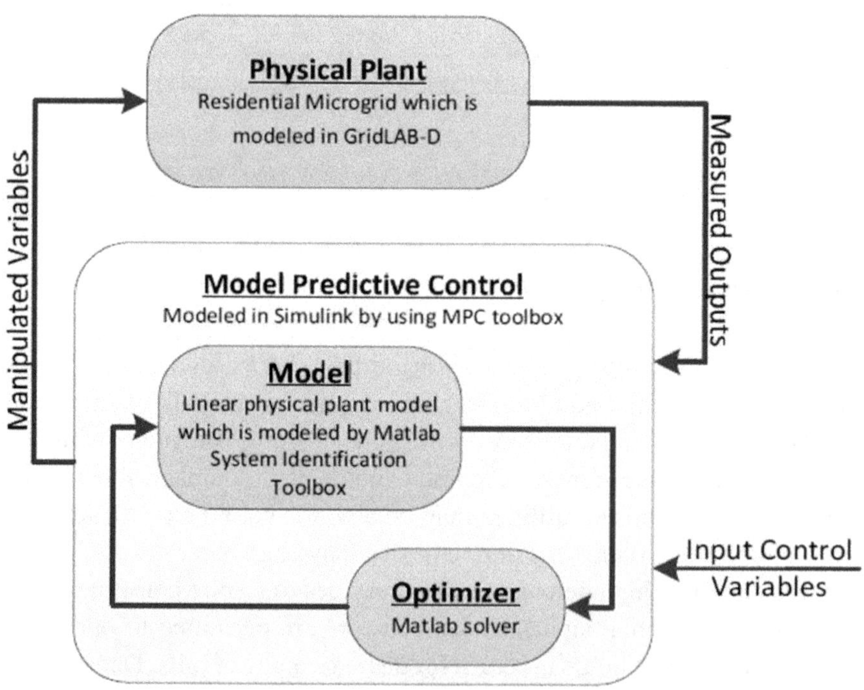

Realability of the Grid

One of the issues in building a reliable residential microgrid is managing voltage drop at the perimeter. Maintaining a specified voltage level within a defined range is a critical auxiliary function in the power system to guarantee dependable operation and minimise harmful impacts on equipment. Recent research, however, suggests that higher need side power management in a housing microgrid may offer supplemental services. These services are often provided on the production side. Voltage is not constant throughout the power system due to resistance in conductors and other sections of the system, and it inclines to drop as we go earlier to the load.

Power regulators and transformer taps are often used in a conventional power system to prevent voltage drop. However, in microgrids, these tactics are useless since adverse weather conditions may reduce the output voltage of distributed renewable energy sources like solar PV. We show how to use the GridMat tool to create an algorithm that minimises demand at lower system voltage to reduce voltage loss in a home microgrid. In this example, smart appliances are used to reduce power consumption caused by a local voltage decrease. Consider the heating coil and drum-turning motor of a dryer. Because the dryer's coil uses a lot of electricity and the majority of users aren't concerned with drying time.

The dryer has three settings: ordinary, maximum, and serious. In standard mode, the drier employs the whole capacity of its coils. The dryer disables the loop and limits its size to 80% at maximum and critical settings. All of the dryer motor operating modes listed above result in normal functioning. Customers may quickly switch from high or critical settings to regular mode to speed up the drying process. When the local voltage falls below 118 volts, the controller enters high mode, and when it falls below 116 volts, it enters critical mode. The voltages necessary to transition from high to normal and critical to high are raised to 118.4V and 116.4V, respectively, with the addition of a 0.4V dead band in the opposite direction. Figure 8 displays the outcomes of a smart grid-based local voltage regulation.

Simulation

After modelling the home microgrid, we created the programme as a C coding using the MATLAB Rooted code creation tool, and then we prepared it for practical deployment. To do this, we ran a Hardware-in-the-Loop simulation using our GridMat tool. The platform developed by GridLAB-GridMat D keeps the home microgrid's physical plant model current and provides an HTTP interface for communication between the controller and the actual plant. During the HILS and GridLAB-D operates in actual mode to replicates behaviour of the actual plant.

Figure 8. Power consumption at the high demand

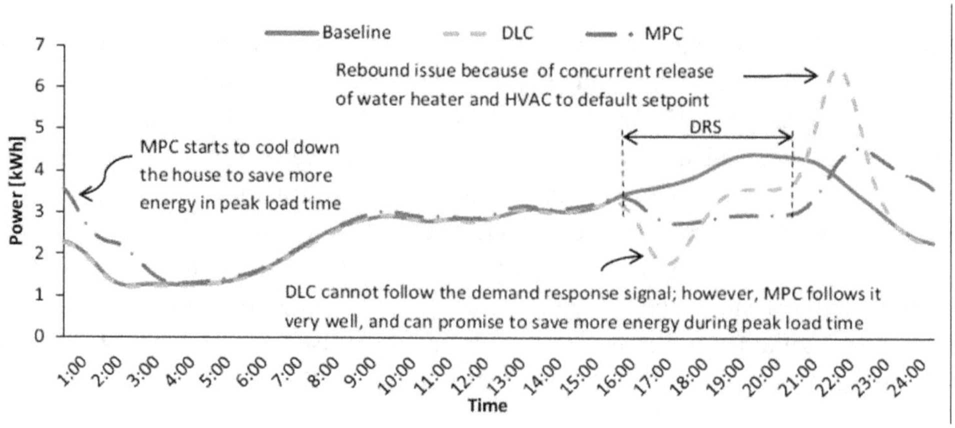

5. CONCLUSION

In conclusion, the research presented in this article presents a novel approach for the design and implementation of Cyber-Physical Energy Systems (CPES) in residential microgrids. The approach utilizes a Model-in-the-Loop Simulation (MILS) environment and multi-tool co-simulation capabilities, which enables the generation of control software that can be tested in both Software-In-the-Loop Simulation (SILS) and Hardware-In-the-Loop Simulation (HILS) settings. The proposed methodology starts with the identification of the needs, components, structure, location, and issues of the residential microgrid, which is then modelled and simulated using the Model-Based Design (MBD) technique. The proposed GridMat tool is a cyber-physical MATLAB package that facilitates the modelling, simulation, analysis, and validation of distribution power systems, such as residential microgrids. The use of GridMat in conjunction with the MBD approach enables the rapid and effective creation of control algorithms for distributed energy resources, demand response, and power system dependability. The results of the simulation and validation demonstrate that the proposed approach can significantly improve the dependability of the residential microgrid power system.

REFERENCES

Balaji, N., Natrayan, L., Kaliappan, S., Patil, P. P., & Sivakumar, N. S. (2022). Annealed peanut shell biochar as potential reinforcement for aloe vera fiber-epoxy biocomposite: Mechanical, thermal conductivity, and dielectric properties. *Biomass Conversion and Biorefinery*, 1–9. doi:10.100713399-022-02650-7

Broy, M., Daembkes, H., & Sztipanovits, J. (2019). Editorial to the theme section on model-based design of cyber-physical systems. *Software & Systems Modeling*, *18*(3), 1575–1576. doi:10.100710270-018-0670-9

Fatimah, Y. A., Widianto, A., & Hanafi, M. (2020). Cyber-physical System Enabled in Sustainable Waste Management 4.0: A Smart Waste Collection System for Indonesian Semi-Urban Cities. *Procedia Manufacturing*, *43*, 535–542. doi:10.1016/j.promfg.2020.02.169

Guan, X., Xu, Z., Jia, Q. S., Liu, K., & Zhou, Y. (2018). Cyber-physical model for efficient and secured operation of CPES or energy Internet. *Science China. Information Sciences*, *61*(11), 3–5. doi:10.100711432-018-9621-8

Herdt, V., & Drechsler, R. (2022). Advanced virtual prototyping for cyber-physical systems using RISC-V: Implementation, verification and challenges. *Science China. Information Sciences*, *65*(1), 1–17. doi:10.100711432-020-3308-4

Jha, A. V., Appasani, B., Ghazali, A. N., Pattanayak, P., Gurjar, D. S., Kabalci, E., & Mohanta, D. K. (2021). Smart grid cyber-physical systems: Communication technologies, standards and challenges. *Wireless Networks*, *27*(4), 2595–2613. doi:10.100711276-021-02579-1

Kim, B. S., Nam, S., Jin, Y., & Seo, K. M. (2020). Simulation framework for cyber-physical production system: Applying concept of LVC interoperation. *Complexity*, *2020*, 1–11. Advance online publication. doi:10.1155/2020/4321873

Kumar, M. S. (2018). Experimental investigations on mechanical and microstructural properties of Al2O3/SiC reinforced hybrid metal matrix composite. []. IOP Publishing.]. *IOP Conference Series. Materials Science and Engineering*, *402*(1), 012123. doi:10.1088/1757-899X/402/1/012123

Kumar, M. S. (2019). Processing and characterization of AA2024/Al2O3/SiC reinforces hybrid composites using squeeze casting technique. *Iran. J. Mater. Sci. Eng*, *16*(2), 55–67.

Meikandan, M., Karthick, M., Natrayan, L., Patil, P. P., Sekar, S., Rao, Y. S., & Bayu, M. B. (2022). Experimental investigation on tribological behaviour of various processes of anodized coated piston for engine application. *Journal of Nanomaterials, 2022*, 1–8. doi:10.1155/2022/7983390

Nagarajan, Rajagopalan, A., Angalaeswari, S., Natrayan, L., & Mammo, W. D. (2022). Combined economic emission dispatch of microgrid with the incorporation of renewable energy sources using improved mayfly optimization algorithm. *Computational Intelligence and Neuroscience, 2022*, 1–22. doi:10.1155/2022/6461690 PMID:35479598

Natrayan, L., & Kumar, M. S. (2020). An integrated artificial neural network and Taguchi approach to optimize the squeeze cast process parameters of AA6061/Al2O3/SiC/Gr hybrid composites prepared by novel encapsulation feeding technique. *Materials Today. Communications, 25*, 101586. doi:10.1016/j.mtcomm.2020.101586

Natrayan, L., & Kumar, M. S. (2020). Optimization of wear behaviour on AA6061/Al2O3/SiC metal matrix composite using squeeze casting technique–Statistical analysis. *Materials Today: Proceedings, 27*, 306–310. doi:10.1016/j.matpr.2019.11.038

Natrayan, L., Merneedi, A., Bharathiraja, G., Kaliappan, S., Veeman, D., & Murugan, P. (2021). Processing and characterization of carbon nanofibre composites for automotive applications. *Journal of Nanomaterials, 2021*, 1–7. doi:10.1155/2021/7323885

Natrayan, L., Senthil Kumar, M., & Chaudhari, M. (2020). Optimization of squeeze casting process parameters to investigate the mechanical properties of AA6061/Al2O 3/SiC hybrid metal matrix composites by Taguchi and Anova approach. In *Advanced Engineering Optimization Through Intelligent Techniques: Select Proceedings of AEOTIT 2018* (pp. 393-406). Springer Singapore. 10.1007/978-981-13-8196-6_35

Natrayan, L., Sivaprakash, V., & Santhosh, M. S. (2018). Mechanical, microstructure and wear behavior of the material AA6061 reinforced SiC with different leaf ashes using advanced stir casting method. *International Journal of Engineering and Advanced Technology, 8*, 366–371.

Pan, Y., & Zhang, L. (2022). Integrating BIM and AI for Smart Construction Management: Current Status and Future Directions. *Archives of Computational Methods in Engineering*. doi:10.100711831-022-09830-8

Paranthaman, V., Sundaram, K. S., & Natrayan, L. (2022). Influence of SiC particles on mechanical and microstructural properties of modified interlock friction stir weld lap joint for automotive grade aluminium alloy. *Silicon, 14*(4), 1617–1627. doi:10.100712633-021-00944-5

Pragadish, N., Kaliappan, S., Subramanian, M., Natrayan, L., Satish Prakash, K., Subbiah, R., & Kumar, T. C. A. (2022). Optimization of cardanol oil dielectric-activated EDM process parameters in machining of silicon steel. *Biomass Conversion and Biorefinery*, 1–10. doi:10.100713399-021-02268-1

Ramaswamy, R., Gurupranes, S. V., Kaliappan, S., Natrayan, L., & Patil, P. P. (2022). Characterization of prickly pear short fiber and red onion peel biocarbon nanosheets toughened epoxy composites. *Polymer Composites*, *43*(8), 4899–4908. doi:10.1002/pc.26735

Sandborn, M., Wright, C., White, J. (2021). CaSTL: Towards cyber-physical authentication of additively manufactured components. *Procedia Computer Science, 198*, 764–769. doi:10.1016/j.procs.2021.12.319

Santhosh, M. S. (2018). Investigation of mechanical and electrical properties of Kevlar/E-glass and Basalt/E-glass reinforced hybrid composites. *International Journal of Mechanical and Production Engineering Research and Development*, *8*(3), 591–598. doi:10.24247/ijmperdjun201863

Sundaram. (2019). Smart clothes with bio-sensors for ECG monitoring. *International Journal of Innovative Technology and Exploring Engineering*, *8*(4), 298–301.

Tripathi, D., Biswas, A., Tripathi, A. K., Singh, L. K., & Chaturvedi, A. (2022). *An integrated approach of designing functionality with security for distributed cyber-physical systems*. Springer US. doi:10.100711227-022-04481-9

Veeman, Shree, M. V., Sureshkumar, P., Jagadeesha, T., Natrayan, L., Ravichandran, M., & Paramasivam, P. (2021). Sustainable development of carbon nanocomposites: Synthesis and classification for environmental remediation. *Journal of Nanomaterials*, *2021*, 1–21. doi:10.1155/2021/5840645

Velmurugan, G., Siva Shankar, V., Kaliappan, S., Socrates, S., Sekar, S., Patil, P. P., Raja, S., Natrayan, L., & Bobe, K. (2022). Effect of aluminium tetrahydrate nanofiller addition on the mechanical and thermal behaviour of luffa fibre-based polyester composites under cryogenic environment. *Journal of Nanomaterials*, *2022*, 1–10. doi:10.1155/2022/5970534

Yogeshwaran. (2015). Mechanical properties of leaf ashes reinforced aluminum alloy metal matrix composites. *International Journal of Applied Engineering Research: IJAER*, *10*(13), 11048–11052.

Chapter 17
New Generation Mobile Cyber Security Threats:
QR Codes and Social Engineering Threats

Aykut Aydın
Ahmet Yesevi University, Turkey

Gurkan Tuna
https://orcid.org/0000-0002-6466-4696
Trakya University, Turkey

ABSTRACT

With the invention of Quick Response (QR) code technology by the Japanese automotive company Denso Wave, the ease of use and the development of image processing receivers are making its use widespread in all sectors day by day. The widespread use of QR code, especially in applications containing personal data such as banking, finance, and messaging applications, has made it the focus of fraudsters and cyber attackers. QR code supported phishing attacks are among the new methods of fraudsters because awareness of the safe use of QR code has not yet been formed. In this context, in this chapter, the authors try to draw attention to the safe use of QR code, and present cyber security risks that QR codes may contain, common usage areas of QR codes, QR code generators, and issues related to QR code awareness.

DOI: 10.4018/978-1-6684-9267-3.ch017

INTRODUCTION

Quick Response (QR) code is a two-dimensional symbol consisting of black patterns and motifs on a white background, invented by Denso, one of the Toyota group companies, in 1994 (Chen et al., 2019; Hung et al., 2020). It was approved in June 2000 as an ISO international standard (ISO/IEC18004) (Soon, 2008). Although QR coded were originally intended to be used in the production control of automotive parts, their use in other areas is becoming widespread over time. Factors such as the fact that it has more features than conventional barcode types, has much higher data density, can be used free of charge by everyone after Denso company made its QR code patent public, and smart mobile devices equipped with cameras that allow QR code to be read have become widespread. makes it attractive. A QR code can basically store a coded link address, with the information capacity it can contain, and enables automatic access to this link address via QR code readers (Soon, 2008). QR Codes actually took place as a matrix symbol that was developed to enable all high-capacity and high-density features.

Using a QR code is actually pretty easy. You only need a smart device with an internet connection to use it. However, with QR code shares; business cards, flyers, advertisements, banking transactions, identity verification stages, restaurant menus are made easily accessible. With the camera directed towards the area with the QR code, the barcode is translated and automatically forwarded to the connection address, thanks to the software hosted on the smart device. QR codes have many common uses such as transportation, banking, service sector, identity verification, based on its benefits such as design elegance, ease of expression and information load capacity it can contain. It is widely used for ticket tracking and control in public transportation vehicles (Finžgar & Trebar, 2011). It is also used for path planning of robots used in areas such as product tracking, storage systems and factory production lines (Teja & Kumaar, 2018). In addition, QR codes are created for the purpose of checking the authentication of paper-based documents using digital signatures (Warasart & Kuacharoen, 2012). In mobile applications, QR codes are also available for authentication purposes such as access with two-factor authentication, multiple login permission, and web-based access (Eminagaoglu et al., 2014). Moreover, it is used for secure authentication in Internet Banking transactions using QR codes (Shamal et al., 2014).

By coding up to 7,089 characters consisting of numbers as numerical data, 4,296 characters as alpha-numeric, 2,953 characters as binary bytes, and 1,871 characters from Kanji expressions can be stored via QR codes (Kieseberg et al., 2010; Bilir, 2020). The amount of data that can be stored for a QR code varies depending on its version, mode and error recovery level. There are 40 different versions for QR codes starting from version 1 to 40 (Tiwari, 2016). In terms of QR code structure, it consists of 5 basic areas, which can be defined as version information, formatting information, coded data, required patterns and silent zone. As the data storage capacity for each code is different, the versions increase as the amount of data they store increases. However, the error correction levels of the codes also differ (Bilir, 2020). The formatting information is located next to the separators and consists of 15 bits of data. Here, the QR code contains the data about the selected masking model, along with the error correction level. The encoded data, the data of the QR code are stored in this area. Data is stored by converting the binary numbers '0' and '1' into black and white cells. The detection pattern consists of three similar squares in all corners of the QR code, except for the lower right corner. These square patterns in the QR code consist of a 3*3 dimensional matrix. These squares are designed to ensure correct recognition and correct orientation of QR code readers. It is surrounded by white separators with one-pixel area, making it easy to recognize and easily distinguish from real data. Changes to be made in the shapes in the detection pattern prevent QR decoders from reading the code (Bilir, 2020).

The alignment pattern allows the QR code reader to correct these distortions in case the area where the QR code is located is bent or curved in any way. It supports decoder to avoid moderate image distortions. There are no alignment patterns used for the purpose described in QR code version 1. However, there is an alignment pattern in QR code version 2 (Bilir, 2020). The timing pattern enables the module width to be determined by the solver, thanks to the alternatively available black and white module. In case of cell spacing or symbol distortion, the data cell is arranged in vertical and horizontal directions to fix the center coordinate. Silent zone is the area surrounding the QR code where there is no data. Thanks to this empty space, the QR code can be read without error (Kiesberg et al., 2010).

While the focus is on the benefits of QR code technology in the context of facilitating life, such as its development, dissemination and bringing it to different areas, due to its ease of use, however, the cyber security risks that QR codes may contain should be studied with the same importance. For this reason, in almost all newly developing technological developments, the cyber security factor should be carried out in coordination with the project, not at a secondary level, and updates should be made together with measures against new threats that may occur after development. With any cyber vulnerability that may occur, the damage to personal information security will be significant.

QR CODE TYPES AND QR CODE GENERATORS

QR code version 1 is a QR code that can encode 1,167 numbers and has a maximum version of 14 (73 x 73 modules) (Bilir, 2020). QR code version 2 was created by developing Version 1. This ensures that the code can be read without any problems even if it is corrupted in some way. The corrupted codes of the QR code, which cannot be fully read by the reader, are efficiently converted thanks to the alignment pattern it contains. The maximum capacity can go up to Version 40, and the QR code of 40 (177x177 modules) can encode 7.089 numbers (Bilir, 2020).

Micro QR code has a direction detection pattern so that it can be written in a small area. Its most important feature is the location detection pattern layout, unlike the standard QR code, which requires serious space. The direction detection pattern is located at three corners in a symbol. A QR code requires at least four module wide margins, while a Micro QR code requires two module wide margins. In micro QR codes, it is possible to print data in much smaller areas (Bilir, 2020).

A iQR code is a matrix type 2 size code that makes it easy to read in terms of its location and size. iQR codes have the capacity to store more data than standard QR codes and Micro QR codes and allow a wide range of code options including large codes. These have application examples in various fields by printing them as rectangle, black and white color inversion or as a dot pattern code (Bilir, 2020).

A logo QR code stands out as a unique type of QR code with the possibility of personalization. Visual richness can be created by blending it with images and texts. Since it is a designable QR code type, it can be easily distinguished from ordinary QR codes.

Encrypted QR codes can be defined as a restricted reading function. They can also be called as a QR code containing encrypted data. In the encrypted QR code system, the data information is encrypted using various encryption techniques and then the encrypted data is applied to the QR code encoder to generate the QR code. Then, the QR code is scanned by the decoder and decoded, then the information is obtained by using decryption techniques on the data (Tiwari, 2016).

QR codes can contain many different types of information. This data is processed and made readable through various reader applications on the smart mobile device. Although many examples are given of the types of data that can be uploaded to the QR code, common usages are contact information, which includes contact information such as phone number, e-mail address, website, address and note, can be shared. A calendar event, event information including event title, start and end date times, time zone, location and description can be presented with a QR code. By uploading a text-formatted content to QR codes, the visuality of content such as reminders, diaries and advertisements can be enriched. A QR code can be created that contains information such as Wi-Fi network configurations, security camera access. There are almost endless examples of options for encoding the link address into the QR code, and link addresses can be added to social media accounts such as Facebook and Instagram, and to content such as a video shared on the YouTube platform (Narayanan, 2012).

QR codes can be created commonly through websites, mobile applications and programs that can be downloaded to computers (Çataloğlu & Ateşkan, 2014). However, the widespread use of it day by day increases the diversity of QR code generators. Therefore, QR code generators should be carefully selected between safe or insecure generators. Although QR codes are perceived as safe in theory and in practice, when people handle them the wrong way, the threats they can pose can be alarming. When creating QR codes, users can choose one the two types of QR code shapes. These can be defined as static and dynamic QR codes. Static QR codes allow users to have one-time QR code generation. They are more secure in many aspects as they cannot be edited. Nevertheless, dynamic QR codes allow users to create and edit their content without the need to create new ones. So this type of QR code can pose a threat to users. However, since QR code generators currently require an account login, only the user has access to modify the data within the dynamic QR code (Baidong & Yukun, 2019).

CYBER SECURITY RISKS AND ATTACK METHODS CARRIED BY QR CODES

The most common attack scenarios that QR codes carry with cyber security risks include phishing attack, SQL command injection, XSS cross-site scripting, browser-based attacks, QR code reader application attacks, malware propagation, barcode-in-barcode attacks, barcode tampering and barcode tampering (Focardi, Luccio, & Wahsheh, 2019).

Phishing Attacks

Phishing attacks can be described as a type of social engineering realized in order to gain unauthorized access to people's sensitive confidential information (Bilir & Ergüner Özkoç, 2020). By stealing such information, the cyber attacker tries to enter the areas where he/she is not authorized. Phishing attacks stand out as the most attractive and effective method for social engineering (Mitnick & Simon, 2011).

In phishing attacks, it is aimed to steal sensitive personal data such as user information, user names, password information or credit card information. One of the prominent tactics in phishing attacks is the ue of phishing e-mail messages and encoded QR codes that contain links to fake websites or websites that host malicious software. Among the most important reasons for the widespread use of encoded QR codes is that the address of the target QR code cannot be read directly. When this QR code is scanned with the help of a reader, the user can only see which link address it is directed to (Bilir, 2020).

Due to the fact that some platforms such as Twitter, which have become widespread recently, impose letter restrictions, and the link addresses are long and cause a bad image in the text, platforms for creating shortened link addresses such as bit.ly have been developed. Thus, it is thought that the probability of the target not gaining access will be low when it sees this shortened, supposedly reliable but masked link address. When the target accesses the abbreviated common link address offered for access permission by scanning the QR code, they are redirected to fake websites, and the user unwittingly enters the mask website, i.e. information belonging to the user, such as user name, password and credit card. (Kapsalis, 2013).

SQL and Command Injection Attacks

SQL and command injection attacks are used by cyber attackers to access automated systems (Bilir & Ergüner Özkoç, 2020). For example, let's assume that a QR code analysis software is connected to a database with the codes it contains. Let's consider the scenario where QR code information is used for the query parameters run in this connected database. Here, if the QR code ('1' OR '1'='1') contains a query as in the command example in parentheses, the reader runs the query without verifying whether it comes from an authenticated source (Kilir, 2020). This information could lead to a potential hacker accessing and displaying the database, which is otherwise intended for an authorized user. However, QR code samples are not yet used to provide a database query (Bilir & Ergüner Özkoç, 2020). But, for such a situation that may occur in the future, it is thought that it can be used to attack such systems via QR codes. Attempts have been made to use the QR code to authenticate users who access Google's Gmail account, and it is already in use (Sharma, 2012).

In a command injection attack, the attacker can inject HTML codes by changing the content of the web page. When the user visits this modified page, the web browser interprets this code and causes the codes of malicious software to run on the mobile device on which the user reads the QR code (Ahuja, 2014).

In other words, it appears as a situation where the input via QR code is used as a command line parameter. Attackers thus exploit sensitive information about the target by executing random commands with the help of QR code. Thus, the attacker can initiate attacks such as rootkits, spywares, Denial of Service (DoS), Distributed DoS (DDoS), or they can connect to a remote computer and access system resources from there (Sharma, 2012). If command injection vulnerabilities are not eliminated, the reader can be exposed to command injection or buffer overflows (Bilir & Ergüner Özkoç, 2020). The attacker can take control of the user's contact information or communication content such as e-mail messages, SMS messages, contacts information, and the system by taking control of the smart mobile device (Kieseberg et al., 2010).

Cross-Site Scripting (XSS) and Browser-Based Attacks

QR codes can be used to run browser-based and cross-site scripting attacks (Bilir & Ergüner Özkoç, 2020). A QR code may contain an encrypted link address and when the encrypted connection address gives a warning message over the browser, if the user accesses the connection address, and malicious malicious software containing the warning can be run and damage the system (Sharma, 2012).

Cross-site scripting attacks by injecting scripts containing malicious software into trusted websites occur when the attacker uses an internet application to deliver malicious code, usually in browser-side script format, to a different end user. Mistakes that make such attacks successful are quite common. It occurs wherever the output generated by any internet application comes from a user and is used as input without applying validation parameters or coding. Here, hackers can use a cross-site command execution attack to send malicious scripts to the user (Bilir & Ergüner Özkoç, 2020). The script is run although the user's internet browser is not able to know that it is inconvenient to trust this script. Considering that the script is coming from a secure source, the script hosting malicious software can access session icons held via the internet browser, all cookie data used on this website or sensitive information. These scripts can also rearrange the content data of the HTML-coded website page (Li et al., 2018).

QR Code Reader Threats

Many QR code reader applications require full permission to access device resources such as user's device location, contact list and photos during the installation process on mobile device. However, in the case of a QR code-triggered security vulnerability, the attacker can capture the user's private data (Kiesberg et al., 2010).

QR Code Related Frauds

With QR codes, changes can be made on the systems in order to commit fraud or fraud (Bilir & Ergüner Özkoç, 2020). For example, by deceiving the installed system, the consumer can be deceived by replacing the QR code of the expensive product X with the QR code of an inexpensive product Y (Kieseberg et al., 2010). Users are deceived in this way by taking advantage of people's weaknesses, such as through social media platforms, fake bank advertisements, fake link addresses in the lottery account, and if you log in to the bank account, it's just a matter of time to win a bonus prize. When the user enters this information, fraudsters make money transfers or shopping from their bank accounts using the information they have obtained (Bilir & Ergüner Özkoç, 2020).

Social Engineering Attacks

As social engineering attacks, more specifically, sharing of content and posters containing QR code enriched with visual richness are used, rather than spear phishing or other attack methods. In fact, in this attack method using classical social engineering principles, the QR code increases the effectiveness of the attack (Kieseberg et al., 2010).

Examples of Attacks Realized With QR Codes

A malicious QR code was identified through Kaspersky Lab in September 2011. In the detected attack method, it was determined that malicious software was installed on the mobile device, with the target reading the QR code with the mobile device and directing it to the link address embedded in the QR code (Bilir & Ergüner Özkoç, 2020). In addition, it was observed that there are QR codes directed to websites with a large number of malicious software for mobile applications and Trojan horse software that send short messages to GSM (Wane & Jamankar, 2013). Especially in fake bank advertisements, which are brought together with the target through tools such as social media platforms, users' information is captured by clicking on the link address. Thus, by clicking on the link address, the data falls on the screen of the fraudster. By using the personal data of the legal user, fraudsters can use this data to access data such as the customer of the bank, make purchases and even transfer money to other accounts.

Thanks to the newly used mobile malware, the victim's mobile device is controlled and the verification messages sent by the bank can be captured by the fraudster. Therefore, it is understood that two-step verification methods for fraudsters have been exceeded recently. While making transactions from bank ATMs with these QR codes and password information captured by malicious people, the victim only receives warning messages. For a while, in order to prevent cyber piracy in China, it was decided to stop QR code payment methods in the banking sector by the Central Bank of the People's Republic of China (Bilir, 2020).

In 2012, a banner with a QR code prepared by the fans was shown in the football match between Karşıyaka and Göztepe sports clubs held at the İzmir Atatürk Stadium football field (Bilir, 2020). It was seen that there were many slang and profanity expressions in the QR code on the banner intended to be read by Göztepe supporters. Karşıyaka supporters who lost the match opened this banner with the QR code and dedicated it to their opponent. Thanks to the security cameras, the people who prepared the banner and opened it in the stadium were identified by the security forces (Bilir, 2020).

In 2016, in order to raise awareness, a banner was prepared by a company named Garrison Information Security Ltd. Sti., and the movements of people were observed for 3 months (Bilir, 2020). QR code contents, which contain harmless content, were presented to the citizens at universities and on the streets in various parts of our country. Over time, users were directed to the website page that contains information thanks to these QR codes read, and statistical information was collected at the same time (Bilir & Ergüner Özkoç, 2020).

MEASURES TO BE TAKEN AGAINST ATTACKS PERFORMED BY QR CODES

Engineering attacks target the weakest link in the security chain, come to mind among the measures that can be taken against the aforementioned and similar QR code attacks. In this sense, users should be careful against social engineering attacks, follow cyber warnings carefully, and be sensitive to unknown sources that they do not trust. Otherwise, despite all the precautions that can be taken, it will be inevitable for cyber attackers to reach their goals thanks to the weakest link, the human. On the other hand, some precautions can be taken for the use of QR codes before they reach the end user. For example, since people cannot see the link address contained in the QR code at once, it is possible to display the content of the link address in order to inform the user (Krombholz et al., 2014).

With the help of a QR code reader, which has been tested for security among QR code readers, the QR code can be passed through a set of reliability test stages. Especially when choosing a QR code reader, it should be taken into account in a sense that it is preferred from known application stores such as the Play Store, which has undergone security tests. There are several security issue handling approaches for QR code. Before the user can access the link address contained in the QR code, he/she must provide the link address, its source and the controls that can be performed manually. If an insecure connection of the redirected website is attempted or if it does not have a valid certification, the QR code should send a warning message to the user (Bilir & Ergüner Özkoç, 2020). In addition, digital signatures should be included in the QR code, and data privacy is ensured by simultaneous encryption methods (Ahuja, 2014).

In a secure QR code system; data integrity should be protected, authentication mechanisms should be included in the firmware (Bilir & Ergüner Özkoç, 2020). In addition, online content should be verified and malicious software should be detected and neutralized (Bani-Hani & Yarub, 2014). Testing the data integrity and verification mechanism of a QR code generator can be a precaution against attacks that can be carried out by cyber attackers. It is important for the integration and standardization of digital signatures for QR code in order to verify the creators of the code, thus ensuring control if the QR code is changed. Such measures significantly complicate QR code attacks, as the digital signature for this purpose requires the checksum information for the attacker and modification for this dependent verification process. The increase in the amount of encoded data reduces the coding area that contains the actual data. On the other hand, QR code reader software should verify these digital signatures and be applied in a way that indicates the success of this verification (Krombholz et al., 2014). Along with QR codes, in banking and finance applications the use of One Time Password (OTP) is recommended (Chu et al., 2013).

In order to test the verification of QR code generators, hashing and encryption algorithms are used together with digital signature algorithms. It is important to test the security of online content such as targeted phishing, malware content, as a result of redirecting to untrusted link addresses, which are the most frequently encountered QR code attack methods. Direct QR code readers are expected to be able to detect whether the application contains fake or real link addresses. Therefore, different techniques were proposed to determine whether a QR code contains malware or not (Bilir & Ergüner Özkoç, 2020). For example, it is known that the application named QRphish API is better than this public blacklist application such as Phistank (David, 2006) with an accuracy parameter of 93.34%, the detection mechanism has the ability to detect 82.9% more connection addresses (Alnajjar et al., 2016). QR code contents should be isolated from the executable codes as well as the commands in order to distance them from the sources of the device the QR code is being read. While isolating the content, it is aimed to prevent many attacks such as privacy violation, obtaining personal data, DDoS, bot net attacks against the reader. The simplest method for this is to prevent the access of applications, including the application with a QR code reader, to the management resources (camera, media contents, phone book information, location data, etc.) of the devices on which QR code reader applications are run. Buffer overflow regarding resource permissions of the QR code reader application should also be prevented from being used to initiate SQL injections.

By default, mobile devices do not have security software installed. However, it would be beneficial to have security software that can be installed on mobile devices. Programs used to provide such security have an important place in eliminating malicious acts of cyber attackers. In QR code applications, except banking applications, user name, password, bank account information, credit card details are not required. If there is a website where this information is requested, no information should be shared without logging in. In these cases, it is better to doubt the link addresses contained in the QR code.

QR Code Reader Software Related Solutions

Test software can be used to prove QR code reliability. For example, Norton Snap QR Code Reader, developed by Symantec, is a security application that runs on Android operating systems. It was designed to protect QR codes from online threats; however, it is no longer supported. In addition, it can control the connection addresses that contain the QR code. Thanks to these security checks, it makes recommendations about the reliability of the website. In case the internet address is not secure, access to this site is blocked by the application (Bani-Hani & Yarub, 2014).

A QR code reader for Android operating systems was developed by Kaspersky Lab. It is one of the smart QR code readers that protects it from dangerous connections without exposing it to attacks such as malicious software and identifies theft traps. Similarly, Trend Micro QR scanner can provide high-quality security controls within the codes by providing the process of being directed to a safe web page away from the dangers of scammers, hackers, and codes containing malicious software. SafeQR application offers effective methods of detecting malicious link addresses against phishing and attack methods that may contain malicious software. In addition, the level of security perception is kept high so that the user can make a better decision by scanning the QR code (Shin & Yao, 2013).

Visual QR Code Sharing/Masking

Visual QR code sharing or masking method is another measure that can be taken against QR code manipulation attacks. A QR code conforming to the specifications exhibits a particular pattern in black and white module distributions, and this pattern is determined by the mask used to determine whether it has changed in relation to the module color (Bilir & Ergüner Özkoç, 2020). The more the deviation from the distribution equation of the black and white module is different, the percentage of QR code change will increase in proportion to the difference. Therefore, the error rate and the change between security should be analyzed in detail and the reader application should be secured (Krombholz et al., 2014).

Authentication Based Solutions

QR codes are useful in many applications, but they are vulnerable to an online attack, called QRLjacking. One of the technical precautions that can be taken against QRLJacking is the ability to log in with a QR code. Requesting confirmation for each session, for example, receiving a confirmation message or sending a notification during login is another measure. Making IP restrictions, restricting authentication parameters on a different network are other measures to reduce attacks (Bilir & Ergüner Özkoç, 2020).

Location-based restrictions can be provided to prevent authentication factors according to different locations. Authentication using voice is among the techniques that mitigate attacks. Thanks to technologies such as this, which are possible to generate unique data with voice-based verification method and reconvert it in its original form, measures can be taken against QRLJacking (Owasp, 2019).

Using Digital Signatures

Digital signatures are among the most effective tools to increase QR code security. QR code integration and standardization are important in verifying the manufacturer of the code and ensuring that the QR code has been replaced with another QR code. The necessity of making changes in the checksum and verification processes with the digital signature method provides complexity against many QR code-based attacks. In fact, the amount of increase in the amount of encoded data can reduce the area to be encoded the actual data. In addition, QR code readers need to be alerted whether they are successful in verifying digital signatures and similar to secure internet connection addresses. In conclusion, it is beneficial to continue to develop digital signatures in the context of integration with QR code (Krombholz et al., 2014).

Adding Watermarks

In parallel with the increase in commercial activities in digital media, the need for security is increasing. Therefore, novel watermark methods are needed (Bilir & Ergüner Özkoç, 2020). In (Zhang & Yoshino, 2008) a novel watermark method to place a digital image inside QR codes was proposed. In the proposed method, by using the watermark based on Discrete Wavelet Transform (DWT), the original image is divided into blocks and embedded into the watermark signals. The proposed method was shown to be robust in JPEG format compression. The embedded information was extracted correctly even if the original image was 11% compressed. Studies have been carried out on the watermarking method with the audio watermark scheme, which is based on QR code, genetic algorithm methods with clipped wave conversion (Bilir & Ergüner Özkoç, 2020). However, still methods are needed to hide data (Poomvichid et al., 2012).

QR Code Visualization

QR codes that contain visual features with colors, letters, pictures or logos are called Visual QR codes. QR code generation, which provides visual appeal, has become widespread recently. However, it is necessary to create a QR code that is visually interesting without losing its readability. For this reason, half-tone QR code approaches are recommended for QR code generation with high quality visuals (Chu et al., 2013).

Recently, QR codes have started to be supported with aesthetic elements and to gain visual richness with the methods carried out by researchers. In this direction, the shape and color information of the modules in QR codes can be changed or the content such as a picture, photograph, symbol, logo can be embedded in the QR code (Lin et al., 2015). This method supports the user with significant signs in detecting QR codes that have been changed by cyber attackers (Bilir & Ergüner Özkoç, 2020). No matter how complex the created theme is, any modified QR code that the attacker can create will be distinguishable even by eye (Krombholz et al., 2014).

Masking Solution

During QR code creation stages, a mask is applied to each separate section and the resulting information is graded. Looking at the grading results, the mask is selected according to this distribution result (Kieseberg et al., 2010). The distribution of black and white modules for the QR code follows a specific pattern if the technical specifications are suitable. Determining the color of the module for the pattern is determined by the mask. The stability provided by Reed Solomon error correction codes does not have any negative effect on the resolution of the QR code with pixel values that may be corrupted. The height of the deviation in the distribution in the black and white module will be a factor in determining the change that can be made in the QR code (Krombholz et al., 2014).

A CASE STUDY

In recent years, cyber attackers have focused heavily on phishing attacks. Although different studies have been carried out about phishing attacks, it seems that they did not focus on QR code-based phishing attacks. But QR code-based phishing attacks have become widespread recently. In the case study designed to be realized in a test environment, firstly a QR code was created. Then, a tool that allows the website to work as online by downloading the directory structure, a database to store the user name and password information to be obtained sequentially, and a PHP editor for coding were used. In the study, Instagram as one of the most popular social media platforms was targeted.

Information was provided for the recovery of the Instagram account seized by cyber attackers. In addition, technical methods such as two-step verification methods, application code, backup code, which must be taken so that hackers cannot easily take over their accounts, such as not sharing passwords for the use of social media platforms and not accessing unsafe source links and content were provided. It was aimed to raise awareness by mentioning the precautions.

Design of the Case Study

A program called "HTTrack Website Copier", which is a website copying tool that allows any website to be completely downloaded to a computer and enables the downloaded website content to be used offline, was used. HTTrack allows downloading HTML, text, images and all other necessary files by copying the directory structure of the site you want to download in the same way. To put it briefly, the "mirrored" website can be opened via any internet browser, allowing navigation between these links as if browsing online, and these processes are performed offline. HTTrack has versions with different graphical interface options. In addition, it can be used as a command line tool. WebHTTrack interface, which is one of the graphical interfaces, it allows to use the program over any internet browser (Firefox, Opera, Chrome etc.).

This interface can be used on any system regardless of the platform. WinHTTrack interface shown in Figure 1, which is another graphical interface designed for Windows operating system, allows to be used as a separate application independent of the internet browser.

As shown in Figure 2, the directories of the website belonging to Instagram were downloaded to the computer for offline use through the program. The Instagram page files downloaded with the WinHTTrack Website Copier program are shown in Figure 3. The main pages of websites are usually located in a file named index.html. The site is used offline by clicking on the home page file among the downloaded files.

As shown in Figure 4, a database was created in order to ensure that user name and password information can be archived and viewed regularly. The database was created using phpMyAdmin. phpMyAdmin is an open source tool written in PHP. Its main use is to provide MySQL database management over the Internet. It stands out as a free software that can perform functions such as creating and deleting databases, adding/modifying/deleting tables, adding/modifying/deleting fields, running SQL queries, managing users, privileges and field keys.

As shown in Figure 5, in the database created with the name "instagramph", variables are defined with the name of id, kadi (username) and password in order to add the user name and password information obtained from the victims to the database, respectively.

Figure 1. Graphical user interface of WinHTTrack website copier

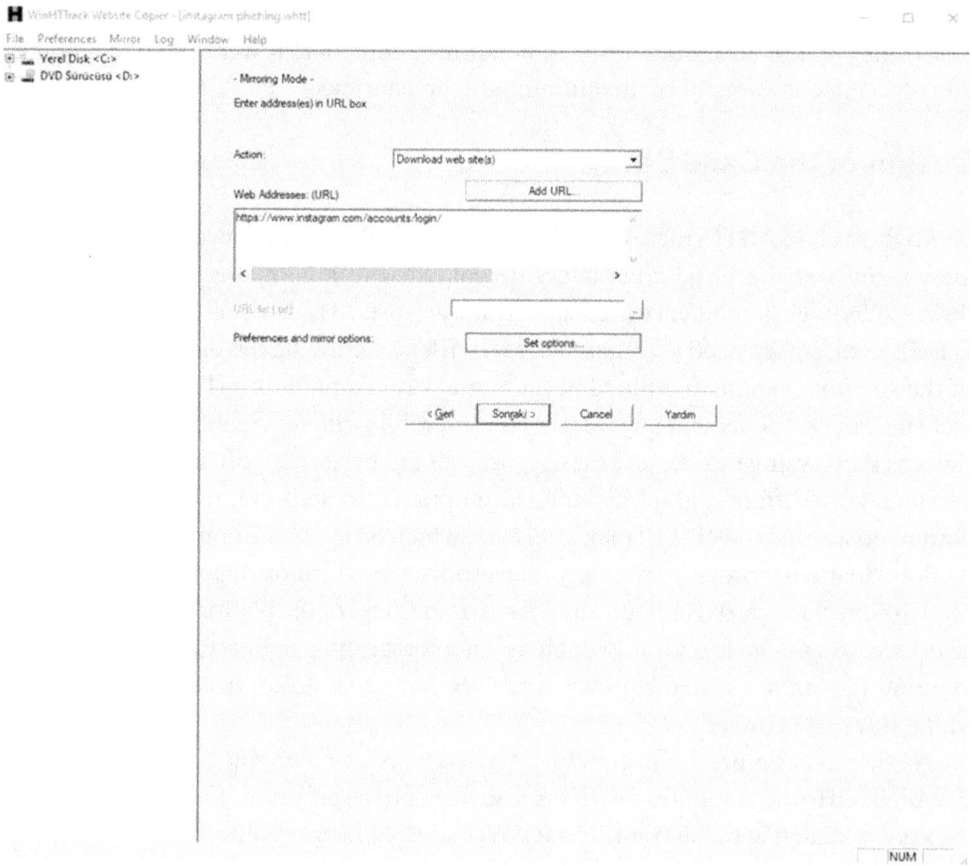

As shown in Figure 6 and Figure 7, the values obtained from the victim, which will be defined in the variable named "name", are transferred to the database. The values of the user name and password information obtained from the victim are assigned to the variables called "kuadi" and "password". Instagram page directories, database and action.php commands are run together on localhost.

After the database design and related coding, as shown in Figure 8 a QR code was created to map the website link address. The QR code created here contains the forwarded link address. Since the application scenario was realized offline, "localhost/instagramphishing" was redirected as the link address. Before performing actions, cyber attackers generally collect information from their victims. For this, they typically design an attractive login page or a page similar to the design of the actual one as shown in Figure 9.

Figure 2. Downloading Instagram for offline use

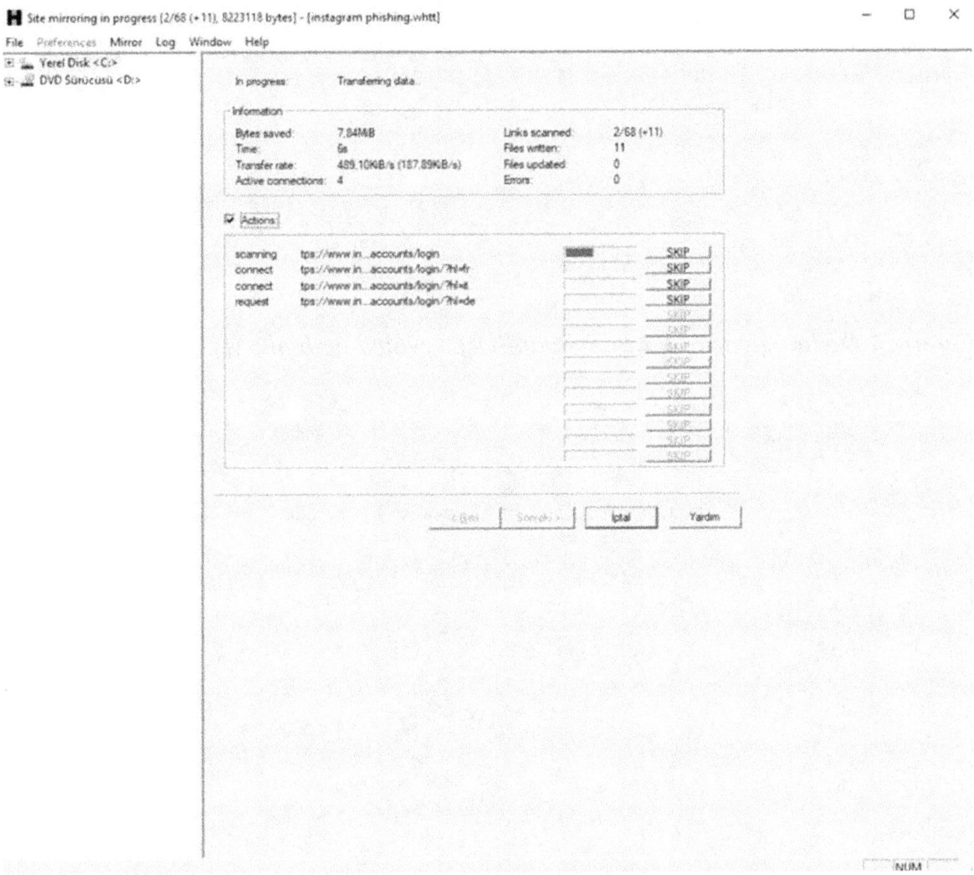

Figure 3. Files belonging to Instagram

Figure 4. Creating a database for Instagram information with phpMyAdmin

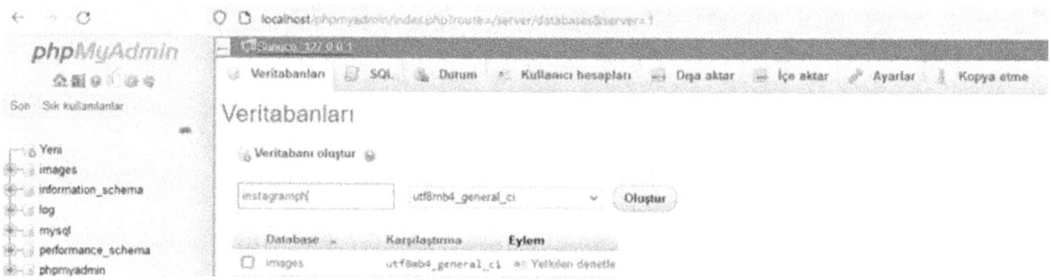

Figure 5. Defining variables in the database created with phpMyAdmin

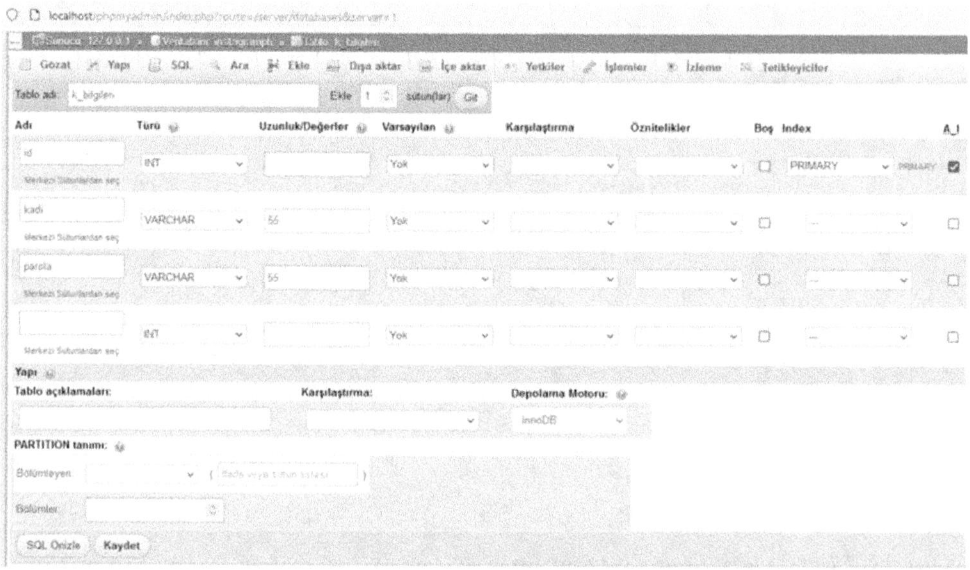

After the victim enters his/her user name and password, they are directed to the original Instagram page or they are informed that the content could not be accessed. Most victims are unaware that their account information is in the hands of the cyber attacker in this situation. As it can be seen in Figure 10, the cyber attacker, on the other hand, has already started to store user information in the database.

Figure 6. Variables to assign user name and password information in Instagram source codes

```
<form class="_ab3b" id="loginForm" method="post" action="action.php"><div class="_ab8w  _ab94 _ab99 _ab9f _ab9m
_ab9p  _abc2 _abcm"><div class="_ab32"><div class="x6s0dn4 x1npaq5j x1c83p5e x1enjb0b x199158v x9f619 x78zum5
x1q0g3np xvs91rp x1n2onr6 xh8yej3 xnz67gz x5n08af xx7zo7k x1fzb3qy xb0nk2e xubunj8"><label class="_aa48"><span
class="_aa4a">Phone number, username, or email</span><input aria-label="Phone number, username, or email"
aria-required="true" autocapitalize="none" autocorrect="off" maxlength="75" name="username" type="text" class="
_aa4b _add6 _ac4d"></label><div class="x6s0dn4 x6umtig x1b1mbwd xaqea5y xav7gou x9f619 x78zum5 x1c4vz4f x2lah0s
xdl72j9 x1q0g3np xk390pu x5yr21d xdj266r x11i5rnm xat24cr x1mh8g0r xexx8yu x1sxyh0 x18d9i69 xkhd6sd x1n2onr6
xxymvpz"></div></div></div><div class="_ab32"><div class="x6s0dn4 x1npaq5j x1c83p5e x1enjb0b x199158v x9f619
x78zum5 x1q0g3np xvs91rp x1n2onr6 xh8yej3 xnz67gz x5n08af xx7zo7k x1fzb3qy xb0nk2e xubunj8"><label class="_aa48
_aa49"><span class="_aa4a">Password</span><input aria-label="Password" aria-required="true" autocapitalize="none"
autocorrect="off" name="password" type="password" class="_aa4b _add6 _ac4d" value="faNHmM4jVEQ4jTt"></label><div
class="x6s0dn4 x6umtig x1b1mbwd xaqea5y xav7gou x9f619 x78zum5 x1c4vz4f x2lah0s xdl72j9 x1q0g3np xk390pu x5yr21d
xdj266r x11i5rnm xat24cr x1mh8g0r xexx8yu x1sxyh0 x18d9i69 xkhd6sd x1n2onr6 xxymvpz"><div class="_ab8w  _ab94
_ab99 _ab9f _ab9m _ab9p  _abb0 _abcm"><button class="_acan _acao _acat _aj1-" type="button">Show</button></div></
div></div></div><div class="_ab8w  _ab94 _ab99 _ab9f _ab9m _ab9p  _abak _abb8 _abbq _abb- _abcm"><button class="
_acan _acap _acas _aj1-" disabled="disabled" type="submit"><div class="_ab8w  _ab94 _ab99 _ab9f _ab9m _ab9p _abcm"
>Log in</div></button></div><div class="_ab39"><div class="x78zum5 x1q0g3np"><div class="x1iyjqo2 xs83m0k xjm9jq1
x1n2onr6 xwtuau4 x11mr3az"></div><div class="x1qjc9v5 x6umtig x1b1mbwd xaqea5y xav7gou x9f619 x1roi4f4 x78zum5
xdt5ytf x1c4vz4f x2lah0s x1nxh6w3 x1s688f x1ly1vsg xdj266r xpdqn1h xat24cr x1sliqq xexx8yu x4uap5 x18d9i69
xkhd6sd x1n2onr6 xtvhhri x11njtxf">or</div><div class="x1iyjqo2 xs83m0k xjm9jq1 x1n2onr6 xwtuau4 x11mr3az"></div>
</div></div><div class="_ab8w  _ab94 _ab99 _ab9f _ab9m _ab9p  _abak _abb8 _abbq _abb- _abcm"><button class="_acan
_acao _acas _aj1-" type="button"><span class=" _a04s _ab36"></span><span class="_ab37">Log in with Facebook</span>
</button></div></div><a class="x1i10hfl xjbqb8w x6umtig x1b1mbwd xaqea5y xav7gou x9f619 x1ypdohk xt0psk2 xe8uvvx
xdj266r x11i5rnm xat24cr x1mh8g0r xexx8yu x4uap5 x18d9i69 xkhd6sd x16tdsg8 x1hl2dhg xggy1nq x1a2a7pz _ab2y _a6hd"
href="https://www.instagram.com/accounts/password/
reset/?next=https%3A%2F%2Fwww.instagram.com%2Flogin%2F%3F__coig_login%3D1" role="link" tabindex="0">Forgot
password?</a></form>
```

Figure 7. Transferring the user name and password to the database with action.php coding

```
action.php
<?php

$kuadi=$_POST['username']
$sifre=$_POST['password']


$servername = "localhost";
$username = "root";
$password = "1";
$dbname = "instagramph";

// Create connection
$conn = new mysqli($servername, $username, $password, $dbname);
// Check connection
if ($conn->connect_error) {
  die("Connection failed: " . $conn->connect_error);
}

$sql = "INSERT INTO kbilgileri (id, kadi, parola)
VALUES ('$id','$kuadi','$sifre')";

if ($conn->query($sql) === TRUE) {
  echo "New record created successfully";
} else {
  echo "Error: " . $sql . "<br>" . $conn->error;
}

$conn->close();
?>
```

Figure 8. Simple QR code with forwarded link address

Figure 9. Entering the user name and password of the target on the site directed by QR code

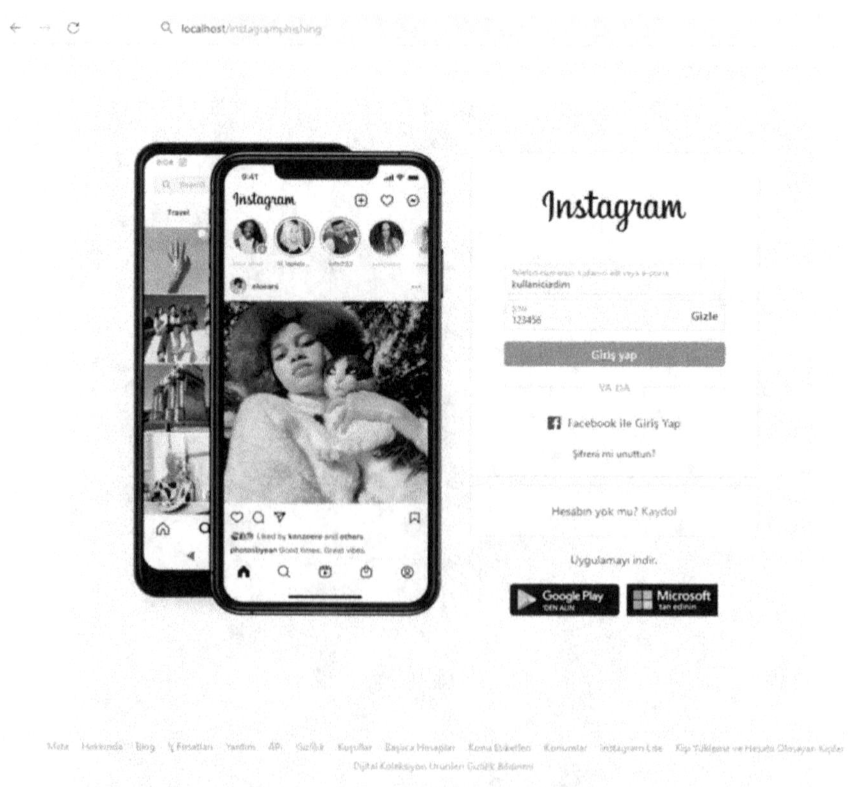

Figure 10. Looking at the user name and password entered by the victim in the database

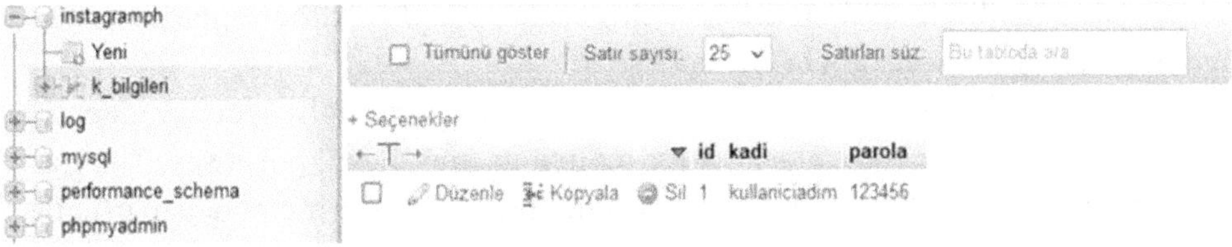

FINDINGS, INTERPRETATION, AND SUGGESTIONS

Ordinary users do not generally pay attention to personal information security, and those who care do not have cyber awareness. Although information about cyber activities is known by people, technical methods and tightening of personal data protection are generally seen as insignificant. However, sensitive personal data obtained by cyber attackers can be used for malicious purposes. In such cases, users who suffered a loss regrets it, but in cases of being late, it may cause irreversible consequences and users may be exposed to cyberbullying with demands such as being forced to pay ransom. Contary to common belief, in general, the awareness level of cyber attacks that may arise from QR codes, especially among young people is quite low. Sometimes, it is seen that QR codes, which are presented with interesting content, especially with social engineering methods, are read, since users are curious to discover even if they are aware that the QR code may contain malicious software. The sense of curiosity unfortunately causes the security sensitivities to be thrown into the secondary plan by the users. Here, it is beneficial to observe the balance of curiosity and security by the users.

With the increase in the use of QR code technology, messaging applications such as WhatsApp, SnapChat and social media platforms such as Instagram are working to secure their users' information and to prevent hackers from stealing their sensitive information. Therefore, QR codes are becoming today's primary solution in combating identity and data theft. In order to ensure the information security of QR codes, two-factor authentication methods are already commonly used in applications. The use of QR codes for malicious purposes and exploiting security vulnerabilities in mobile devices can allow cyber attackers to perform different actions such as creating new contact lists on the mobile devices, accessing existing contacts on the mobile devices, sending messages to people, making calls, tracking the users' locations, uploading unwanted content to the mobile devices, and accessing content and images. The cyber attackers can send e-mails by accessing the users' e-mail accounts, hijack their social media accounts, transfer money and make purchases by accessing bank and financial accounts, and request ransom by encrypting important content on the mobile devices. It is expected that phishing attacks via e-mail, which is used as a classical method, will be replaced or used together with attacks made via QR codes.

Username and password information obtained from victims by social engineering methods are used by cyber attackers for malicious and illegal purposes. The main purpose is mostly to obtain large amounts of money by ransom method. In particular, the method of demanding ransom by hijacking accounts on the Instagram platform is among the most widely used. After obtaining the user name and password, as well as trying to demand ransom, cyber attackers can send slang messages, and even get involved in crime through the victim's account. Up until this point, most victims are unaware of the situation. In case of a hacked account, the best thing to do is to inform the Instagram support that the account has been stolen and a recovery is requested by sharing the detail information with the device and browser that the victim has previously accessed.

CONCLUSION

With the widespread use of the internet and mobile devices, QR codes are used in almost every area of our lives in social media and messaging applications, finance and banking applications, official documents, advertisements, business cards, advertisements and product tracking. On the other hand, most users are unaware of cyber activities associated with QR codes and information stored in QR codes cannot be checked immediately. Therefore, QR codes can lead to significant risks. QR code attacks include SQL and command injection, phishing, fraud, spyware, spread of malware such as viruses. The purpose of such attacks is to access the personal data of internet users, to use the resources requested by QR code reader devices without permission, to allow other attacks without the users being aware of this situation, which can lead to material and moral losses. Widely used QR code generators and readers with high user ratings should be preferred. The basic measure against QR code related attacks is that before accessing a link address directed with the QR code, the link address should be copied and it should be examined whether there is a typo.

It is extremely important to realize that scanning a random, unknown QR code and accessing the forwarded link address is quite similar to accessing the unknown unknown domain. It is very difficult and incomprehensible to verify that any link address is malicious just by looking at its spelling. In the first step, efforts should be made to raise awareness of users not to have QR codes read by readers of unknown origin, and not to share their personal data with unverified sources when they give up access to invisible link addresses. It is considered that it will be beneficial to increase the security levels in terms of constantly installing the updated versions of the web browsers and operating systems of the mobile devices used, finding the authentication systems of the QR code generators, verifying the online content by ensuring data integrity, and isolating the malicious software that may be contained in the QR code.

One of the main limitations of this study is that the study was carried out in an offline test environment. However, it is considered that testing of QR code related attacks in real world scenarios by informing the necessary authorities and obtaining permissions will increase awareness. Regarding QR codes, users can be informed by common access of QR codes or link addresses that contain maliciously directed link addresses. As it is widely accepted, training users is the most effective method to overcome cyber security problems. However, it is expensive and time-consuming.

ACKNOWLEDGMENT

Some parts of this chapter were produced from the first author's master's project. This research received no specific grant from any funding agency in the public, commercial, or not-for-profit sectors.

REFERENCES

Ahuja, S. (2014). QR codes and security concerns. *International Journal of Computer Science and Information Technologies*, *5*(3), 3878–3879.

Alnajjar, A. Y., Anbar, M., Manickam, S., Elejla, O., & El-Taj, H. (2016). QRphish: An Automated QR Code Phishing Detection Approach. *Journal of Engineering and Applied Sciences (Asian Research Publishing Network)*, *11*(3), 553–560.

Baidong, H., & Yukun, Z. (2019). Research on Quickpass Payment Terminal Application System Based on dynamic QR code. *Journal of Physics: Conference Series*, *1168*(3), 032059. doi:10.1088/1742-6596/1168/3/032059

Bani-Hani, R., Wahsheh, Y. A., & Al-Sarhan, M. B. (2014). Secure QR code system. *2014 10th International Conference on Innovations in Information Technology* (pp. 1-6). 10.1109/INNOVATIONS.2014.6985772

Bilir, M. O. (2020). *Qr Kod Güvenlik Farkindaliği Üzerine Ankara Ilinde Bir Araştirma*. [Unpublished Master's Thesis. Baskent University, Ankara, Türkiye].

Bilir, M. O., & Ergüner Özkoç, E. (2020). QR Kod Güvenlik Farkındalığı Üzerine Ankara İlinde Bir Araştırma. *Journal of Internet Applications and Management, 11*(2), 113–129. doi:10.34231/iuyd.763399

Çataloğlu, E., & Ateşkan, A. (2014). Use of QR codes in education with examples. *Elementary Education Online, 13*(1), 5–14.

Chen, R., Yu, Y., Xu, X., Wang, L., Zhao, H., & Tan, H.-Z. (2019). Adaptive Binarization of QR Code Images for Fast Automatic Sorting in Warehouse Systems. *Sensors (Basel), 19*(24), 5466. doi:10.339019245466 PMID:31835866

Chu, H. K., Chang, C. S., Lee, R. R., & Mitra, N. J. (2013). Halftone QR codes. *ACM Transactions on Graphics, 32*(6), 217. doi:10.1145/2508363.2508408

David, U. (2006). *OpenDNS*. Phistank. https://www.phishtank.com/

Eminagaoglu, M., Cini, E., Sert, G., & Zor, D. (2014). A two-factor authentication system with QR codes for web and mobile applications. *2014 Fifth International Conference on Emerging Security Technologies* (pp. 105-112). IEEE. 10.1109/EST.2014.19

Finžgar, L., & Trebar, M. (2011). Use of NFC and QR code identification in an electronic ticket system for public transport. *SoftCOM 2011, 19th International Conference on Software, Telecommunications and Computer Networks* (pp. 1-6). IEEE.

Focardi, R., Luccio, F. L., & Wahsheh, H. A. (2019). Usable security for QR code. *Journal of Information Security and Applications, 48*, 102369. doi:10.1016/j.jisa.2019.102369

Hung, S. H., Yao, C. Y., Fang, Y. J., Tan, P., Lee, R. R., Sheffer, A., & Chu, H. K. (2020). Micrography QR Codes. *IEEE Transactions on Visualization and Computer Graphics, 26*(9), 2834–2847. doi:10.1109/TVCG.2019.2896895 PMID:30716038

Kapsalis, I. (2013). *Security of QR Codes*. [Unpublished master's thesis, Norwegian University of Science and Technology, Trondheim, Norway].

Kieseberg, P., Leithner, M., Mulazzani, M., Munroe, L., Schrittwieser, S., Sinha, M., & Weippl, E. (2010). QR code security. *8th International Conference on Advances in Mobile Computing and Multimedia* (pp. 430-435). ACM. 10.1145/1971519.1971593

Krombholz, K., Frühwirt, P., Kieseberg, P., Kapsalis, I., Huber, M., & Weippl, E. (2014). QR code security: A survey of attacks and challenges for usable security. In *International Conference on Human Aspects of Information Security, Privacy, and Trust* (pp. 79-90). Springer, Cham. 10.1007/978-3-319-07620-1_8

Li, L., Fan, M., & Wang, G. (2018). LWSQR: Lightweight Secure QR Code. *International Conference on Frontiers in Cyber Security* (pp. 241-255). Springer. 10.1007/978-981-13-3095-7_19

Mitnick, K. D., & Simon, W. L. (2011). *The art of deception controlling the human element of security*. John Wiley & Sons.

Narayanan, A. S. (2012). QR codes and security solutions. *International Journal of Computer Science and Telecommunications*, *3*(7), 69–72.

Owasp. (2019). *Index.* Owasp. https://www.owasp.org/index.php/Qrljacking

Poomvichid, T., Patirupanusara, P., & Ketcham, M. (2012). The QR code for audio watermarking using Genetic algorithm. *International Conference on Machine Learning and Computer Science* (pp. 171-174). IEEE.

Sharma, V. (2012). A study of malicious QR codes. *International Journal of Computational Intelligence and Information Security*, *3*(5), 21–26.

Shin, D., & Yao, H. (2013). A User Study of Security Warnings for Detecting QR Code Based Attacks on Android Phone. *Journal of Wireless Mobile Networks, Ubiquitous Computing and Dependable Applications*, *4*(4), 49–64.

Soon, T. J. (2008). QR code. *Synthesis Journal*, *3*, 59–78.

Teja, P. R., & Kumaar, A. N. (2018). QR Code based Path Planning for Warehouse Management Robot. *2018 International Conference on Advances in Computing, Communications and Informatics* (pp. 1239-1244). IEEE. 10.1109/ICACCI.2018.8554760

Tiwari, S. (2016). An introduction to QR code technology. *2016 International Conference on Information Technology* (pp. 39-44). IEEE. 10.1109/ICIT.2016.021

Wane, A. R., Jamankar, S. P., & Chandure, O. V. (2013). An effective mechanism for ensuring security of QR code. *International Journal of Advanced Research in Computer Science*, *4*(6), 175–179.

Warasart, M., & Kuacharoen, P. (2012). Paper-based document authentication using digital signature and QR code. *4th International Conference on Computer Engineering and Technology*. Springer.

Zhang, S., & Yoshino, K. (2008). DWT-Based Watermarking Using QR Code. *Science Journal of Kanagawa University*, *19*, 3–6.

ADDITIONAL READING

Aktaş, C. (2017). *The evolution and emergence of QR codes*. Cambridge Scholars Publishing.

Diogenes, Y., & Ozkaya, E. (2019). *Cybersecurity - Attack and Defense Strategies* (2nd ed.). Packt Publishing.

Ozkaya, E. (2018). *Learn Social Engineering: Learn the art of human hacking with an internationally renowned expert*. Packt Publishing.

Winter, M. (2011). *Scan me: everybody's guide to the magical world of qr codes--barcodes mobile devices and hyperlinking the real to the virtual* (1st print). Westsong Pub.

KEY TERMS AND DEFINITIONS

App Store: It means an online store where customers can purchase and download various software applications. Generally, apps sold through app stores are designed for mobile devices. Examples of common application stores include Google Play Store, Apple Store, Huawei Store, Samsung Galaxy Store.

Phishing: It is a type of online attack in which scammers send emails to random user accounts. E-mails are sent by pretending to be known internet pages or the user's bank account, internet service provider.

Quick Response (QR): It is a two-dimensional barcode type consisting of different patterns and drawings in square shape, usually black on a white background.

Social Engineering: In the context of information security, it is the psychological manipulation of people to perform actions or disclose confidential information.

Social Media Platforms: They are online platforms used to establish social relationships. They are used to share people's personal or career interests, activities, backgrounds or real life connections. Social networking services differ in form and number of features. Examples of the most commonly used ones are Facebook and Instagram.

Chapter 18
Prediction and Development of Healthcare Systems Through Cyber Physical Systems Using Cloud Computing and Big Data

R. Kavitha
Velammal College of Engineering and Technology, India

S. Selvakanmani
https://orcid.org/0000-0001-6469-6084
Department of Information Technology, RMK Engineering College, India

A. Anbumani
Department of Computer Science Engineering, Velammal Institute of Technology, India

P. Praveen
Department of Mechanical Engineering, Graphic Era University (Deemed), India

Ramya Maranan
Department of Research and Innovation, Saveetha School of Engineering, SIMATS, India

ABSTRACT

The integration of modern information technology in healthcare has brought about a significant improvement in healthcare technologies. However, the increasing volume of healthcare data generated poses a significant challenge in managing and analysing the data. Moreover, big data, which is rapidly produced from multiple sources, has become a major concern in healthcare. To address these challenges, this

DOI: 10.4018/978-1-6684-9267-3.ch018

study proposes a Health-CPS cyber-physical system, which uses cloud computing and big data analytics to improve healthcare service and create a more favourable healthcare environment. The proposed system consists of a data management stage for circulated storing and similar processing, an information-oriented service stage, and a uniform data collection layer. This research aims to demonstrate the potential of big data and cloud technologies to enhance healthcare system efficiency and provide patients with access to a wide range of intelligent healthcare applications and services.

1. INTRODUCTION

Over the past two decades, there has been a significant increase in the use of information technology in medicine, resulting in improved data availability, traceability, and liquidity in areas such as biomedical databases, electronic health records (EHRs), and public health (Devarajan and Ravi 2019). However, this increase in data has also led to challenges in data management, storage, and processing. One major factor contributing to this growth in healthcare data is the development of medical information technology, predominantly the implementation of clinic evidence systems. Additionally, the proliferation of wearable health devices has also played a role in the rapid generation of medical data (Sethuraman et al. 2020).

This data is often complex and varied, coming in various formats such as text, images, audio, or video, and is generated in real-time by medical equipment, particularly wearable technology. This makes it necessary to manage and process data quickly, particularly in emergency situations (Kaiser et al. 2021). Despite these challenges, there is still a lot of value to be found in healthcare data, such as personalized health counseling and public health warnings. (Sood and Mahajan 2018).

However, there are still obstacles to overcome, such as heterogeneous data fusion and the lack of universal standards and methods for data access and analysis. This research aims to address these challenges by proposing a healthcare CPS based on cloud computing and big information technologies. The suggested system comprises a cloud-based and information-determined platform for the preservation and examination of various and diverse healthcare data, as well as a healthcare application service cloud with a shared API for programmers and a streamlined crossing point for users (Dey et al. 2018). The key advancement in this study is the introduction of a consolidated data gathering layer for integrating public medical resources. This will enable the smooth storage, examination and retrieval of various and diverse healthcare data, enabling the extraction of more profound understanding and value from the data (Gupta and Singh 2022).

The use of cloud computing and big data technologies in the proposed system will enable the effective administration and processing of large amounts of data, and provide the necessary infrastructure for information storing and examination. The proposed healthcare request service cloud will provide a unified interface for developers and users, making it easier for healthcare providers to access and analyze data (Vangipuram et al. 2021).

Figure 1. Application of health care in various fields

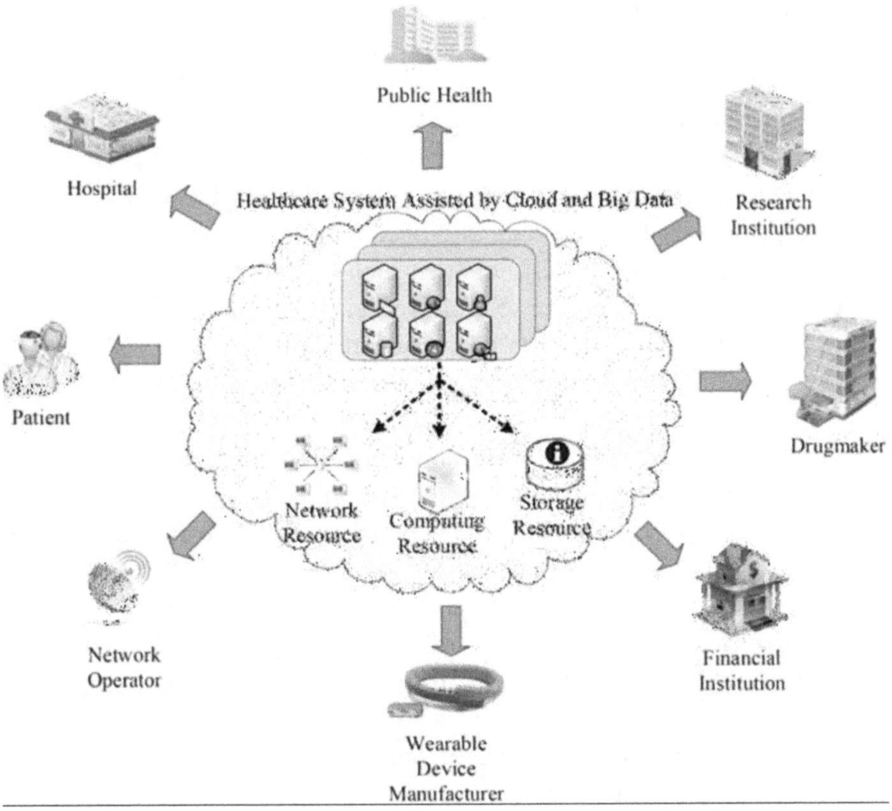

Additionally, the proposed system will also address the issue of data security and privacy, as it will include measures to ensure the safe and secure storage and transmission of sensitive patient data. In conclusion, the suggested healthcare CPS that relies on cloud and big data innovations aims to solve the issues of data management, storage, and processing in the healthcare sector by offering a consolidated platform for the integration of public therapeutic capitals and individual healthiness plans. (Das et al. 2022). This will enable the efficient storage, analysis, and retrieval of multisource heterogeneous healthcare data, and provide the necessary infrastructure for data storage and analysis. The proposed system will also address the issue of data security and privacy, by ensuring the safe and secure storage and transmission of sensitive patient data resources and personal health devices. Figure 1 shows the healthcare sector functions.

2. ARCHITECTURE OF THE PROPOSED SYSTEM

2.1 Issues Related to the Design

Cloud computing and big data have become essential technologies for the healthcare industry, and are increasingly becoming the driving force behind innovation in the sector. The ability to collect, store, and analyze large amounts of data is becoming increasingly important in medicine, and cloud-based medical information sharing has become commonplace. Patients are also becoming more involved in their own healthcare, thanks to the use of technologies such as smartphones, and wireless sensors. This shift has changed the role of physicians from one of sole decision maker to that of a consultant and helper in decision making (Kandasamy et al. 2020).

The healthcare industry is undergoing a significant transformation, with the introduction of new participants and technologies in the ecosystem, as shown in Fig. 1. To address the challenges of this new ecosystem, it is necessary to develop a healthcare system that is better adapted to dealing with these new challenges. One important aspect of this adaptation is the development of unified data management standards for multisource heterogeneous data. With the diversity of data types and variability of similar information, health care information can be difficult to utilize. To make the system scalable, it must support a wide range of medical equipment devices and use universal values to recover the effectiveness of data storing, query, recovery, dealing out, and examination. This will allow for the efficient analysis of diverse healthcare data, including organized, semi-structured, and unstructured data. The use of a unified programming interface will also decrease system complexity while improving access and development effectiveness. Finally, a platform for application services with a northbound common interface is also necessary. This platform will provide a wide range of applications and services for a wide range of occupations, and will be critical for resource efficiency, technical help, and data interchange. In conclusion, cloud storage and data storage of huge files are essential technologies for the healthcare industry, and are driving innovation in the sector (Wang et al. 2022). The proposed healthcare system must be adapted to deal with the challenges of this new ecosystem, by developing unified data management standards, diversified data analysis modules, a unified programming interface, and a platform for application services with a northbound common interface.

2.2 Architecture of Cloud and Big Data

The suggested Health-CPS design features three tiers: gathering information, managing data, and service provision. These tiers are depicted in Figure 2.

Figure 2. Various areas of data collection

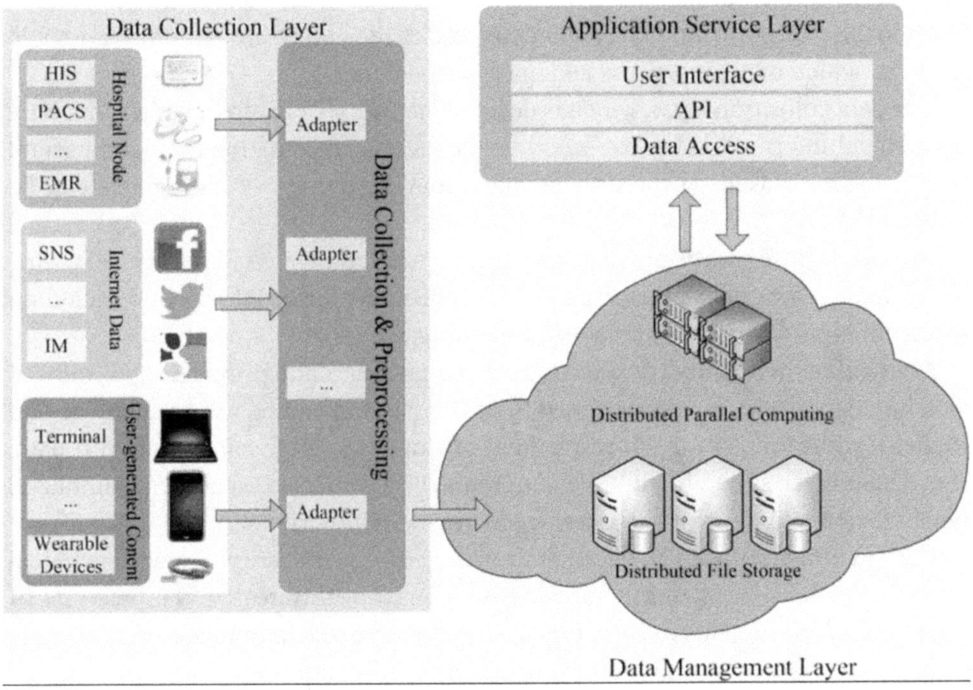

The data acquisition tier, which is composed of data points and connectors, delivers a uniform system access interface for diverse and varied data from various sources such as medical institutions, the web, and user-generated content. The DFS component, aided by big data-related technologies, provides efficient data storage and input/output for diverse healthcare data. (Fayad et al. 2022; Pattepu et al. 2022).

The application service tier enables consumers to access straightforward visual information analysis findings and also provides an exposed united API for programmers to build user-centric apps that offer comprehensive, expert, and customized healthcare services.

In summary, the proposed Health-CPS architecture is divided into three tiers: data acquisition, data management, and application service. The data acquisition tier provides a uniform system access interface for diverse data, while the information management tier comprises of a DFS component and a DPC component for effectual data storage and processing. The application service tier provides visual data analysis findings and an open API for building user-centric healthcare apps.

3. COLLECTION OF DATA

Figure 3 illustrates how various forms of healthcare data are collected by data nodes in the data collection layer and delivered to the cloud through programmable adapters, which may preprocess and encrypt the data.

The data collection layer, which is designed to be user-centered, gathers not only old medical information but too everyday behaviour information that is important for healthcare. Based on the data sources, four categories of data nodes can be constructed:

Research information: Pharmaceutical and additional scientific examine organizations have collected a substantial amount of investigate information, including data from high-throughput screening and clinical studies.

Medical expenditure data: Medical behaviour data provides a wealth of spending data that is not often linked to healthcare but can be used to analyze and predict healthcare costs, such as medical insurance reimbursement and bill data. This data is often stored in multiple geographically dispersed databases maintained by healthcare organizations that use standardized data formats.

Patient information: The information gathered by healthcare providers for clinical diagnosis such as ECG and therapeutic imaging. To fully realize the potential of medical data mining, this information can be standardized, administered, and made available to scientists while preserving patient confidentiality.

Personal movement and mood information: Although this type of information is produced outside of the healthcare industry, it is still applicable to individual health. For sample, individual trade ingesting data can reveal lifestyle choices that can be used to measure a person's health risk and tailor a health plan to that individual. Information shared on social media also allows for the collection of personal emotion data that can be used in affective computing and mental health evaluation. A physician can adjust a patient's treatment plan based on their mental state, particularly during recovery. An emotionally responsive wellness service encourages the integration of modern medicine with humanistic care.

Figure 3. Collection of healthcare data

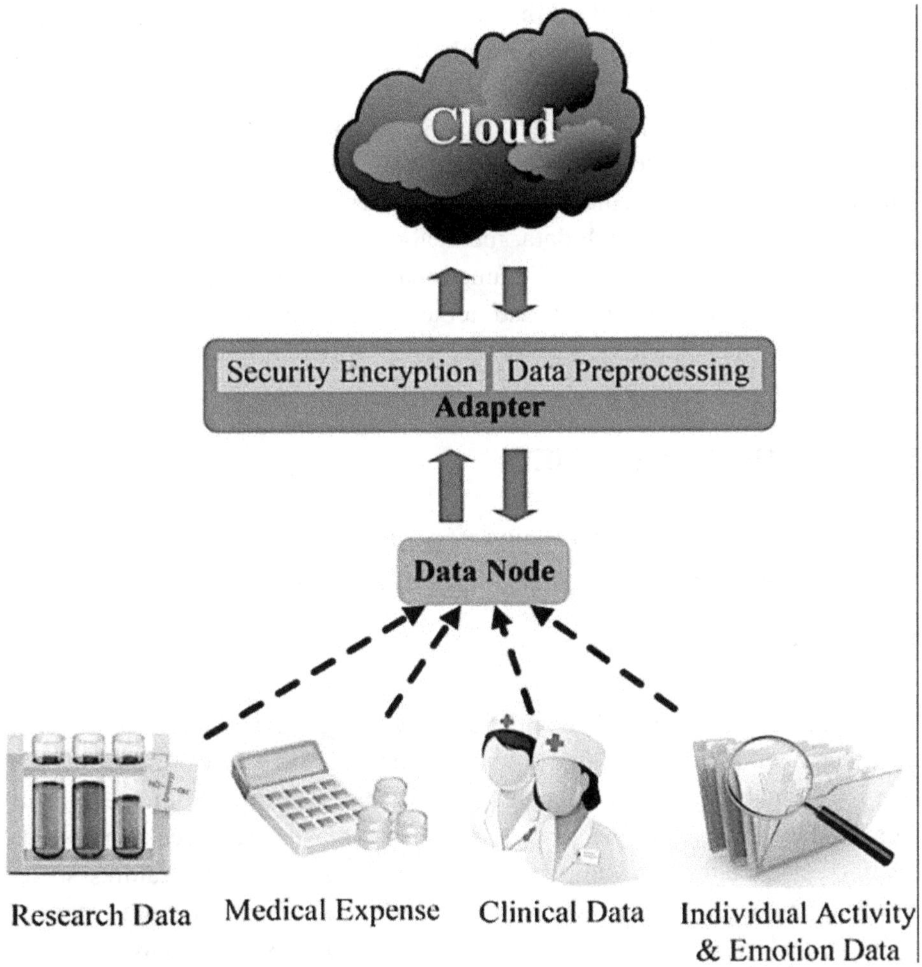

The adapters in the data collection layer act as middleware that allow data nodes to connect to the system. They also act as preprocessors and encryptors for the raw data. In addition to data format translation, the preprocessing module provides data cleaning, redundancy elimination, and compression. The functional unit of an adapter can be adjusted to enhance the system's scalability. The following conditions must be met before the adapter's relevant modules can be updated online:

Data variation: If a data point is updated or substituted, and the format of the new device differs from the original, the functional components will not function correctly. The server then logs the type of changed data point and re-validates the encryption module online, while the connector requests that the pre-processing module be reconfigured to be compatible with the new one.

Data standard library expansion: The data standard library is predictable to be enlarged and delivered to the suitable module for informing whenever a novel kind of expedient with no structure-definite information standard gains access to the system.

In summary, Figure 3 illustrates how healthcare data is collected by data nodes in the data collection layer and delivered to the cloud through programmable adapters, which preprocess and encrypt the data. Four categories of data nodes can be constructed based on data sources: research data, medical expenditure data, clinical data, and personal activity and mood data. The functional unit of an adapter can be adjusted to enhance the system's scalability and updates are made to the adapter's relevant modules online when data node variation or data standard library expansion occurs.

4. DATA MANAGEMENT TIER

As illustrated in Figure 4, the information management tier is made up of components to support efficient management and analysis of diverse data.

DFS Component

Distributed File Storage (DFS) is a technology that enables the efficient storage, management, and retrieval of large amounts of data across a network of computers. Unlike traditional file storage systems, which store data on a single server, DFS distributes data across multiple servers, allowing for faster data access, higher data availability, and better data fault tolerance. DFS also provides features such as high-throughput data upload and download, multitenant user management, access control list, and rapid data retrieval and exchange. DFS is commonly used in big data analytics, cloud computing, and other applications that require the management of large amounts of data. It is implemented using various distributed file system protocols such as HDFS, GlusterFS, Ceph and so on.

To uniformly manage diverse and varied data, the subsequent 3 units are intricate in creating the data morals:

Data Description: Data description is a process of providing information about data that is stored in a system. Data description is important for ensuring data availability, reliability, and traceability. Data availability refers to the ability to access the data when needed. Reliability means that the data is accurate and consistent. Traceability refers to the ability to trace the origin of data and the changes made to it over time.

Figure 4. Layers of management of data

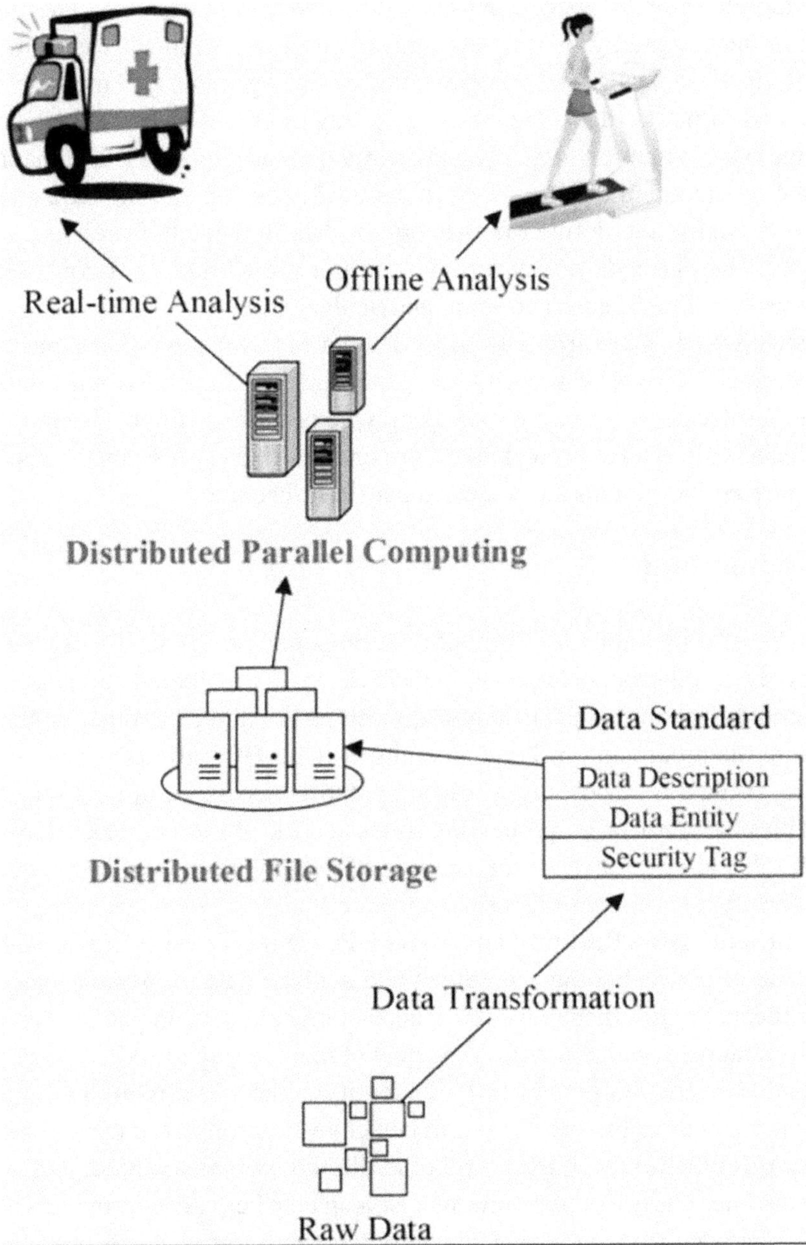

Data Objects: Data objects are standardized definitions of information entities, which include qualities, kind, rate variety, and the associations with other information. They are used to support the integration of data in the same category and are vital for data alteration. Data objects are used to ensure that data from different sources can be integrated and understood in a consistent and meaningful way. They provide a common understanding of the data and its context, making it easier to share, access, and analyze. Data objects are used in data management systems to define the structure and format of data. They provide a way to organize and describe data in a consistent and structured way. They also provide a way to access and manipulate data in a consistent and efficient way. In data warehousing and big data analytics, data objects are used to define the structure of data in the data warehouse and the data marts. They provide a way to organize and access data in a consistent and efficient way, making it easier to share and analyze data.

Security Label: A security label, also known as a security tag, is a mechanism used to define the level of security and authorization associated with data. It is used to control access to data by specifying who can view, edit, or delete the data. Security labels are used to protect data from unauthorized access and to ensure that only authorized individuals can access sensitive information.

DPC Component

The DPC (Distributed Parallel Computing) component is a type of software that is used to analyze and process large amounts of data in a distributed and parallel way. It is commonly used in big data analytics, machine learning, and other applications that require the processing of large amounts of data. DPC allows for the efficient processing of data by distributing the computation across multiple computers or nodes. This allows for faster processing times and the ability to handle larger data sets. DPC also allows for real-time analysis of data by providing the capability to process data in real-time, as it is being generated.

Real-time analysis: Real-time analysis of DPC (Direct Patient Care) in healthcare refers to the use of technology to collect and analyze data on patient interactions with healthcare providers in real-time. This can include data on patient symptoms, diagnosis, treatment, and outcomes. The goal of real-time analysis is to improve the quality and efficiency of patient care by identifying and addressing issues as they arise, rather than after the fact. This can be achieved through the use of electronic health records (EHRs), wearables, and other digital tools that collect and transmit data in real-time. The use of real-time analysis can help healthcare providers to make more informed decisions, improve patient outcomes, and reduce costs.

Offline analysis: Offline analysis of DPC (Direct Patient Care) in healthcare refers to the process of analyzing data on patient interactions with healthcare providers after the fact. This can include data on patient symptoms, diagnosis, treatment, and outcomes, which is collected and stored in electronic health records (EHRs) or other databases. The goal of offline analysis is to identify patterns and trends in patient care, such as common diagnoses or treatment outcomes, that can be used to improve the quality and efficiency of care.

Offline analysis can be done using various methods such as statistical analysis, data mining, machine learning, and natural language processing. This can help healthcare providers to identify areas for improvement, such as inpatient care, and provide more accurate diagnoses and treatments (Natrayan L & Kumar MS. 2020). Additionally, offline analysis can help in identifying the risk factors for certain diseases, which can help in preventative care. Offline analysis can also help to reduce costs by identifying inefficiencies and unnecessary procedures and treatments, and identifying ways to improve the coordination of care among different healthcare providers (Veeman et al., 2021).

In summary, the data management tier in Health-CPS is designed to provision efficient administration and examination of diverse data. The DFS component is responsible for providing efficient storage, high data upload and download speeds, high data fault tolerance, multitenant user management, access control list, and quick data retrieval and exchange. The DPC component is responsible for analyzing and processing data from DFS, and provides both real-time and offline analysis capabilities using various data mining and machine learning algorithms. This allows for the discovery of valuable insights and predictions that can be used to improve the overall healthcare system (Natrayan L et al., 2021).

5. APPLICATION LAYER

An Application Service Layer (ASL) is a software layer that sits between the application software and the underlying infrastructure in a Cyber-Physical System (CPS). In the context of healthcare, a CPS is a system that integrates physical devices, such as medical equipment, with software and communication networks to provide improved patient care (Natrayan L et al., 2018).

The ASL in a healthcare CPS serves as an intermediary between the applications (such as electronic health records, medical imaging systems, and telemedicine platforms) and the underlying infrastructure (such as servers, networks, and storage devices). The ASL provides a number of services to the applications, including:

Data management: The ASL can manage the storage, retrieval, and sharing of patient data across different applications and systems.

Security: The ASL can provide security services, such as authentication and encryption, to protect patient data from unauthorized access.

Communication: The ASL can facilitate communication between different applications and systems, such as sending alerts or notifications to healthcare providers.

Integration: The ASL can integrate different applications and systems to allow them to work together seamlessly, providing a unified view of the patient's health status (Pragadish et al., 2022).

Scalability: The ASL can allow the system to scale up and down as the number of users and devices increases or decreases (Balaji et al., 2022).

The ASL in a healthcare CPS can play a critical role in improving the quality, efficiency, and safety of patient care by enabling the integration of various healthcare applications and devices, and providing services such as data management, security, and communication (Ramaswamy et al., 2022).

6. PROPOSED TESTBED ARCHITECTURE FOR HEALTHCARE

The use of cyber physical systems (CPS) in healthcare has been gaining traction in recent years as a way to improve the delivery of healthcare services. The layers of healthcare service are illustrated in figure 5.

Figure 5. Various service layer

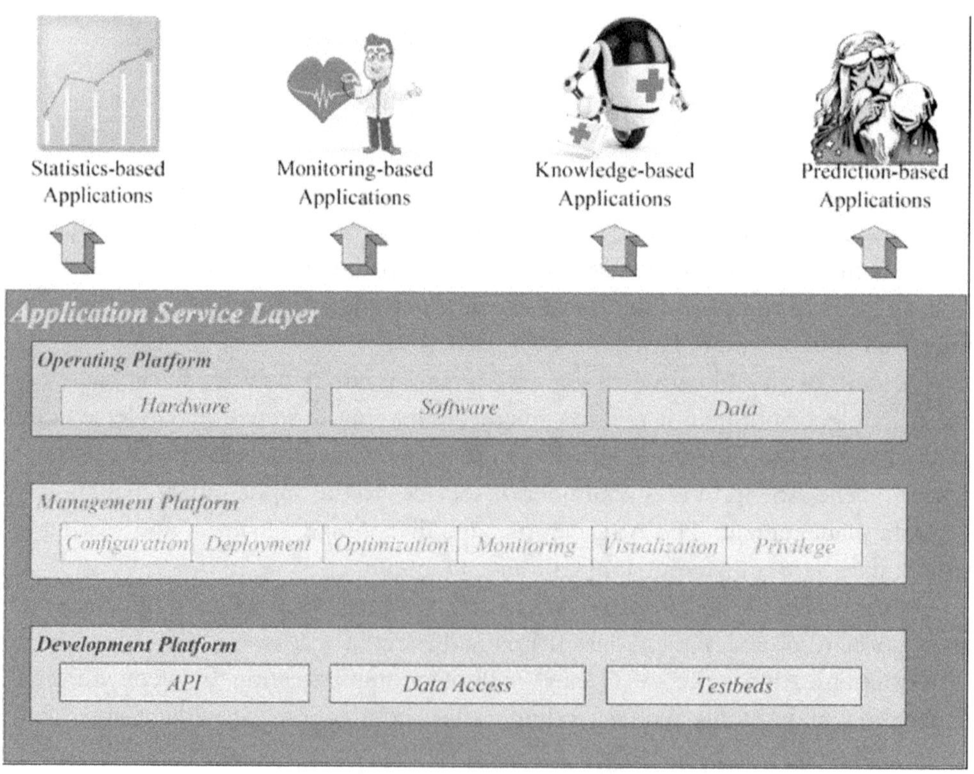

Researchers at the Entrenched and Persistent Computing Laboratory developed a testbed for CPS-assisted healthcare services that utilizes a robotic platform (Yogeshwaran et al., 2015). The testbed is composed of several components including a robot hardware, embedded software, and a cloud-based healthcare system (Kumar MS et al., 2018). The robot hardware includes mechanical devices, environment sensors, biosensors, and cameras, while the embedded software includes motion control, voice recognition, and sensory data pre-processing. The cloud-based healthcare system provides networking, processing, and storage resources for sensory information examination and well-being demonstrating (Santhosh MS et al., 2018). The testbed allows for both homebased and isolated users to operate the robot using a smartphone, and the robot's sensors measure vital signs such as temperature, humidity, heart rate, and ECG (Velmurugan et al., 2022). The data collected by the robot is processed and stored in the cloud for analysis and to aid in treatment decision making. In conclusion, the use of cyber physical systems in conjunction with cloud computing and big data has the potential to revolutionize the healthcare industry by allowing for the efficient management and analysis of large amounts of healthcare data (Natrayan, L., & Kumar, M. S. 2020). This research proposes a healthcare CPS that utilizes cloud and big data technologies to establish a stage for the storing and investigation of different -source varied healthcare information, as well as a unified request software design interface for designers and a united boundary for users (Nagarajan et al., 2022). The system is able to collect and process a wide range of healthcare data, including data from research, medical expenditures, clinical proof, and personal activity and mood data. By using big data-related technologies such as distributed file storage and distributed parallel computing, the system is able to effectively manage and analyze the large amounts of data, ultimately discovering new knowledge that can be used to improve healthcare outcomes.

CONCLUSION

In conclusion, the use of advanced analytics technology has greatly improved healthcare by allowing for patient-centered prevention, prediction, and therapy. Big data technology is becoming increasingly prevalent in healthcare and is expected to be used in a variety of ways, including as a guide for lifestyle choices, a tool for decision making, and a source of innovation. This paper described a smart health system that utilizes cloud computing and big data, which includes three main components: a unified data collection layer, a cloud-enabled platform for data storage and analysis, and a unified interface for developers and users. By utilizing this system, healthcare can overcome traditional challenges such as centralized resources, information silos, and patient passivity. In the future, the focus will be on developing a variety of applications using this system to improve healthcare outcomes.

REFERENCES

Balaji, N., Natrayan, L., Kaliappan, S., Patil, P. P., & Sivakumar, N. S. (2022). Annealed peanut shell biochar as potential reinforcement for aloe vera fiber-epoxy biocomposite: Mechanical, thermal conductivity, and dielectric properties. *Biomass Conversion and Biorefinery*, 1–9. doi:10.100713399-022-02650-7

Das, D., Ghosal, S., & Mohanty, S. P. (2022). CoviLearn: A Machine Learning Integrated Smart X-Ray Device in Healthcare Cyber-Physical System for Automatic Initial Screening of COVID-19. *SN Computer Science*, *3*(2), 1–11. doi:10.100742979-022-01035-x PMID:35132394

Devarajan, M., & Ravi, L. (2019). Intelligent cyber-physical system for an efficient detection of Parkinson disease using fog computing. *Multimedia Tools and Applications*, *78*(23), 32695–32719. doi:10.100711042-018-6898-0

Dey, N., Ashour, A. S., & Fong, S. J. (2018). Medical cyber-physical systems. *Survey (London, England)*, 1–13. PMID:29525900

Fayad, M., Mostefaoui, A., Chouali, S., & Benbernou, S. (2022). Toward a design model-oriented methodology to ensure QoS of a cyber-physical healthcare system. *Computing*, *104*(7), 1615–1641. doi:10.100700607-022-01058-5

Gupta, A., & Singh, A. (2022). *An Intelligent Healthcare Cyber Physical Framework for Encephalitis Diagnosis Based on Information Fusion and Soft-Computing Techniques*. Ohmsha. doi:10.100700354-022-00175-1

Kaiser, F. K., Wiens, M., & Schultmann, F. (2021). Use of digital healthcare solutions for care delivery during a pandemic-chances and (cyber) risks referring to the example of the COVID-19 pandemic. *Health and Technology*, *11*(5), 1125–1137. doi:10.100712553-021-00541-x PMID:33875933

Kandasamy K, Srinivas S, Achuthan K, Rangan VP (2020) IoT cyber risk: a holistic analysis of cyber risk assessment frameworks, risk vectors, and risk ranking process. *Eurasip Journal on Information Security 2020*(1). doi:10.1186/s13635-020-00111-0

Kumar, M. S. (2018). Experimental investigations on mechanical and microstructural properties of Al2O3/SiC reinforced hybrid metal matrix composite. []. IOP Publishing.]. *IOP Conference Series. Materials Science and Engineering*, *402*(1), 012123. doi:10.1088/1757-899X/402/1/012123

Nagarajan, Rajagopalan, A., Angalaeswari, S., Natrayan, L., & Mammo, W. D. (2022). Combined economic emission dispatch of microgrid with the incorporation of renewable energy sources using improved mayfly optimization algorithm. *Computational Intelligence and Neuroscience*, *2022*, 1–22. doi:10.1155/2022/6461690 PMID:35479598

Natrayan, L., & Kumar, M. S. (2020). An integrated artificial neural network and Taguchi approach to optimize the squeeze cast process parameters of AA6061/Al2O3/SiC/Gr hybrid composites prepared by novel encapsulation feeding technique. *Materials Today. Communications*, *25*, 101586. doi:10.1016/j.mtcomm.2020.101586

Natrayan, L., & Kumar, M. S. (2020). Optimization of wear behaviour on AA6061/Al2O3/SiC metal matrix composite using squeeze casting technique–Statistical analysis. *Materials Today: Proceedings*, *27*, 306–310. doi:10.1016/j.matpr.2019.11.038

Natrayan, L., Merneedi, A., Bharathiraja, G., Kaliappan, S., Veeman, D., & Murugan, P. (2021). Processing and characterization of carbon nanofibre composites for automotive applications. *Journal of Nanomaterials*, *2021*, 1–7. doi:10.1155/2021/7323885

Natrayan, L., Sivaprakash, V., & Santhosh, M. S. (2018). Mechanical, microstructure and wear behavior of the material AA6061 reinforced SiC with different leaf ashes using advanced stir casting method. *International Journal of Engineering and Advanced Technology*, *8*, 366–371.

Pattepu, S., Mukherjee, A., Routray, S., Mukherjee, P., Qi, Y., & Datta, A. (2022). Multi-antenna relay based cyber-physical systems in smart-healthcare NTNs: An explainable AI approach. *Cluster Computing*, *0*. doi:10.100710586-022-03632-0

Pragadish, N., Kaliappan, S., Subramanian, M., Natrayan, L., Satish Prakash, K., Subbiah, R., & Kumar, T. C. A. (2022). Optimization of cardanol oil dielectric-activated EDM process parameters in machining of silicon steel. *Biomass Conversion and Biorefinery*, 1–10. doi:10.100713399-021-02268-1

Ramaswamy, R., Gurupranes, S. V., Kaliappan, S., Natrayan, L., & Patil, P. P. (2022). Characterization of prickly pear short fiber and red onion peel biocarbon nanosheets toughened epoxy composites. *Polymer Composites*, *43*(8), 4899–4908. doi:10.1002/pc.26735

Santhosh, M. S. (2018). Investigation of mechanical and electrical properties of Kevlar/E-glass and Basalt/E-glass reinforced hybrid composites. *International Journal of Mechanical and Production Engineering Research and Development*, *8*(3), 591–598. doi:10.24247/ijmperdjun201863

Sethuraman, S. C., Vijayakumar, V., & Walczak, S. (2020). Cyber Attacks on Healthcare Devices Using Unmanned Aerial Vehicles. *Journal of Medical Systems*, *44*(1), 29. doi:10.100710916-019-1489-9 PMID:31838588

Sood, S. K., & Mahajan, I. (2018). A Fog Assisted Cyber-Physical Framework for Identifying and Preventing Coronary Heart Disease. *Wireless Personal Communications*, *101*(1), 143–165. doi:10.100711277-018-5680-y

Vangipuram, S. L. T., Mohanty, S. P., & Kougianos, E. (2021). CoviChain: A Blockchain Based Framework for Nonrepudiable Contact Tracing in Healthcare Cyber-Physical Systems During Pandemic Outbreaks. *SN Computer Science*, *2*(5), 1–16. doi:10.100742979-021-00746-x PMID:34179827

Veeman, Shree, M. V., Sureshkumar, P., Jagadeesha, T., Natrayan, L., Ravichandran, M., & Paramasivam, P. (2021). Sustainable development of carbon nanocomposites: Synthesis and classification for environmental remediation. *Journal of Nanomaterials*, *2021*, 1–21. doi:10.1155/2021/5840645

Velmurugan, G., Siva Shankar, V., Kaliappan, S., Socrates, S., Sekar, S., Patil, P. P., Raja, S., Natrayan, L., & Bobe, K. (2022). Effect of aluminium tetrahydrate nanofiller addition on the mechanical and thermal behaviour of luffa fibre-based polyester composites under cryogenic environment. *Journal of Nanomaterials*, *2022*, 1–10. doi:10.1155/2022/5970534

Wang, B., Zhou, H., Yang, G., Li, X., & Yang, H. (2022). Human Digital Twin (HDT) Driven Human-Cyber-Physical Systems: Key Technologies and Applications. [English Edition]. *Chinese Journal of Mechanical Engineering*, *35*(1), 11. doi:10.118610033-022-00680-w

Yogeshwaran. (2015). Mechanical properties of leaf ashes reinforced aluminum alloy metal matrix composites. *International Journal of Applied Engineering Research: IJAER*, *10*(13), 11048–11052.

Chapter 19
Recognition and Implementation of Cyber Physical Systems in the Development of Smart Factories in Industry 4.0 Through Optimization Techniques

L. Natrayan
https://orcid.org/0000-0003-1001-3422
Department of Mechanical Engineering, Saveetha School of Engineering, SIMATS, India

ABSTRACT

This chapter investigates the recognition and implementation of cyber physical systems (CPS) in the development of smart factories in Industry 4.0. With the integration of physical and digital components, CPS allows for the seamless integration of control systems, communication networks, and sensor systems leading to the optimization of factory processes. The use of optimization techniques such as machine learning and control theory is also discussed in the context of CPS implementation in smart factories. The data collected and analyzed in real-time through the integration of CPS technology leads to improvements in efficiency, reduction of downtime, and increase in overall production. The study concludes that the recognition and implementation of CPS in smart factories can lead to significant improvements in productivity and competitiveness. The research also provides insights for further research in the field of CPS implementation in smart factories and Industry 4.0.

DOI: 10.4018/978-1-6684-9267-3.ch019

1. INTRODUCTION

The fourth industrial revolution, also known as Industry 4.0, has brought about a new era of intelligent and connected systems in manufacturing. Cyber Physical Systems (CPS) play a critical role in this revolution by enabling the integration of physical and digital systems in the manufacturing process. Smart factories leveraging CPS to optimize production processes are at the forefront of Industry 4.0. (Busygin et al. 2018) The recognition and implementation of CPS in developing smart factories are vital for advancing manufacturing and for companies to stay competitive in the global market. In this research article, we will examine the current state of CPS in smart factories and their potential for optimization through advanced techniques such as machine learning, artificial intelligence, and data analytics (Park et al. 2013). Additionally, we will identify the challenges that must be overcome to fully realize the potential of CPS in smart factories and propose solutions for addressing these challenges. This research aims to provide insights on how to design and implement CPS in smart factories to optimize production processes, and to serve as a guide for companies looking to adopt CPS in their manufacturing operations (Mueller et al. 2017).

The integration of CPS in smart factories has the potential to significantly improve the efficiency, flexibility, and scalability of production processes. Real-time monitoring and control enabled by CPS allows factories to respond quickly to changes in demand and optimize production in real-time (Alguliyev et al. 2021). Additionally, CPS can collect and analyze large amounts of data from production processes, enabling factories to identify bottlenecks and inefficiencies and make data-driven decisions to improve production processes (Wang et al. 2016).

One of the key ways CPS can be used to optimize production is through implementing machine learning and artificial intelligence techniques. These techniques can be used to analyze data from production processes, identify patterns, and predict future production needs (Verma 2022). This can help factories to optimize production schedules, reduce waste, and improve overall production efficiency. Additionally, data analytics can identify trends and patterns in production data, enabling factories to make data-driven decisions to improve production processes. However, the implementation of CPS in smart factories also poses certain challenges. One major challenge is the integration of CPS into existing production processes and systems, which can be complex and time-consuming. Additionally, there are concerns about the security of CPS systems and the potential for cyber attacks on factories (AlZubi et al. 2021).

Companies must have a clear strategy for implementing CPS in smart factories to overcome these challenges. This should include a plan for integrating CPS into existing systems and processes and a strategy for addressing security concerns. Additionally, companies should invest in developing the skills and expertise needed to design, implement, and maintain CPS systems. In conclusion, CPS is crucial in developing smart factories in Industry 4.0 (He et al. 2020; Wan et al. 2021). The integration of CPS in smart factories has the potential to significantly improve the efficiency, flexibility, and scalability of production processes. However, implementing smart factories also poses certain challenges that must be addressed. This research article aims to provide insights on how to design and implement CPS in smart factories to optimize production processes and to serve as a guide for companies looking to adopt CPS in their manufacturing operations (Parvin et al. 2013; Bordel et al. 2018; Priyadarshini et al. 2021; Wu and Li 2021).

2. PROPOSED DESIGN

The Smart-CPMS is an innovative manufacturing system capable of rapidly and appropriately responding to changes in the industrial environment without needing external intervention. Figure 1 illustrates the smart cyber-physical manufacturing method paradigm. Individualized goods are a major trend in industry, and customers demand more personalized end products, minimum product life cycles, highest superiority, and minimising pricing. To meet these criteria, manufacturing systems must be furnishhavethat allow them to adapt to production changes.

Figure 1. Cyber-physical system enabled manufacturing environment

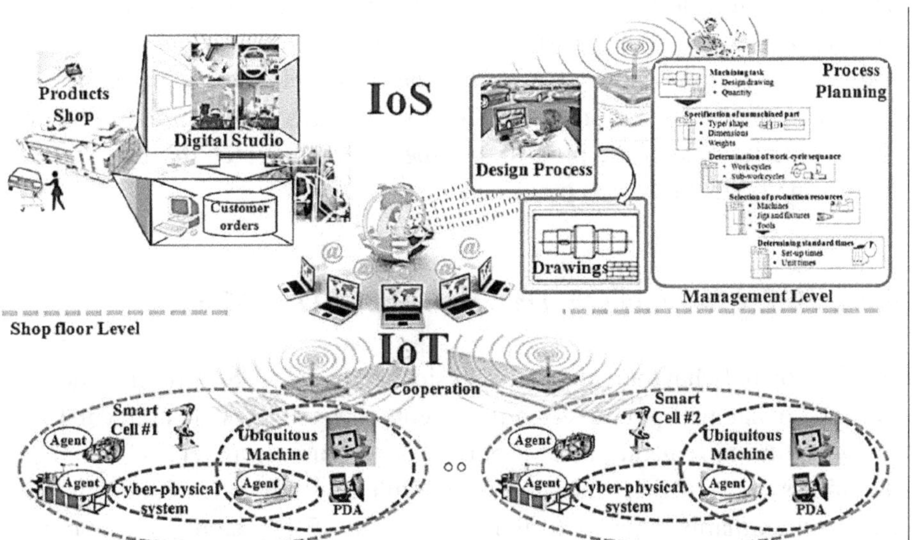

In the digital studio, customers use head-mounted displays with automated suggestions to select product attributes and make their purchases. The design and process planning department then uses this product information to create a production schedule for the unfinished goods on the shop floor. To optimize this process planning, biologically encouraged approaches such as genetic algorithms and optimization method of ant colony are employed. Other technologies, such as agent-based dynamic scheduling and holonic execution manufacturing systems, have also been used to address optimization issues at the management level. The Smart-CPMS utilizes the internet of services (IoS) for efficient task sharing from design to development planning.

On the production area, the Smart-CPMS is seen as an Internet of Things (IoT) collaboration between smart cells for work completion. To every cellEverys CPS are selfd entities that can adapt to changes and collaborate with supplementary CPSs to complete development planning. A CPS comprises a measured physical device and a cyber section, and is ablcanonomously and intelligently. The ability for CPSs to independently perceive and act is made possible by cognitive agent technology and artificial intelligence. Additionally, machines are equipped with pervasive technology to allol-time monitoring and control.

Human-machine interfaces (HMI), smartphones, and personal digital assistants are use the machine to the workforce. The CPS can record its status and connect with other CPSs through the execution system, planning of enterprise approach, wireless fidelity network, RFID approach, and sensor communication. A computer program with consciousness is referred to as a cognitive agent. A CPS, as deAs Figure 2, includes coma CPS ponents for awareness, choice-making, information, communiqué, and control. The awareness module receives and analyzes sensor data to determine the status of the physical component, and the decision-making module uses this information, anowledge-based reasoning, and artificial intelligenceto make decisions. New plans, such as needs for interfacing with other CPSs or new machining limitations for the process, are then communicated through the control module to the HMI, PDA, and other CPSs.

3. IMPLEMENTATION OF THE PROPOSED MODEL

The evaluation of behaviours in cyber-physical systems (CPS) is often done using a testbed model that incorporates cognitive agents. The model is illustrated in Figure 3 and involves the implementation of cognitive agents on a computer (PC) connected to a machining system (MES). The MES is a complex system that includewithponents, such as programmable logic controllers (PLCs), an RFID system, a communication system, and a production implementation system. To enable communication between

Figure 2. Proposed model block diagram

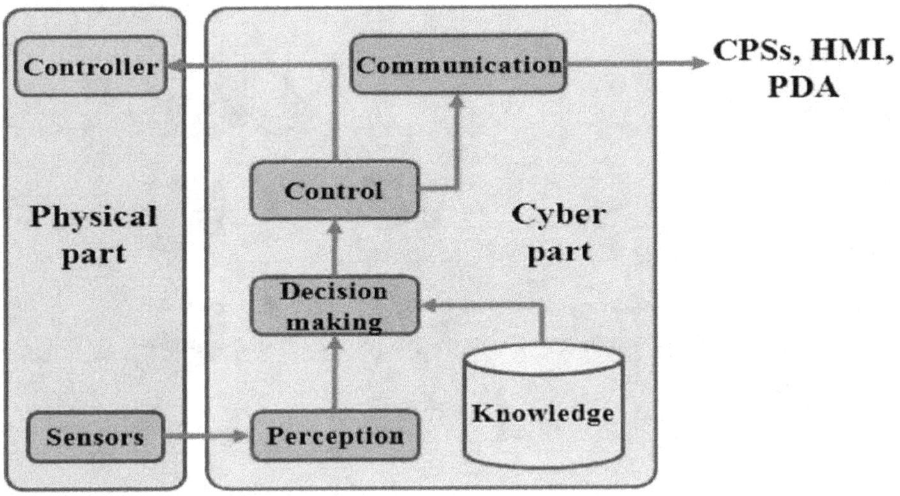

the cognitive agents on the PC and the various components of the MES, a BuffaloTM Ethernet converter hardware is used. This allows for seamless communication and data transfer between the two systems. Additionally, the testbed model includes the use of three switches as a means of introducing disruptions to the system. These disruptions are used to test the resilience and robustness of the CPS in the face of unexpected events. Overall, this testbed model provides a useful framework for evaluating the behaviours of CPS systems and assessing their performance in real-world scenarios.

One example of a cyber-physical system (CPS) is shown in detail in Figure 4. This system uses an RFID device to read and write information about a workpiece connected to a programmable logic controller (PLC) through an RS232 cable. The communication between the RFID reader and the PLC is facilitated through a message architecture that employs a byte Cyclic Redundancy Check (CRC) as a means of error detection. This architecture is also depicted in Figure 4. The PLC and PC communicate with each other via an internet connection, allowing for efficient and reliable data transfer. This CPS is also connected wirelessly to other CPSs through a BuffaloTM Ethernet converter, enabling seamless integration and coordination with other systems. When the workpiece information is directed to the process agent on the computer, the PLC typically activates a green light as a signal that the information has been received and processed successfully. However, if a disturbance is introduced by turning on a switch, the system will respond by activating a red light and alarm, signalling that a problem has occurred. In this situation, the machine agent will generate a new action strategy or coordinate with

Figure 3. Testbed of the cyber-physical system

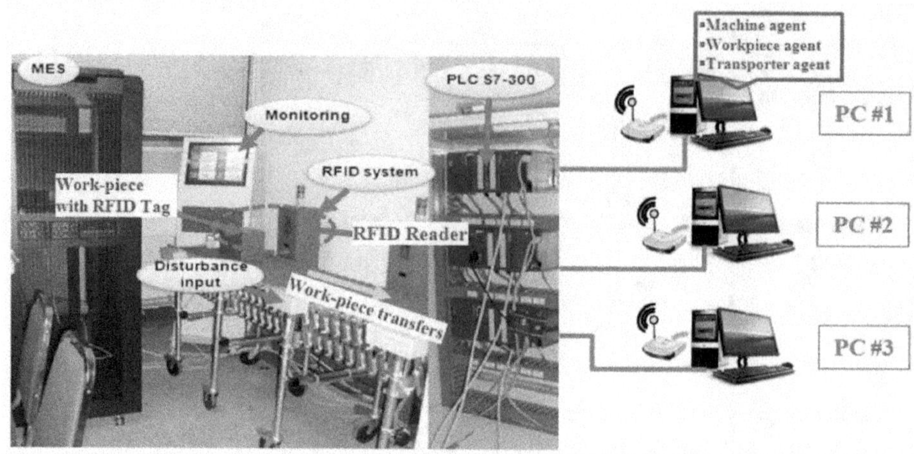

further agents in the system to mitigate the problem. The testbed for this CPS was activated using PLC scripts. Its performance was evaluated using a machining process scenario, which included turning, drilling, and milling operations performed on three machines with processing times of 5, 3, and 8 minutes, respectively. Figure 5 illustrates the system's performance under normal conditions. The control software for this CPS was developed using the I/O diagram and operational sequence of the PLCs, ensuring that it is tailored to the system's specific requirements and its machining process scenario.

Figure 6 illustrates the system's operating mechanism and PLC program when a disturbance condition occurs. The testbed for this system includes three "disturbance input" devices specifically designed to introduce different types of disturbances to the system. Each device can generate five different types of disturbances, resulting in 15 scenarios that can be evaluated. The system's PLC program is programmed to respond to these disturbances by initiating pre-determined actions. These actions may include shutting down certain operations, re-routing workpieces, or activating alarms to signal the occurrence of a disturbance. The PLC program also can analyze the disturbance and determine the appropriate course of action, considering factors such as the type of disturbance, its severity, and its potential impact on the system's overall performance (Natrayan L & Kumar MS. 2020).

Using these disturbance input devices allows for a comprehensive evaluation of the system's performance in the face of a wide range of potential disturbances. This is important as it allows for identification of potential weaknesses and vulnerabilities in the system, and the development of strategies to mitigate them. Overall, this testbed provides a robust framework for evaluating the system's ability to handle disturbances and ensures that it can operate effectively and efficiently under all conditions.

Figure 4. Smart cell architecture

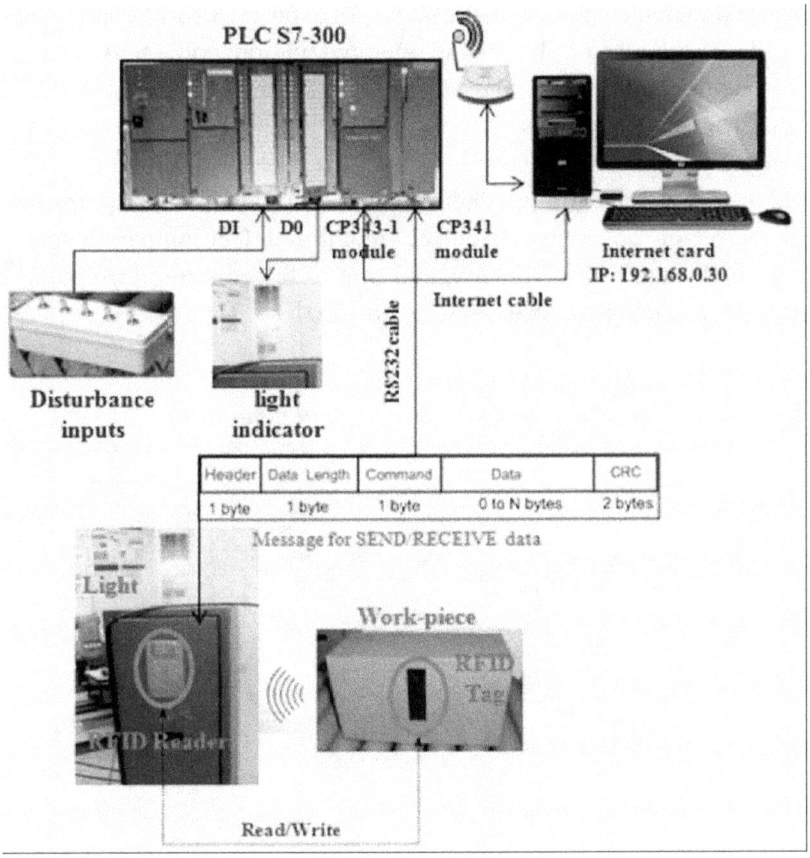

4. RESULT AND DISCUSSION

Monitoring the state of the machining process is crucial for ensuring the quality of the final product throughout the manufacturing process. This study applied the proposed CPS concept to a specific application in the manufacturing industry using a machining condition on a milling machine (Veeman et al., 2021). External disruptions, such as changes in orders or issues with suppliers, as well as internal disruptions related to the physical system, process, and service, such as tool wear and errors in procedures, were identified by their symptoms and used to construct rules or reasoning processes for the CPS to predict and identify disturbances. A CPS database, including information on resources, disturbances, and recovery strategies for each disturbance, was created based on the categorization of disturbances. Each machine in the system was meticulous by its corresponding machine agent. Data on the shop floor was gathered using KEP Server ExOPC to implement the perception function. Pattern recognition methods, such as artificial neural networks, were employed to detect certain disruptions. An experiment was conducted using a contour milling with a end mill tool and operated without coolant as dry machining on a high-speediness vertical center as shown in figure 7, using AISI 1043 carbon steel as the workpiece material (Natrayan L et al., 2021). Cutting tool wear was found to directly impact surface roughness during the machining process, making it a crucial aspect in determining whether the final product met requirements. As a result, the degree of tool wear was closely monitored. Cutting parameters, such as machining depth, machining velocity, and rate of fefeed rateimized using a static technique before machining (Natrayan L et al., 2018).

Figure 5. Performance of the system under normal condition

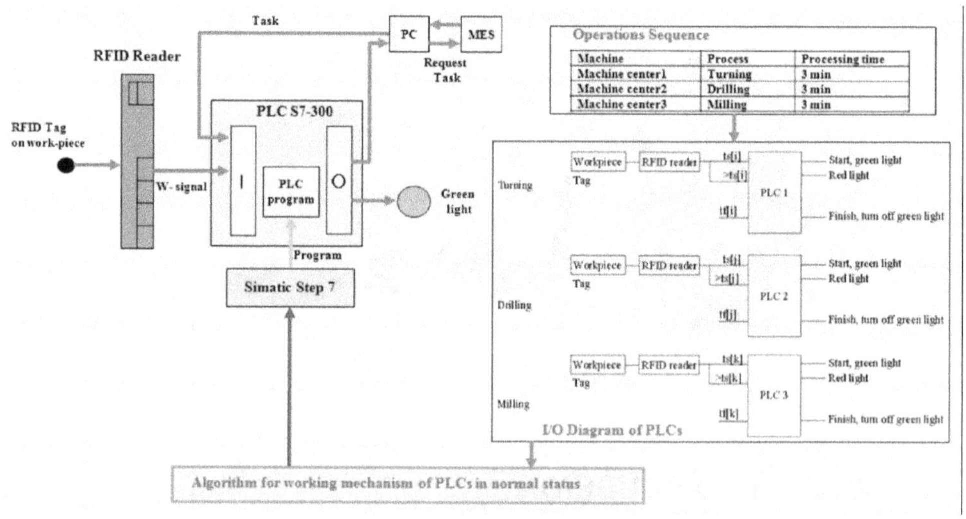

The approach used in this study for evaluating the behaviors of cyber-physical systems (CPS) in a machining process has the limitation of not considering the kinetics of the cutting process (Pragadish et al., 2022). To address this limitation, this study recommends using dynamic optimization, which allows for creating new machining conditions while considering the present condition of the cutting process, such as wear of the tool (Balaji et al., 2022). This method allows new cutting conditions to be produced and communicated to the cutting tool during progression, providing greater control and flexibility in maintaining product quality. Overall, this study suggests implementing CPS control throughout the process to improve the superiority of the final product Yogeshwaran et al., (2015).

Figure 6. Performance of the system under disturbed condition

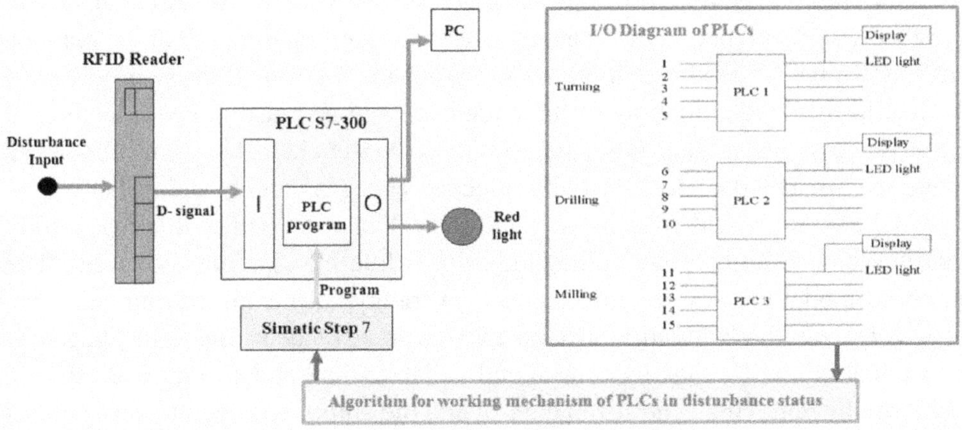

Figure 7. Machining tool controlled by the CPS enabled system

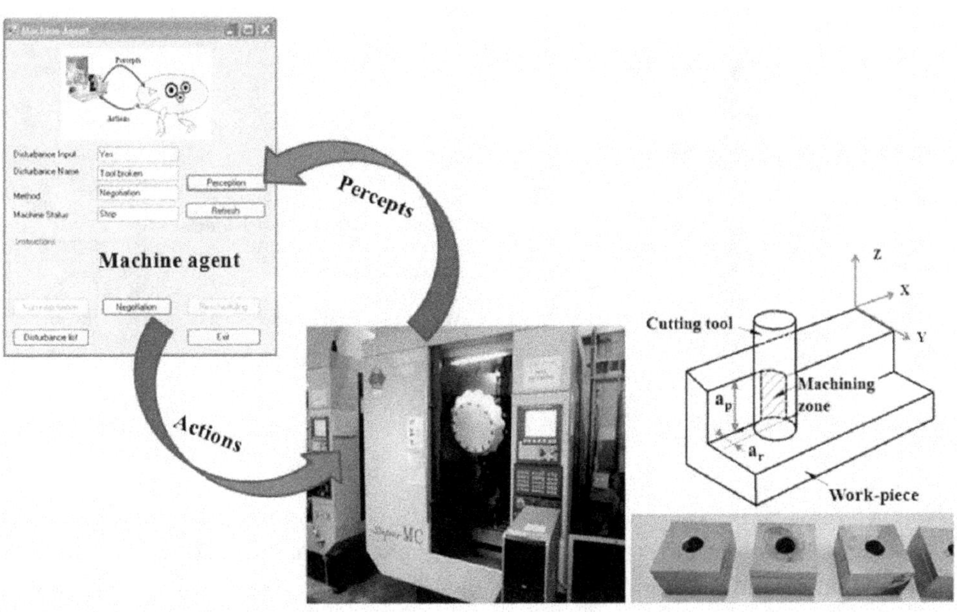

The ability of a production system to quickly respond and recover autonomously to disruptions is essential for keeping the manufacturing process running smoothly and preventing shut downs (Ramaswamy et al., 2022). However, traditional methods such as the lean manufacturing and computer aided manufacturing lack thcannot to the complex and active industrial atmosphere. In case of a disruption, these systems must be halted to reprogram and replan the automated operations using pre-installed programs, limiting system flexibility and increasing downtime. To address this issue, a testbed was developed in the workroom to prove the functionality of the planned method (Kumar MS et al., 2018). By using swarm intelligence, the system enhanced the intelligence and autonomy of the components in an existing FMS, leading to the advance of cps at the cyber equal (Santhosh MS et al., 2018) . This allows the suggested Smart-CPMS to adjust to production changes quickly and instantly.

A CPS was created to achieve steady excellence by self-regulating cutting conditions on a high-speed milling machine while considering tool wear. The proposed model has several advantages over traditional manufacturing concepts, such as increased adaptability through CPSs with advanced capabilities like self-choice making, self-adjusting control and collaboration, and an increased system resilience by removing centralized control and the ability to integrate autonomous behaviours through flexible decision making.

Figure 8. Model for optimal feed

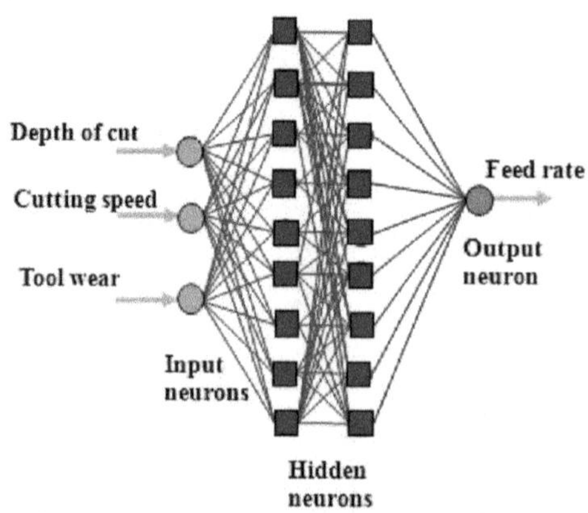

Network configuration and training parameters

Network configuration	3-9-9-1
Number of hidden layer	2
Number of neurons in each hidden layer	9
Activation function	Sigmoid
Learning rate	0.55
Momentum coefficient	0.7
Number of epochs	2000
RMSE (Root Mean Square Error)	0.292

To guarantee the superiority of the final part is maintained, a system was implemented to monitor and manage tool wear. This system utilizes predictive algorithms to anticipate when the cutting tool may reach its wear limit, at which point it is replaced with a new one. This proactive approach ensures that the tool is replaced before it begins to negatively impact the quality of the final product. Another artificial neural network calculates new input conditions when the tool wear is still within the allowable limit. These limitations are used to optimize the cutting process and improve production efficiency. The system architecture for determining the optimal rate of feed is illustrated in Figure 8. The network's input layer includes various settings such as the cut depth, machining velocity, and rate of wear of the tool. The network processes these inputs, which then outputs the recommended feed rate. By constantly monitoring tool wear and adjusting the cutting parameters as needed, this system helps to ensure that the final machined part meets the desired quality standards while also improving the efficiency of the production process. The Root mean square error from the neural network is just 0.292 percentage which shows the effectiveness and accuracy of the approach.

CONCLUSION

This study introduces a novel methodology for emerging a smart manufacturing system that incorporates cyber-physical systems (CPS) to adapt to production changes autonomously and intelligently. The CPS is equipped with cognitive agent technology, enabling it to mimic human cognitive processes and independently perceive, communicate and control. Artificial intelligence algorithms such as neural networks and ant colony optimization empower the CPS to make decisions, learn, and improve its knowledge. The CPS performance was demonstrated through a testbed. Its application was exemplified in a practical scenario of self-regulating cutting conditions on a high-swiftness milling machine while considering the tool quality.

REFERENCES

Alguliyev, R., Imamverdiyev, Y., & Sukhostat, L. (2021). Hybrid DeepGCL model for cyber-attacks detection on cyber-physical systems. *Neural Computing & Applications*, *33*(16), 10211–10226. doi:10.100700521-021-05785-2

AlZubi, A. A., Al-Maitah, M., & Alarifi, A. (2021). Cyber-attack detection in healthcare using cyber-physical system and machine learning techniques. *Soft Computing*, *25*(18), 12319–12332. doi:10.100700500-021-05926-8

Balaji, N., Natrayan, L., Kaliappan, S., Patil, P. P., & Sivakumar, N. S. (2022). Annealed peanut shell biochar as potential reinforcement for aloe vera fiber-epoxy biocomposite: Mechanical, thermal conductivity, and dielectric properties. *Biomass Conversion and Biorefinery*, 1–9. doi:10.100713399-022-02650-7

Bordel, B., Alcarria, R., de Rivera, D. S., & Robles, T. (2018). Process execution in Cyber-Physical Systems using cloud and Cyber-Physical Internet services. *The Journal of Supercomputing*, *74*(8), 4127–4169. doi:10.100711227-018-2416-4

Busygin, A. G., Konoplev, A. S., & Zegzhda, D. P. (2018). Providing Stable Operation of Self-Organizing Cyber-Physical System via Adaptive Topology Management Methods Using Blockchain-Like Directed Acyclic Graph. *Automatic Control and Computer Sciences*, *52*(8), 1080–1083. doi:10.3103/S0146411618080059

He, F., Zhuang, J., & Rao, N. S. V. (2020). Discrete game-theoretic analysis of defense in correlated cyber-physical systems. *Annals of Operations Research*, *294*(1–2), 741–767. doi:10.100710479-019-03381-1

Kumar, M. S., Natrayan, L., Hemanth, R. D., Annamalai, K., & Karthick, E. (2018, August). Experimental investigations on mechanical and microstructural properties of Al2O3/SiC reinforced hybrid metal matrix composite. [). IOP Publishing.]. *IOP Conference Series. Materials Science and Engineering*, *402*(1), 012123. doi:10.1088/1757-899X/402/1/012123

Mueller, E., Chen, X. L., & Riedel, R. (2017). Challenges and Requirements for the Application of Industry 4.0: A Special Insight with the Usage of Cyber-Physical System. [English Edition]. *Chinese Journal of Mechanical Engineering*, *30*(5), 1050–1057. doi:10.100710033-017-0164-7

Natrayan, L., & Kumar, M. S. (2020). Optimization of wear behaviour on AA6061/Al2O3/SiC metal matrix composite using squeeze casting technique–Statistical analysis. *Materials Today: Proceedings*, *27*, 306–310. doi:10.1016/j.matpr.2019.11.038

Natrayan, L., Merneedi, A., Bharathiraja, G., Kaliappan, S., Veeman, D., & Murugan, P. (2021). Processing and characterization of carbon nanofibre composites for automotive applications. *Journal of Nanomaterials*, *2021*, 1–7. doi:10.1155/2021/7323885

Natrayan, L., Sivaprakash, V., & Santhosh, M. S. (2018). Mechanical, microstructure and wear behavior of the material AA6061 reinforced SiC with different leaf ashes using advanced stir casting method. *International Journal of Engineering and Advanced Technology*, *8*, 366–371.

Park, S. O., Xiao, B., Leung, V., & Jeong, Y. S. (2013). Advanced technologies and applications for Highly-Reliable Cyber Physical System (HRCPS). *The Journal of Supercomputing*, *66*(1), 1–3. doi:10.100711227-013-1027-3

Parvin, S., Hussain, F. K., Hussain, O. K., Thein, T., & Park, J. S. (2013). Multi-cyber framework for availability enhancement of cyber physical systems. *Computing*, *95*(10–11), 927–948. doi:10.100700607-012-0227-7

Pragadish, N., Kaliappan, S., Subramanian, M., Natrayan, L., Satish Prakash, K., Subbiah, R., & Kumar, T. C. A. (2022). Optimization of cardanol oil dielectric-activated EDM process parameters in machining of silicon steel. *Biomass Conversion and Biorefinery*, 1–10. doi:10.100713399-021-02268-1

Priyadarshini, I., Kumar, R., Tuan, L. M., Son, L. H., Long, H. V., Sharma, R., & Rai, S. (2021). A new enhanced cyber security framework for medical cyber physical systems. *Software-Intensive Cyber-Physical Systems*, *35*(3–4), 159–183. doi:10.100700450-021-00427-3

Ramaswamy, R., Gurupranes, S. V., Kaliappan, S., Natrayan, L., & Patil, P. P. (2022). Characterizing prickly pear short fiber and red onion peel biocarbon nanosheets toughened epoxy composites. *Polymer Composites*, *43*(8), 4899–4908. doi:10.1002/pc.26735

Santhosh, M. S., Sasikumar, R., Natrayan, L., Kumar, M. S., Elango, V., & Vanmathi, M. (2018). Investigation of mechanical and electrical properties of Kevlar/E-glass and Basalt/E-glass reinforced hybrid composites. *International Journal of Mechanical and Production Engineering Research and Development*, *8*(3), 591–598. doi:10.24247/ijmperdjun201863

Veeman, D., Shree, M. V., Sureshkumar, P., Jagadeesha, T., Natrayan, L., Ravichandran, M., & Paramasivam, P. (2021). Sustainable development of carbon nanocomposites: Synthesis and classification for environmental remediation. *Journal of Nanomaterials*, *2021*, 1–21. doi:10.1155/2021/5840645

Verma, R. (2022). Smart City Healthcare Cyber Physical System: Characteristics, Technologies and Challenges. *Wireless Personal Communications*, *122*(2), 1413–1433. doi:10.100711277-021-08955-6 PMID:34462622

Wan, B., Xu, C., Mahapatra, R. P., & Selvaraj, P. (2021). Understanding the Cyber-Physical System in International Stadiums for Security in the Network from Cyber-Attacks and Adversaries using AI. *Wireless Personal Communications*, *127*(2), 1207–1224. doi:10.100711277-021-08573-2

Wang, Y., Lin, Z., Liang, X., Xu, W., Yang, Q., & Yan, G. (2016). On modeling of electrical cyber-physical systems considering cyber security. *Frontiers of Information Technology and Electronic Engineering 17*(5), 465–478. doi:10.1631/FITEE.1500446

Wu, G., & Li, Z. S. (2021). Cyber—Physical Power System (CPPS): A review on measures and optimization methods of system resilience. *Frontiers of Engineering Management*, *8*(4), 503–518. doi:10.100742524-021-0163-3

Yogeshwaran, S., Prabhu, R., Natrayan, L., & Murugan, R. (2015). Mechanical properties of leaf ashes reinforced aluminum alloy metal matrix composites. *International Journal of Applied Engineering Research: IJAER*, *10*(13), 11048–11052.

Chapter 20
Securing Decentralized Finance Through Blockchain Technologies

C. V. Suresh Babu
https://orcid.org/0000-0002-8474-2882
Hindustan Institute of Technolgy and Science, India

Abhinaba Pal
https://orcid.org/0009-0004-1727-7628
Hindustan Institute of Technology and Science, India

ABSTRACT

Decentralized finance (DeFi) has become one of the most successful applications of blockchain and is widely regarded as the future of finance. DeFi's enormous success is based on real-world data that is not directly available on the blockchain. Furthermore, since a blockchain is deterministic, it cannot directly obtain in-deterministic data from the outside world (off-chain). As a result, oracles have emerged as a viable solution for supplying off-chain data to blockchain applications. This study conducts a thorough examination of DeFi oracles, starting with introducing DeFi oracle application scenarios, then moves onto characterizing them into several types based on their design factors. Finally, it analyzes and address the security aspects of oracles in DeFi and explores the risks associated with oracles, such as data inaccuracy, centralization, manipulation, and oracle failure, and proposes solutions to mitigate these risks by focusing on the Chainlink network.

DOI: 10.4018/978-1-6684-9267-3.ch020

INTRODUCTION

An oracle is a third-party service or technology that helps integrate off-chain data with on-chain smart contracts in DeFi. Oracles are necessary to enable DeFi protocols to interact with the outside world because the blockchain is only capable of processing and verifying on-chain data and cannot access off-chain data sources. For example, Price Feeds: a limit order in a cryptocurrency trade, in which a person writes a smart contract to automatically sell an Ethereum (a cryptocurrency) token when the price reaches a certain target level. This contract requires access to real-time Ethereum prices in order to function. If the data obtained is inaccurate or not obtained in a timely fashion, it could lead to huge losses for a trader. Now ideally, this could have been easily solved by querying the REST API of some financial exchange. But the problem is in maintaining determinism, which is important so that different nodes that execute the contract can come to a consensus. In this case, allowing the smart contract to query the price off of the internet would result in disparities in values across nodes, making consensus impossible. The solution: an Oracle is made to provide accurate and reliable price data from external sources such as cryptocurrency exchanges as an external entity and post the data on-chain. Once the data is on the blockchain, smart contracts can access and use it. Here we suggest a potential trust architecture, a few promising building blocks for trustworthy DeFi oracles, and a set of metrics for designing security-centric oracles which aim to address its associated risks by focusing on security and reliability as their primary design goals.

There are several different types of oracles used in DeFi, including centralized oracles, decentralized oracles, and hybrid oracles that combine elements of both, all of which face the following common issues:

a) Centralized oracles rely on a single source of data and can be subject to a single point of failure or manipulation and hence is subject to a substantial cybersecurity threat landscape.
b) Whereas, decentralized oracles use multiple sources of data and consensus mechanisms to provide more secure and reliable data feeds. This dependence on a broad spectrum of data providers again introduces the possibility of an adversarial attack, in which an adversary bribes existing nodes or establishes nodes to produce a corrupt report. Collusion among Oracle nodes could also be problematic.
c) Oracle Failure: This can happen for a number of reasons, including network congestion, hardware failure, or human error. Such failures can lead to incorrect data being sent to smart contracts, resulting in significant financial losses.

d) Lack of Standards: The lack of standards in the DeFi ecosystem makes it difficult for developers to implement secure and reliable oracle systems. The absence of clear guidelines on how oracles should be integrated with smart contracts and data sources can lead to vulnerabilities.

RELATED WORKS

Blockchain oracles are middleware entities that connect smart contracts to external data sources, such as market prices, weather information, sports outcomes, and so on. Oracles are essential for enabling smart contracts to interact with the real world and execute complex logic based on various inputs. However, oracles also introduce challenges and risks for the security and reliability of smart contracts, especially in decentralized finance (DeFi) applications that involve large amounts of value.

One of the main challenges of blockchain oracles is how to ensure the trustworthiness and quality of the data they provide. Depending on the source and type of data, oracles may be subject to manipulation, corruption, or inconsistency. For example, a malicious oracle may feed false data to a smart contract to trigger an undesired outcome, such as a liquidation or a payout (Caldarelli & Ellul, 2021). Alternatively, an oracle may be compromised by an attacker who gains access to its data feed or its private key. Moreover, different oracles may provide different data for the same query due to latency, network congestion, or aggregation methods.

To address these challenges, various solutions have been proposed and implemented in the blockchain space. Some of these solutions include:

- Using multiple oracles to provide redundancy and consensus on the data. For example, Chainlink is a decentralized oracle network that allows smart contracts to access data from multiple independent node operators who are incentivized to provide accurate and timely data.
- Using cryptographic techniques to verify the authenticity and integrity of the data. For example, Town Crier is a system that leverages trusted hardware (Intel SGX) to provide secure and confidential data delivery for smart contracts (Priem, 2020).
- Using game-theoretic mechanisms to align the incentives of the oracle providers and the users. For example, Augur is a prediction market platform that uses a reputation system and a dispute resolution process to ensure honest reporting of real-world events. One study provides a systematic and up-to-date overview of the state-of-the-art in blockchain oracles for DeFi and discusses various use cases of DeFi, such as lending, trading, and insurance, and their potential impact on the traditional financial system (Bartoletti et al., 2020;

Beniiche, 2020; Foxley, 2020; Junejo et al., 2021; Liu et al., 2020; Zetzsche et al., 2020). Another paper proposes the use of multi-sourced oracles, where data is fetched from multiple sources, and a consensus mechanism is used to determine the final value. This approach can help to mitigate the risk of data manipulation and collusion attacks by minimizing the reliance on a single source (Al Breiki et al., 2020; Filippi & Mannan, 2020; Kumar et al., 2020; Priem, 2020; Yu & Zichichi, 2021). It also suggests the use of reputation-based mechanisms to ensure the reliability of oracles. The authors propose a reputation-based framework where oracle providers are ranked based on their past performance and reliability. This approach can incentivize oracle providers to maintain high-quality data feeds and deter them from engaging in malicious activities.

Alternatively, one study also highlights the several challenges that need to be addressed to enable mainstream adoption of DeFi, such as scalability, interoperability, and user experience (Zhao, 2022) providing insights into potential solutions to these challenges, such as layer 2 scaling solutions, cross-chain interoperability protocols, and user-friendly interfaces. The technical contribution of this paper lies in its focus on the challenges and opportunities of building DeFi applications on Ethereum, such as scalability issues due to its limited capacity to process transactions. Ethereum has a limited gas limit for each block, which puts a cap on the number of transactions that can be processed in a single block. This has led to high gas fees and slow transaction times, making it difficult for DeFi applications to scale and handle high volumes of users.

Also, the gas price is the amount of ether that users must pay to execute a transaction on the Ethereum network. Gas prices are determined by the market (Zetzsche et al., 2020), and during times of high demand, prices can become very expensive, making it prohibitively expensive for some users to use the network. The paper by Siddiqui and Kumar (Bartoletti et al., 2020) is unique in its approach as it not only provides a detailed analysis of the existing literature on DeFi but also offers insights into the future of this field. They suggest that the use of blockchain-based identity systems, such as those based on self-sovereign identity, can help increase access to financial services for under-served communities. The authors note that the governance models of DeFi systems play a crucial role in their success. They suggest that the development of decentralized governance models, such as those based on prediction markets and quadratic voting, can help ensure the long-term sustainability of these systems.

Identified risks/limitations of modern Oracle services that have been reviewed:

(a) Relying on erroneous centralized oracle models for financial market information (e.g stock price data) may lead to misreporting exploits (Foxley, 2020) where costs are different from a commodity's actual market value. Lending protocols face risks of accumulating harmful debt and reaching a state of insolvency if they fail to promptly and effectively liquidate positions that have insufficient collateral. Additionally, users' funds can be unfairly converted into cash due to manipulated price fluctuations, leading to financial losses.

(b) Schelling Point oracles (Junejo et al., 2021) rely on the wisdom of the crowd by collecting and aggregating information from participants. However, they can be susceptible to manipulation if a large number of participants collude or provide inaccurate data intentionally. This is particularly catastrophic for synthetic assets where liquidity providers are at risk of losing their assets if positions relying on manipulated prices are used to settle payments or generate problematic loans.

(c) The decentralized oracle network- Chainlink (Breidenbach, 2021) is currently still relatively limited in terms of the number of nodes it has (~332 nodes), especially when compared to a blockchain platform like Ethereum, which has thousands of nodes (Foxley, 2020). It implies that further decentralization of Chainlink's network could enhance its redundancy and security. Also, The native token of the Chainlink network, (LINK), has not fared as well as tokens from other high-performing projects, like as Ethereum (ETH), despite the network's evident success.

Identified theoretical solutions to the "integrity problem" of Oracles

(a) In the conventional sense, reputation gets built over time based on an entity's behavior. However, the EtherScore protocol seeks to emulate this idea by consolidating on-chain activities linked to addresses. Its primary objective is to provide decentralized applications (DApps) with a means to recognize and incentivize users who possess favorable reputations. This is accomplished through the implementation of non-fungible token (NFT) badges, which empower applications to establish loyalty programs and grant rewards to contributors without depending on external platforms for off-chain analytics. By incorporating user analytics validations directly into smart contracts (Al Breiki et al., 2020), a comprehensive on-chain fidelity program can be developed.

(b) A Human Protocol is a technological framework that establishes a connection between human and machine workers while validating their contributions. Its overarching goal is to cultivate a vast labor pool, enabling humans and

> machines to reach higher levels of potential and comprehension. The protocol employs tokenization to represent work and establishes a linkage between projects and participants worldwide. It provides blockchain infrastructure for recording, verifying, and rewarding contributions. This technology has found applications in data labeling markets and human verification solutions, such as Proof of HUMANity, which safeguards Dapps against bot interference. The protocol is meticulously designed to ensure seamless integration of jobs with the appropriate components, including APIs, Validators, Job Exchanges, oracles, and blockchains, thereby optimizing their execution. Requesters have the option to choose from pre-defined templates or create custom job types, generating a job specification that is intelligently fragmented into numerous tasks. These tasks are then distributed among supportive Job Exchanges for efficient processing.

There has also been a rapid push in the suggested development of decentralized decision-making mechanisms, such as DAOs (decentralized autonomous organizations), which can help ensure that the interests of all stakeholders are represented in the governance process (Tolmach et al., 2021). These are digital entities that operate through a decentralized network of computers, using smart contracts to execute decisions without the need for human intervention. A DAO consists of four major components: (a) the smart contract code, which defines the rules and governance structure of the organization; (b) the token, which represents ownership in the organization and allows members to vote on proposals; (c) the voting mechanism, which allows members to participate in decision-making and determine the direction of the organization; and (d) the treasury, which holds the organization's funds and is managed by the members through the voting process. DAOs are designed to operate in a trustless and transparent manner, allowing for decentralized decision-making and ownership of digital assets. One of the main drawbacks of current DAO systems is their susceptibility to hacking and security breaches, such as the security incident in 2016 involving "The DAO" - a decentralized investment fund built on top of the Ethereum blockchain. The DAO was designed to allow members to propose and vote on investment opportunities using Ethereum's smart contract technology. However, a flaw in the smart contract code allowed a hacker to exploit a vulnerability and siphon off approximately one-third of the fund's total assets, totaling around $50 million worth of Ether at the time. The incident highlighted the potential security risks associated with DAOs and led to a contentious hard fork of the Ethereum blockchain to recover the stolen funds, ultimately resulting in the creation of Ethereum (ETH) and Ethereum Classic (ETC) as separate cryptocurrencies. Another issue is the lack of legal clarity and regulatory oversight, which can create uncertainty for DAO participants and limit their ability to operate within existing legal frameworks. Additionally, the governance structures of DAOs can be complex and difficult to manage, requiring active participation and engagement from members to ensure that decisions are made fairly and transparently.

Other significant emerging trends in the DeFi space include the growth of decentralized exchanges (DEXs), the use of flash loans (Wang et al., 2021), and the rise of algorithmic stablecoins (Bartoletti et al., 2020; Caldarelli & Ellul, 2021; Tolmach et al., 2021; Wang et al., 2021; Zhao, 2022), which in relation to oracles to play out in the following ways:

(a) In the case of DEXs, oracles are used to provide real-time price feeds for assets traded on the exchange. These price feeds are critical for ensuring that traders receive accurate and up-to-date pricing information when executing trades. In a DEX, smart contracts handle the trading of assets, but they are not able to access real-world market data on their own. Oracles are used to provide this data in a trustless and decentralized manner, ensuring that the prices reflected on the DEX are accurate and not subject to manipulation.
(b) Stablecoins, on the other hand, rely on oracles to maintain their peg to a specific asset or basket of assets. For example, a stablecoin pegged to the US dollar would require an oracle to provide real-time price feeds of the USD/ETH exchange rate, which would be used to determine the value of the stablecoin. Without an oracle, stablecoins would not be able to maintain their peg, leading to volatility and instability.

The paper "Enhancing Trust in Trust Services: Towards an Intelligent Human-input-based Blockchain Oracle (IHiBO)" proposes a novel approach to address the limitations of traditional oracle systems by incorporating intelligent human inputs into the blockchain-based oracle system (Yu & Zichichi, 2021). The IHiBO system utilizes a combination of AI-based algorithms and human inputs to provide accurate and reliable data for smart contracts. The system includes a reputation-based mechanism that ensures the honesty and accuracy of the human oracles, as well as a machine learning component that optimizes the selection of human oracles based on their expertise and reliability. The system also incorporates a feedback mechanism that allows for continuous improvement and refinement of the human oracle selection process.

One potential advantage of the IHiBO system is its ability to provide more accurate and reliable data inputs for smart contracts, particularly in situations where automated data sources are insufficient or unreliable. The use of human oracles may also provide a more nuanced and context-specific interpretation of data, allowing for more sophisticated smart contract executions. However, the IHiBO system also has several potential drawbacks. The reliance on human inputs introduces a degree of subjectivity and potential bias into the system, which can be difficult to mitigate. Additionally, the reputation-based mechanism may be vulnerable to manipulation or collusion among participants. The use of machine learning algorithms may also introduce potential vulnerabilities, such as algorithmic bias and exploitation.

The Chainlink 2.0 paper (Breidenbach, 2021) describes the new features and improvements of Chainlink 2.0, including hybrid smart contracts, VRF (Verifiable Random Function), Threshold Signatures, and Off-Chain Reporting (OCR). Chainlink in essence is a fully decentralized oracle network whose architecture allows for a high degree of customization and flexibility. The proposed upgrades aim to enhance the scalability, security, and reliability of the network, enabling it to support more complex and data-rich smart contracts.

Figure 1. Social reduction via conflict resolution

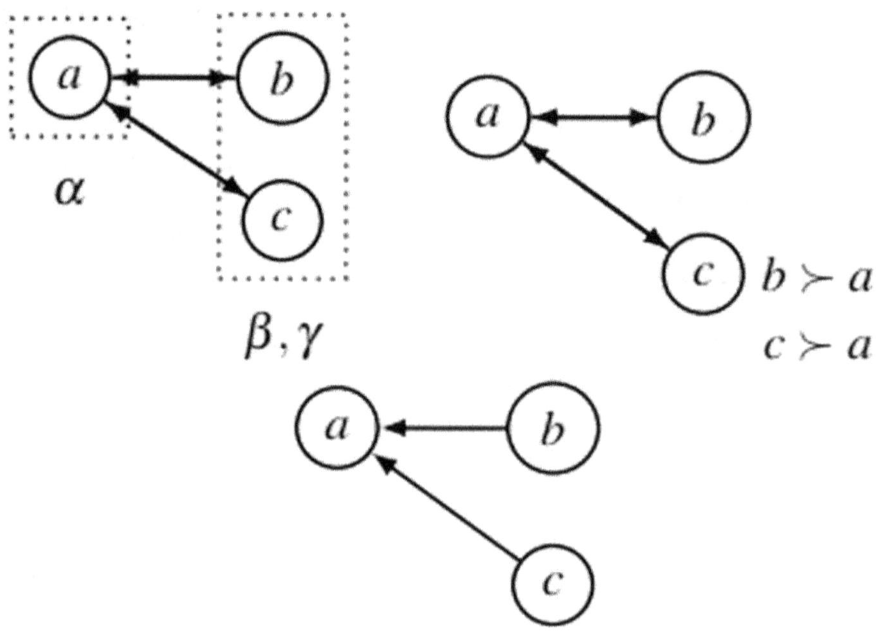

In terms of quantitative analysis, the paper provides some technical details and specifications of the new features, such as the VRF specification, OCR benchmarking data, and the threshold signatures mathematical formula which can be referred to as more of a technical roadmap and specification documention.

Figure 2. The "TEF schematic" shows how transactions are processed in a system. In this example, the transactions go through a "mempool" (a temporary storage area for transactions) in the "MAINCHAIN" (a blockchain network) and then go to the "MS" (a node or server) before being processed by the "DON" (another node or server)

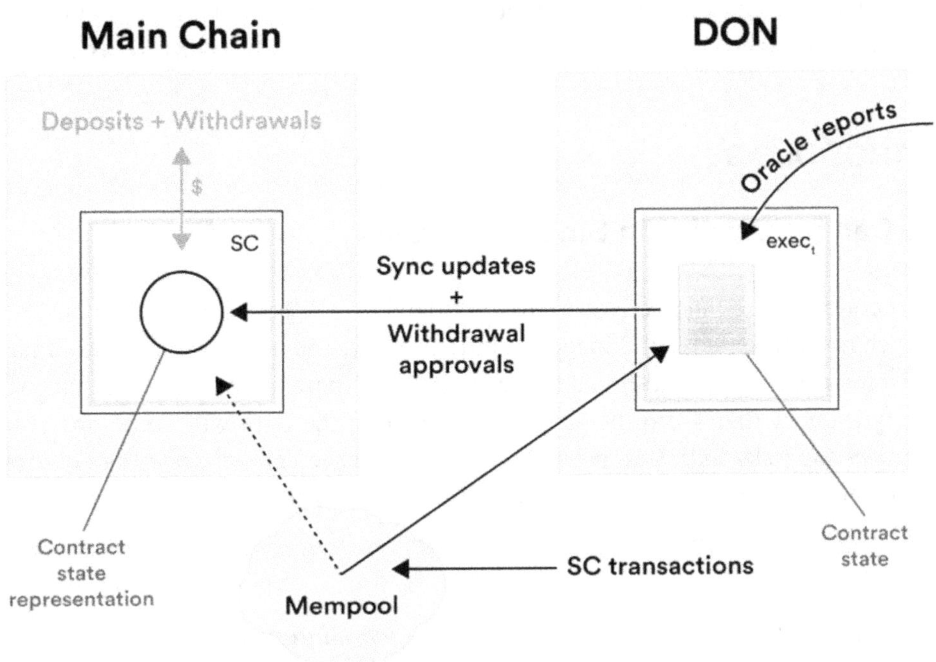

The concept of verified price oracles involves using the burn() method in the smart contract (Kumar et al., 2020), where the provider is free to remove funds from the token reserves. The tokens they lend to the liquidity pool are returned to them along with the interest they earned from the transaction fees. A liquidity provider's share of the token reserve determines how many liquidity tokens are issued for that provider. This is computed using a number of formulas. The formula below is used by the Uniswap protocol to calculate the total number of tokens that will be issued:

In this case, the term s_{minted} refers to the quantity of liquidity tokens that were created in relation to the amount of $x_{deposited}$ tokens that were placed into the pool that contained the tokens $x_{starting}$ and $s_{starting}$, which represented contributions to that liquidity pool. The aforementioned calculation is invalid if liquidity is placed into buffers initially because ($x_{starting} = 0$). The quantity of liquidity tokens issued in this instance is determined by:

$$s_{minted} = \sqrt{x_{deposited} \cdot y_{deposited}}$$

Pair reserves that the depositor loans to the contract are indicated by $x_{deposited}$ and $y_{deposited}$, respectively. The Uniswap protocol also levies a 0.05% protocol fee in addition to the net 0.3% transaction fee paid by merchants. It is based on the Automated Market Maker (AMM) model to facilitate trading without relying on a centralized order book, specifically the, Constant Product Market Maker (CPMM).

METHODOLOGY

The Call for Oracles in Smart Contracts

Smart contracts are intended to run independently on different network nodes.

Each node repeatedly and independently executes these activities as they query their information from additional sources. Because there is no assurance that each node will query data from the same source or that the data will be comparable, this raises the risks indicated in the preceding sentence. Threat actors can quickly take advantage of this weakness to introduce erroneous data into the network and destabilize the blockchain.

Figure 3. Shows how a DApp and an Oracle network interact

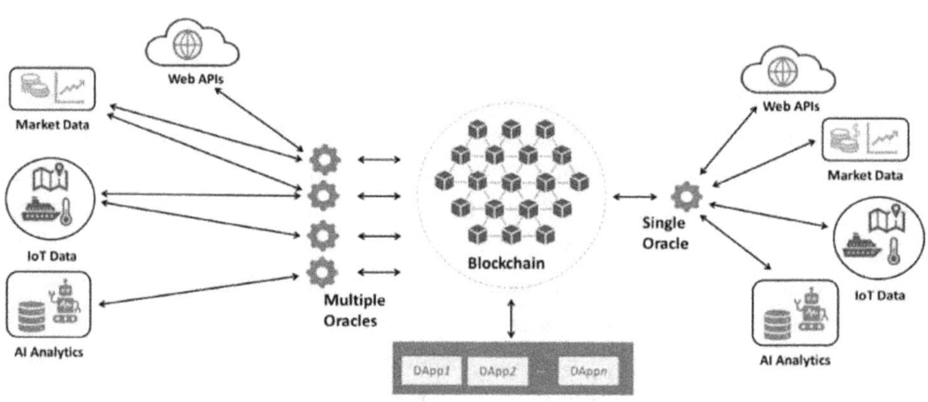

The Archetypes of Oracles

Software oracles: possibly the most popular kind of oracle, they link to the Internet and draw information from servers, databases, APIs, and other online resources. Exchange rates, digital asset values, real-time flight information, etc. are some examples of the information sent by software oracles.

Hardware oracles: are tangible structures that allow smart contracts to interact with the physical environment. Some smart contracts may require data from IoT devices, electronic sensors, automata, and other data collection equipment. A hardware oracle might consist of a device with sensors that "translates" real-world events into a numeric format that smart contracts can use to make choices. This form of oracle is especially useful in the logistics chain and automation industries.

Inbound and outbound oracles: Inbound oracles collect data from outside entities, whilst outbound oracles transmit data off-chain to communicate with the actual world. A smart lock is an example of inbound and outbound oracles collaborating. This can be understood in the context of weather-based insurance contracts. An inbound oracle could track the weather conditions in a particular area and provide real-time data on temperature, precipitation, and other relevant factors. This data would be used to determine whether a triggering event, such as a hurricane or tornado, has occurred, and whether a payout is warranted under the terms of the insurance contract.

Whereas, an outbound oracle could then initiate the payment process, transferring funds from the insurer's account to the policyholder's account automatically, without the need for manual intervention. This would enable faster and more efficient claims processing, reducing the administrative burden on both the insurer and the policyholder.

Centralized & Decentralized Oracles: Centralized oracles refer to oracle systems that rely on a centralized entity or a small group of entities to provide the necessary data inputs to the smart contract. In other words, a centralized oracle relies on a single source of data, which can introduce a single point of failure and potential vulnerabilities. Centralized oracles can be subject to manipulation, censorship, and other types of interference.

Decentralized oracles, on the other hand, use a network of independent data sources to provide the necessary inputs to the smart contract. Decentralized oracles are designed to be trustless, meaning that they do not rely on any single entity to provide accurate data. Instead, they use a consensus mechanism to aggregate data from multiple sources and verify its accuracy before it is used to trigger smart contract executions. Decentralized oracles are considered more secure and resilient than centralized oracles, as they are less vulnerable to manipulation, censorship, and other forms of interference.

Human Oracles: Human oracles are individuals who provide data inputs to smart contracts based on their subjective interpretation of real-world events. Human oracles are used in situations where it is difficult or impossible to obtain reliable data through automated means, such as in the case of sports betting, political elections, and other types of subjective events.

In a human oracle system, individuals are incentivized to provide accurate data inputs by offering rewards for correct predictions or assessments. The system may also use a reputation-based mechanism to encourage honest reporting and discourage fraudulent or malicious behavior.

However, human oracles have their own set of limitations and challenges, such as the potential for bias, manipulation, and collusion among participants. The use of human oracles also requires a significant level of trust in the individuals providing the data inputs, which can be difficult to establish in a trustless and decentralized system. As a result, human oracles are not commonly used in DeFi applications, where automated, trustless oracles are preferred.

Security Concerns/Implications to contemplate while Architecting components of Defi

Front Running by Messenger: The data carrier has the ability to be aware of the outcome prior to transmitting it on the blockchain (Al Breiki et al., 2020). However, the smart contract used by the user must ensure that having knowledge of the accurate response prior to making it public does not compromise the impartiality of the system.

Mirroring: In this attack, an attacker manipulates an oracle to report the value of an asset as a false value that mirrors the value of another asset. For example, suppose a DeFi system uses an oracle to get the price of Ether (ETH) in USD. The attacker can manipulate the oracle to report the value of another asset, say Bitcoin (BTC), as the value of ETH. If the DeFi system uses this manipulated value to calculate the value of a token, the attacker can make a profit by buying the token at a lower price and then selling it at a higher price when the manipulation is reversed.

Data Corruption: this can happen in various ways, including manipulation of the data source, interception of data during transmission, and tampering with the oracle itself. If a smart contract in DeFi relies on such corrupt data, it can lead to incorrect or fraudulent execution of the contract, which can result in financial loss or harm to the ecosystem.

Freeloading: It refers to a situation where a user of a smart contract obtains the data from an oracle without paying the appropriate fee or cost for that data. Oracles are typically operated by third-party providers who charge fees for their services. These fees compensate the providers for the cost of collecting, processing, and transmitting the data to the smart contract or DeFi system (Al Breiki et al., 2020). However, some users may attempt to obtain the data without paying the appropriate fee, which is known as freeloading. This can have several negative impacts on the DeFi ecosystem such as, it can reduce the revenue and profit of the oracle providers, which can affect the quality and reliability of their services. It can also lead to increased costs for legitimate users who are paying the fees, as the providers may need to raise their prices to compensate for the loss of revenue.

Selective Publishing: Selective publishing can also refer to the practice of publishing information only to certain blockchain networks or specific users. This can be done to ensure that certain users have access to information that is relevant to them or to limit the amount of information that is available to users who do not need it. While this can be useful in certain situations, it can also raise concerns about transparency and fairness in DeFi. If certain users or networks have access to information that is not available to others, it can create an uneven playing field and potentially lead to market manipulation or insider trading.

Majority Attacks: Let's say that a DeFi system uses an oracle that collects price data from 10 different data sources or nodes to determine the price of a particular cryptocurrency. Each of the data sources or nodes provides a price for the cryptocurrency, and the oracle calculates the median price to determine the final price that the DeFi system uses. In a majority attack, a malicious actor would aim to control at least 6 of the 10 data sources or nodes used by the oracle. Once they have control of a majority of the data sources or nodes, they can manipulate the prices reported by those sources in their favor.

Oracle Design Requirements

The design of an oracle can vary depending on the specific application and the nature of the data being requested. There are multiple approaches that can be taken (Beniiche, 2020):

Request-Response: Whenever we have excessive data to store in a smart contract and just a tiny fraction of the data is required at any given moment, request-response oracles are utilized. In this example, an application would query the oracle, the oracle would gather pertinent data from off-chain sources and certify it to ensure its legitimacy, and the oracle would then transmit the data to the application, either directly or via an oracle contract. This approach varies from immediate-read in that a smart contract does not instantaneously take up the data; instead, the decentralized application performs its independent processing after receiving a response from the oracle.

Immediate-Read: Immediate-read Oracles have a straightforward design and offer the information required for a smart contract to reach a quick conclusion, such as identifying whether a person has a history of criminal record or getting a person's social security number. This kind of oracle transmits non-changing data and saves it in its contract storage, which is transparent to all other smart contracts.

Publish-Subscribe: Publish-subscribe oracles extend the functionality of immediate-read oracles by functioning as a transmission service. It keeps being refreshed with novel data, which can be subscribed to by parties (smart contracts, decentralized-apps, or the very distributed ledger). It is also often used to provide data that is expected to vary, such as stock index, temperature forecasts, travel times, and so on.

Proposed Solutions and Recommendations for Oracles That Are Authoritative and Reliable

It has been already established that some centralized oracle solutions are capable of delivering high-performance and secure data transmission by using advanced cryptographic techniques to establish a trusted connection between the data source and the blockchain. However, even though centralized oracles may be appropriate for certain use cases, they have a significant drawback. They can act as a single point of failure within the Ethereum network, even if the network itself is decentralized. This vulnerability can be mitigated by using decentralized oracle solutions that aim to secure the data-source-to-oracle pipeline. Several such projects have been proposed to address this issue.

The following discussion will focus on a very crucial decentralized oracle system known as Chainlink, which sees the most number of implementations due to its efficacy from a qualitative standpoint. We will explore how both centralized and decentralized solutions can be used in conjunction to enhance the overall performance of oracle systems and make them more viable.

***Chainlink*:** Starting of with Chainlink 1.0, it is a decentralized oracle network built on the Ethereum blockchain which consists of a network of nodes that securely and transparently connect smart contracts to off-chain data sources, APIs, and payment systems. The system allows smart contracts to securely and reliably access data and off-chain resources, making them more versatile and functional. Chainlink 1.0 introduced a number of innovations, including the concept of "oracles as a service" and the use of multiple independent oracles to provide more secure and reliable data. Overall, Chainlink aims to address the limitations of traditional oracle solutions, such as centralized oracles, and provide a more decentralized and secure oracle network for use in blockchain-based applications.

Figure 4. The diagram shows the positive feedback loop created by the Chainlink staking system. As more users pay fees to use the oracle network, the network grows and becomes more secure. This increased security leads to economies of scale that reduce the average cost of securing the network. As the cost of securing the network decreases, it becomes more affordable for users, which stimulates more demand for oracle services, thus continuing the cycle.

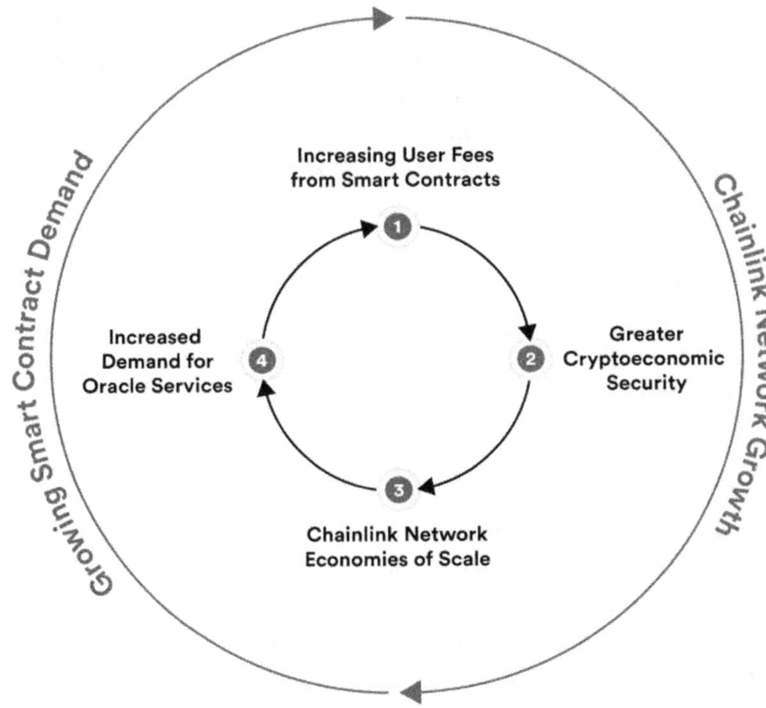

Nodes in the Chainlink network participate by staking deposits of LINK tokens. Staking deposits in a Proof of Stake (PoS) blockchain are analogous to this method. Staking LINK tokens enables nodes to do tasks that require collateral. To ensure that nodes behave honestly, if a node is chosen to fulfil a contract, its LINK is attached to the contract until it is accomplished. If a node fails to produce consistent data, it loses its staked LINK. The rules governing staking and contract fulfilment incentivize nodes to perform honestly and consistently. For effectively operating a node and obtaining data, nodes are also compensated in LINK. If a node is not chosen for a contract, they forfeit their invested

<u>Protocol and architectural considerations:</u> Chainlink's architecture involves its nodes interacting with both the blockchain and external sources. To comprehend its architecture, we must distinguish between on-chain and off-chain components. The purple block in the diagram shown in Chainlink's white paper represents on-chain components, while the green block represents off-chain components.

Figure 5. Using ChainLink to transport data between off-chain sources and on-chain smart contracts entails multiple steps. It begins with the user initiating an on-chain request, which the ChainLink smart contract logs as an event. The event is subsequently obtained by the ChainLink core, which allocates the task to an adapter. To collect the required data, this adapter sends a request to an external API. The ChainLink adaptor then processes the response and returns it to the core. The data is then reported by the ChainLink core to the ChainLink smart contract, which aggregates the responses and returns them to the user's smart contract as a single response.

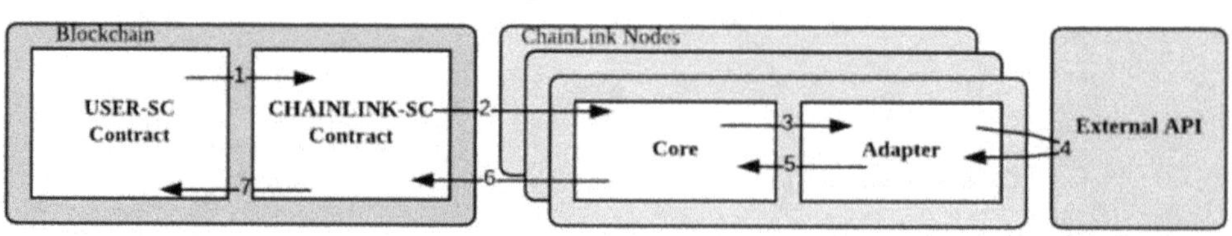

The primary premise of Chainlink is based on the following: The first sort of contract, known as "user contracts" or "USER-SC," is produced by users for decentralised applications. These contracts define the data requirements from external data sources and execute the on-chain logic required for the decentralised application to operate. The second type, known as "Chainlink contracts" or "CHAINLINK-SC,"

is utilised to ensure trust and security through consolidation functions and repute assessment procedures (Breidenbach, 2021).

<u>*Why is it challenging to find an optimal solution to the oracle problem:*</u> The ideal oracle is difficult to produce because it necessitates a completely trustworthy and reliable data source. However, there is no such thing as a completely reliable information source. Data can be compromised or altered as a result of malfunctioning online resources, deceptive vendors, or even genuine errors. This implies that even if the Oracle functions exactly like a Trusted Third Party (TTP), it cannot ensure total security.

Figure 6. Parameters for the development of DeFi Oracles (Zhao, 2022)

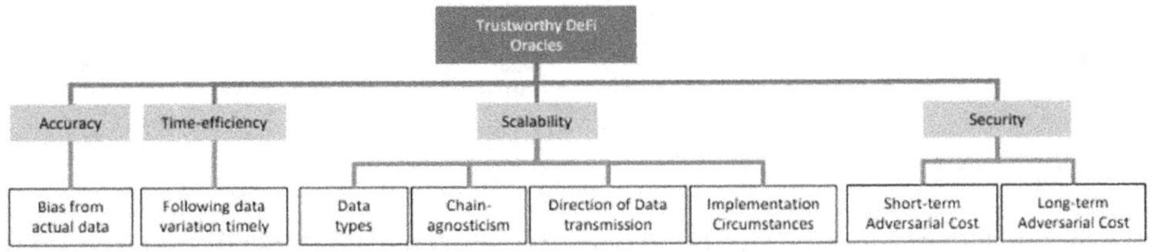

ChainLink's purpose is to construct a real-world oracle system that is as close to ideal functionality as feasible while still operating under realistic trust assumptions. As a result, CHAINLINK-SC is a method of representing the complete ChainLink system as a single entity with a more simplified and streamlined interface. This abstraction helps the authors to simplify the debate and concentrate on the system's overall on-chain functioning rather than individual contracts.

(iii) <u>*On-chain Data:*</u> The CHAINLINK-SC consists of three crucial agreements: the credibility agreement, the matching agreement, and the combining agreement. The credibility agreement monitors the oracle-service providers' performance based on various standards. The matching agreement selects the most suitable oracle for a specific data request, while the combining agreement gathers the feedback from the chosen oracles and delivers the final outcome for the Chainlink inquiry. These agreements are developed as separate components that can be revised or substituted as necessary to stay up-to-date with the changing landscape.

Figure 7. In order to maintain confidentiality on the blockchain regarding the report and the asset that underlies the binary option, the oracle uses a mechanism called Switch. It sends only the binary value switch(r) to the smart contract SC, thereby protecting the privacy of the report and asset. This way, the binary option can be executed without revealing sensitive information on the blockchain (Breidenbach, 2021).

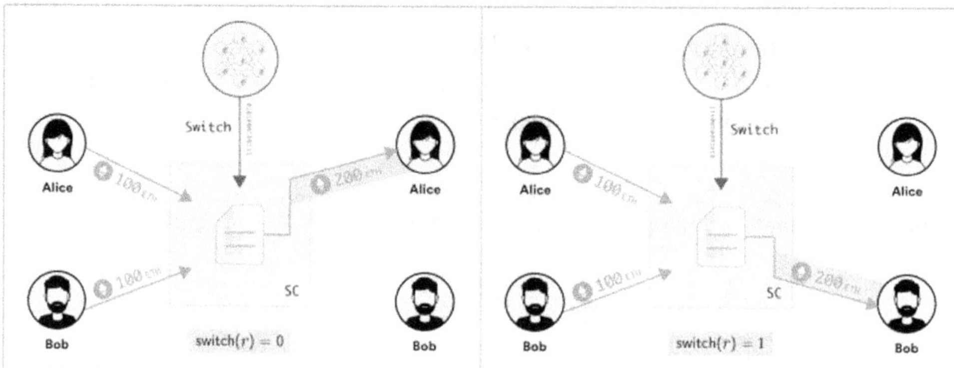

When a data request or query is received by a Chainlink node, it undergoes the following stages to generate a concluded aggregate outcome:

- <u>Nomination of Oracle:</u>

A customer of Oracle services suggests a service level agreement (SLA), which contains specifics such as query parameters, the quantity of Oracles necessary and the wished-for credibility of the Oracles. Once this SLA has been put into place, authorised Oracles will be invited to participate in an auction mediated by the order-matching contract. The reputation contract interfaces with the order-matching contract and selects bids based on the oracles' performance metrics, only evaluating offers that meet the SLA criteria. There is usually an allocated period for bidding within which the order-matching contract will recognise bids.

- <u>Output Assemblage:</u>

A regrouping contract receives the outcomes that the oracles reveal. The regrouping contract calculates a weighted result by adding up the findings from every oracle that completed the task. It also recognizes anomalies or likely incorrect assumptions using the oracles' consensus pool and communicates these authenticity ratings to the reputation contract. The user receives the weighted result in the smart contract at the end.

- Acquiring and publishing of data:

The off-chain oracles carry out the SLA and deliver a report on-chain when an oracle has been chosen using the procedure described above.

(iii) *Off-chain Data:* Off-Chain Reporting (OCR) is a Chainlink oracle network construct that enhances the network's efficiency and decentralization. It does this by allowing the oracles to aggregate data off-chain and submit it on-chain, as opposed to the on-chain aggregation process used in earlier versions.

In OCR, every oracle node retrieves data from various sources and creates a report by processing and aggregating the data off-chain. The report is then signed by the oracle and passed to the network, where it is aggregated with other reports. This aggregation process happens off-chain, and only the final aggregated result is submitted to the blockchain. OCR uses a number of components to accomplish its objectives, including Chainlink Core, which coordinates data requests and report submissions between nodes; the Chainlink Reputation contract, which stores the reputational score of each node; and the Chainlink Order-Matching contract, which matches oracles with requesters based on their SLAs.

- Outboard Adapters:

These are software components that provide a way to integrate external data sources and ser vices with the Chainlink node. Adapters are typically developed as separate services that run on a server or a cloud platform and communicate with the Chainlink node through a RESTful API.

An Adapter typically performs the following tasks:

- Receives a request from a Chainlink node for data from an external data source or service.
- Retrieves the data from the external data source or service using its own protocol and authentication mechanism.
- Processes the data and transforms it into a format that can be consumed by the Chainlink node.
- Sends the processed data back to the Chainlink node in the required format.

- Core:

Since it connects with the distributed ledger and allocates work among its numerous services, the core node application is essential to the Oracle system. Each individual operation is referred to as an assignment, and each assignment is made up of sub-tasks that are executed in a manner similar to a queue. It is designed to be highly scalable, fault-tolerant, and efficient, allowing Chainlink nodes to handle a large volume of data requests and process them quickly and accurately.

FUTURE RESEARCH DIRECTIONS

There is still much to achieve in this line of work. A recurrent weakness of several of the most commonly used Oracle architectures which is often exploited is- not able to deliver information on time (Caldarelli & Ellul, 2021) (a delay in the acquisition of real-time information). This can lead to an exploit called a "time-based attack" where an attacker tries to manipulate the outcome of a smart contract by taking advantage of a time delay in the oracle's response. For example, if the smart contract is set up to execute an action only if a particular condition is met within a certain time frame, an attacker could delay the oracle's response to prevent the contract from executing. This can allow the attacker to manipulate the outcome of the contract in their favor.

Another type of exploit that can occur is called a "data manipulation attack." In this type of attack, an attacker could manipulate the data provided by the oracle to change the outcome of the smart contract. For instance, an attacker could use a "man-in-the-middle" attack to intercept the data being transmitted between the oracle and the smart contract and modify it before sending it to the contract. This can cause the contract to execute incorrectly, leading to undesired outcomes.

This can be potentially resolved in brute-force way by using multiple oracles to provide information to a smart contract; if one oracle fails to provide information on time, the contract can still obtain information from other oracles. Another solution could be to implement super-fast caching mechanisms to store previously obtained data, reducing the need to request the same information repeatedly. This is ultimately still a stopgap solution which can improve response times by reducing the number of requests made to the oracle.

CONCLUSION

There is certainly evidence that blockchain technologies are capable of far more than just monetary transfers. With smart contracts, there are countless opportunities in every imaginable field. As part of our research, we have elaborated upon a decentralized framework for the supply of information by data originators in a trust-less environment.

For distributed ledger networks to connect with external entities, it is crucial to put funds in order to build an infrastructure of reliable oracle solutions. By eliminating potential single point of failure through redundancy, it is rudimentary to shield the consumer-facing side of the service from an antagonistic data suppliers. We also need to ensure that the Oracle system will give results void of error, fraudulent replication, or sabotage as long as a significant proportion of nodes adhere by the guidelines. A number of these approaches, led by Chainlink, have been created over the past few years to keep pace with the platform's exponential growth. The main takeaway from this whitepaper is the importance of combining these innovative solutions, in order to create a more secure and reliable platform to cater to its ever increasing user base.

REFERENCES

Al Breiki, H., & Rehman, M., & Salah, K., & Svetinovic, D. (2020). Trustworthy Blockchain Oracles: Review, Comparison, and Open Research Challenges. *IEEE Access*. . doi:10.1109/ACCESS.2020.2992698

Bartoletti, M., Chiang, J., & Lluch-Lafuente, A. (2020). *SoK: Lending Pools in Decentralized Finance*. Cornell University., doi:10.48550/arxiv.2012.13230

Beniiche, A. (2020). *A Study of Blockchain Oracles*. Cornell University. arxiv.org/pdf/2004.07140.pdf.

Breidenbach, L. (2021). Chainlink 2.0: Next Steps in the Evolution of Decentralized Oracle Networks. *Chainlink Labs*, *15*(April), 1–34.

Caldarelli, G., & Ellul, J. (2021). The Blockchain Oracle Problem in Decentralized Finance—A Multivocal Approach. *Applied Sciences (Basel, Switzerland)*, *11*(16), 7572. doi:10.3390/app11167572

Filippi, P., & Mannan, M. (2020). Blockchain as a confidence machine: The problem of trust & challenges of governance. *Technology in Society*, *62*, 101284. doi:10.1016/j.techsoc.2020.101284

Foxley, W. (2020). *Chainlink to Provide Data for Farming Insurance Startup Arbol*. CoinDesk. www.coindesk.com/chainlink-to-provide-data-for-farming-insurance-startup-arbol

Junejo, A. Z., Hashmani, M. A., & Memon, M. M. (2021). Empirical Evaluation of Privacy Efficiency in Blockchain Networks: Review and Open Challenges. *Applied Sciences (Basel, Switzerland)*, *11*(15), 7013. doi:10.3390/app11157013

Kumar, M., Nikhil, N., & Singh, R. (2020). Decentralising Finance using Decentralised Blockchain Oracles. *2020 International Conference for Emerging Technology (INCET)*, Belgaum, India. 10.1109/INCET49848.2020.9154123

Liu, B., Szalachowski, P., & Zhou, J. (2020). A first look into defi oracles. Computer Science, Cryptography and Security. https://arxiv.org/abs/2005.04377

Priem, R. (2020). Distributed Ledger Technology for Securities Clearing and Settlement: Benefits, Risks, and Regulatory Implications. *Financial Innovation*, *6*(1), 11. doi:10.118640854-019-0169-6

Tolmach, P., Li, Y., Lin, S.-W., & Liu, Y. (2021). *Formal Analysis of Composable DeFi Protocols*. Cornell University., doi:10.1007/978-3-662-63958-0_13/ arxiv.2103.00540

Wang, D., Wu, S., Lin, Z., Wu, L., Yuan, X., Zhou, Y., Wang, H., & Ren, K. (2021). Towards a First Step to Understand Flash Loan and Its Applications in DeFi Ecosystem. *Proceedings of the Ninth International Workshop on Security in Blockchain and Cloud Computing*. ACM. 10.1145/3457977.3460301

Yu, Y., & Zichichi, A. (2021). Enhancing Trust in Trust Services: Towards an Intelligent Human-input-based Blockchain Oracle (IHiBO). *Proceedings of the 54th Hawaii International Conference on System Sciences,* (pp. 1-10). Scholar Space. 10.24251/HICSS.2021.001

Zetzsche, D. A., Arner, D. W., & Buckley, R. P. (2020). Decentralized Finance. *Journal of Financial Regulation*, *6*(2), 172–203. doi:10.1093/jfr/fjaa010

Zhao, Y. (2022). Towards Trustworthy DeFi Oracles: Past, Present and Future. arXiv preprint arXiv:2201.02358

Chapter 21
Recognition of Cyber Physical Systems in Smart Manufacturing Systems With Wireless Connectivity Through Deep Learning Techniques in Industrial Processes

H. Mohammed Ali
Department of Mechanical Engineering, SRM Institute of Science and Technology, India

S. Socrates
Department of Mechanical Engineering, Velammal Institute of Technology, India

K. Balachander
Department of Computer Science Engineering, Velammal Institute of Technology, India

Ramya Maranan
Department of Research and Innovation, Saveetha School of Engineering, SIMATS, India

ABSTRACT

The integration of internet of things (IoT) technologies has been a key aspect in the advancement of modern smart factories, separating them from traditional automated factories. Smart factories use IoT to control automated manufacturing equipment and defect detection tools, leading to improved production efficiency and quality. In this study, the authors propose a concept of a smart factory that integrates deep learning algorithms into a defect detection system and outlines the network architecture for the proposed smart factory. The proposed smart factory system employs equipment for manufacturing, functional testing and problem detection while utilizing checkpoints to identify faulty goods. Defect detection technology is widely used throughout the production, packaging, and functional testing processes, resulting in reduced costs and increased yield of final products. This leads to a more

DOI: 10.4018/978-1-6684-9267-3.ch021

1. INTRODUCTION

In this research article, we propose a method for recognizing cyber physical systems in a smart manufacturing system using wireless connectivity and deep learning techniques in industrial processes. Cyber physical systems (CPS) are a combination of physical and cyber systems that interact with each other in real time, and are becoming increasingly important in the field of manufacturing. The integration of CPS in smart manufacturing systems can improve efficiency and flexibility, but it also poses new challenges in terms of communication and data management. Our proposed method utilizes wireless connectivity and deep learning techniques to overcome these challenges and improve the recognition of CPS in smart manufacturing systems. The results of this research have the potential to enhance the efficiency and flexibility of industrial processes and contribute to the advancement of smart manufacturing technology (Galambos et al. 2018; Liu et al. 2021).

In recent years, the integration of cyber physical systems (CPS) in smart manufacturing systems has gained significant attention due to the potential improvements in efficiency and flexibility. CPS are a combination of physical and cyber systems that interact with each other in real time, and they have the ability to improve the monitoring, control, and decision-making capabilities of manufacturing systems (Cui 2021; Martinez et al. 2022).

One of the main challenges in integrating CPS in smart manufacturing systems is the communication and data management between the physical and cyber systems. Wireless connectivity has been proposed as a solution to this challenge, as it allows for real-time communication and data transfer between the CPS and the smart manufacturing system.

Several researchers have proposed the use of wireless technologies, such as WiFi and Zigbee, for the communication between CPS and smart manufacturing systems. In a study by (Xie et al. 2021), the authors proposed a wireless communication system based on WiFi for the monitoring and control of industrial robots in a smart manufacturing system. The system was able to achieve a high data transfer rate and low latency, demonstrating its potential for use in industrial environments.

Another study (Schneider et al. 2019) proposed the use of Zigbee technology for the communication between CPS and smart manufacturing systems. The authors developed a wireless sensor network based on Zigbee for the monitoring and control of a machining process. The system was able to improve the efficiency and flexibility of the machining process by providing real-time data on the machining conditions. Deep learning techniques have also been proposed as a solution to the challenges of CPS recognition in smart manufacturing systems. Deep learning is a type of machine learning that is able to learn from data and make decisions based on that data. It has been shown to be effective in image and speech recognition, and has been proposed for use in industrial processes as well.

In a study by (Streitz 2019), the authors proposed the use of deep learning for the recognition of CPS in a smart manufacturing system. The authors developed a deep neural network for the recognition of industrial robots in a manufacturing system. The system was able to achieve a high recognition rate and low computational complexity, demonstrating its potential for use in industrial environments.

Another study (Liu et al. 2020) proposed the use of deep learning for the recognition of CPS in a smart manufacturing system. The authors developed a deep convolutional neural network for the recognition of machines and equipment in a manufacturing system. The system was able to achieve a high recognition rate and low computational complexity, demonstrating its potential for use in industrial environments.

Overall, the literature suggests that the integration of CPS in smart manufacturing systems can improve efficiency and flexibility, but it also poses new challenges in terms of communication and data management. Wireless connectivity and deep learning techniques have been proposed as solutions to these challenges. In particular, wireless technologies such as WiFi and Zigbee have been shown to be effective in providing real-time communication and data transfer between CPS and smart manufacturing systems. Deep learning has also been proposed as a solution for CPS recognition in smart manufacturing systems, with studies demonstrating its potential for use in industrial environments (Kathiravelu et al. 2019; Alaghbari et al. 2022; Mishra et al. 2022).

However, despite the promising results from these studies, there is still a need for further research in this field. One area that has not been explored in depth is the combination of wireless connectivity and deep learning techniques for the recognition of CPS in smart manufacturing systems. The proposed research aims to fill this gap by developing a method for recognizing CPS in a smart manufacturing system using wireless connectivity and deep learning techniques in industrial processes. The results of this research have the potential to enhance the efficiency and flexibility of industrial processes and contribute to the advancement of smart manufacturing technology.

2. PROPOSED METHODOLOGY

The illustration in Figure 1 illustrates a potential layout for a smart factory that incorporates a deep learning-based fault detection system. The design suggests that the smart factory should consist of three main components: the industrial process, information management, and profit analysis. During the industrial process, production equipment transforms raw materials into finished goods. To detect faults in semi-finished products, deep learning-based detection technologies are employed. These systems take imageries of semi-finished goods and use deep learning to identify any flaws. Fault detection equipment and manufacturing equipment are placed in alternating positions throughout the operation. At the end of the manufacturing process, functional testing tools are used to confirm that the product is fully functional.

Production equipment records manufacturing parameters and other data, which is then transmitted to the data management division through a network. Similarly, the defect detection equipment records images and findings, which are also sent to the data management division via the network. The functional testing results are also transferred to the data management department via the network. The factory serves as the location where production techniques and data management are housed. Yield analysis can be conducted during and after the manufacturing process. This involves using a microscope to identify faults and investigate their causes. The yield analysis department examines all system data and statistics. The proposed deep learning-based defect detection system is shown in Figure 2.

To create the deep learning algorithm, the architecture must be established. This can include convolutional neural networks (CNN), recurrent neural networks (RNN), and long-short term memory neural networks (LSTM). CNN uses convolutional and pooling layers to enhance pattern identification and data connections (Natrayan L et al., 2020). RNN can switch effort data with varying timeframes because it can maintain previous states. LSTM has three gates: an input gate, a forget gate, and an output gate, which sets it apart from RNN. Each gate has a weight that is determined by input data and used to turn the gate on or off (Paranthaman et al., 2022).

In the second step, the desired learning goal must be defined. This often entails creating a performance indicator and evaluating the success of the built-in deep learning algorithm, also known as the appropriateness function or goal job (Kumar MS et al., 2019). A tailored deep learning procedure that meets our objectives can be achieved during the starting process when the culture goal is clearly defined. Mutual learning objectives include cross-entropy and mean square error (MSE).

In the third stage, the algorithm is trained using a numerical method to identify the optimal weight combination in the proposed deep learning procedure (Nagarajan et al., 2022). Figure 3 displays the proposed network architecture for a deep learning-based defect detection system in a smart factory (Sundaram et al., 2019). Network Unit 1 connects functional testing equipment to the internet by connecting the indication to the IP partaker (Natrayan, L., & Kumar, M. S. 2020). Furthermore, for a good connection, production and defect detection equipment must have distinct, independent IP addresses. The fourth step involves testing and evaluating the deep learning system using images of partly completed products with and without faults (Meikandan et al., 2022). The dataset is used to train the algorithm to detect and

Figure 1. Proposed architecture for smart factory

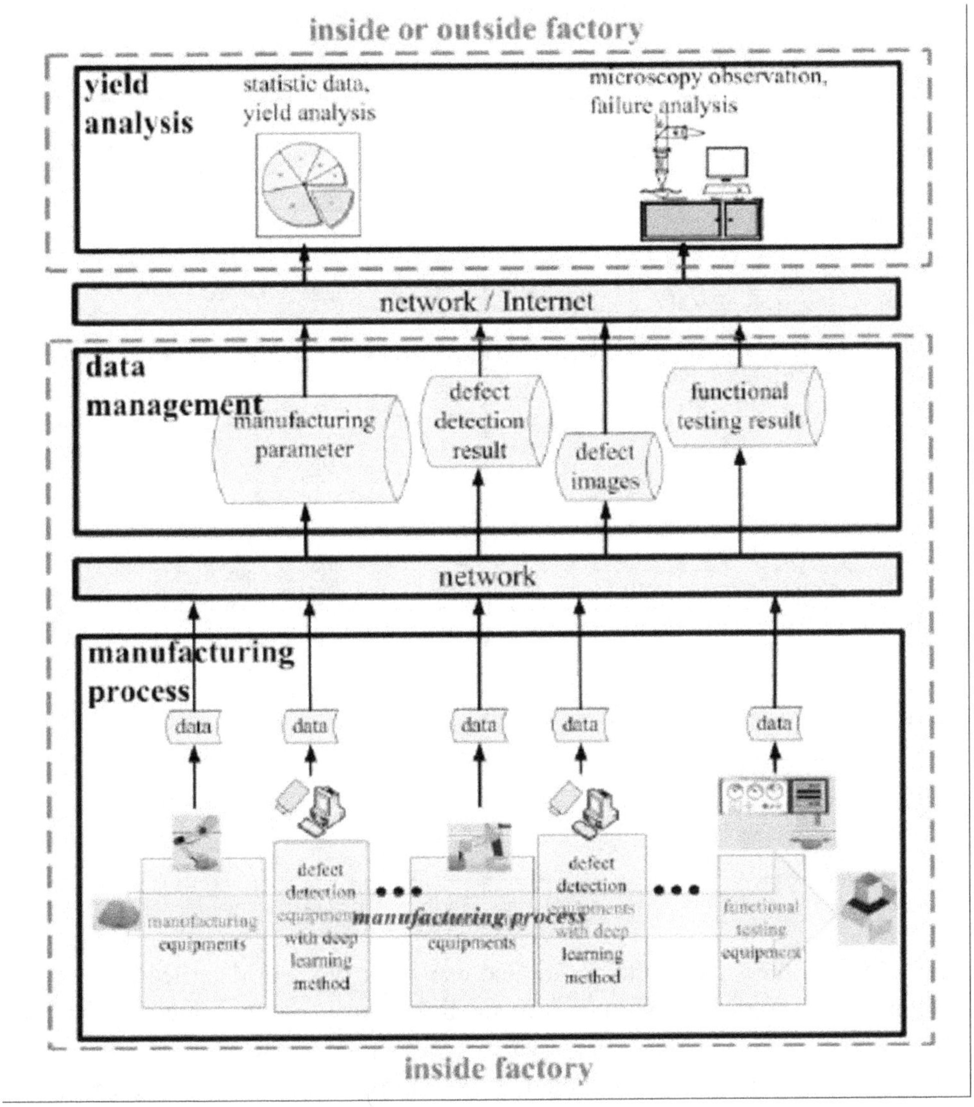

classify issues. The algorithm's performance is measured using metrics such as accuracy, precision, recall, and F1 score (Velmurugan et al., 2022).

Finally, in the fifth and final step, the deep learning-based defect detection system is integrated into the overall smart factory architecture. This involves integrating the system into the production equipment, data management systems, and yield analysis department. The system can then be continuously monitored and optimized to ensure that it is working effectively and efficiently.

Figure 2. Flow chart of the proposed algorithm

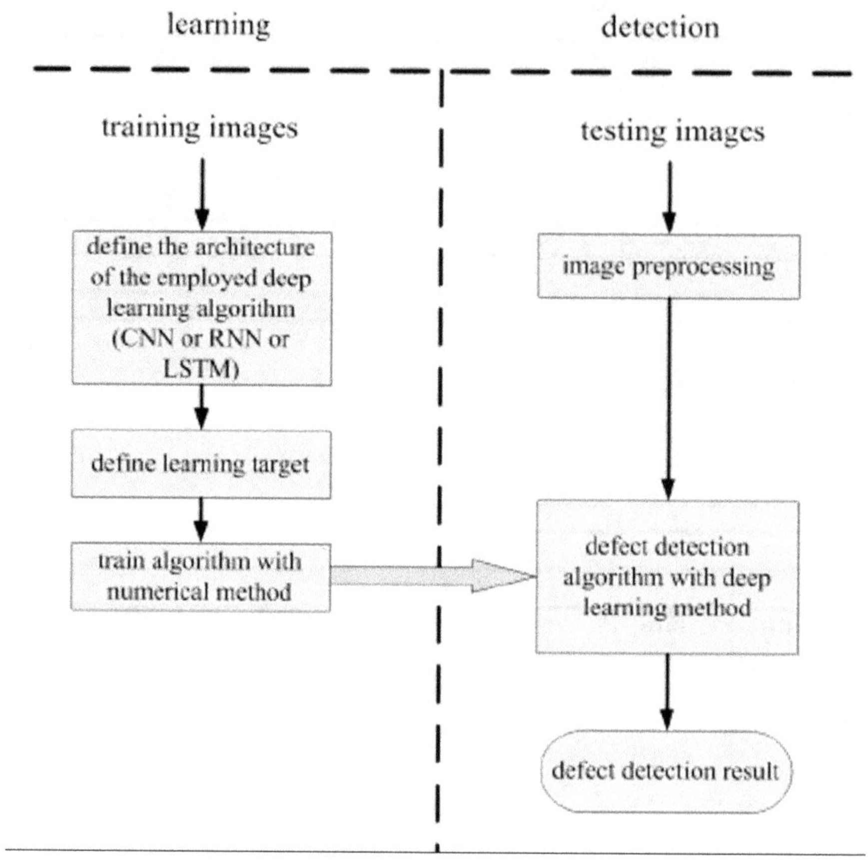

In addition, the smart factory can also use other types of deep learning-based systems such as predictive maintenance, quality control, and resource management (Santhosh MS et al., 2018). These systems work together to provide a complete and integrated smart factory solution that can improve efficiency, reduce costs, and increase productivity (Kumar MS et al., 2018).

In summary, the proposed smart factory design with a deep learning-based fault detection system consists of five main steps: creating the deep learning algorithm architecture, defining the learning goal, training the algorithm, testing and evaluating the system, and integrating the system into the overall smart factory architecture. This design is intended to improve the efficiency and effectiveness of the manufacturing process, while also reducing costs and increasing productivity (Yogeshwaran et al., 2015).

Figure 3. Deep learning architecture

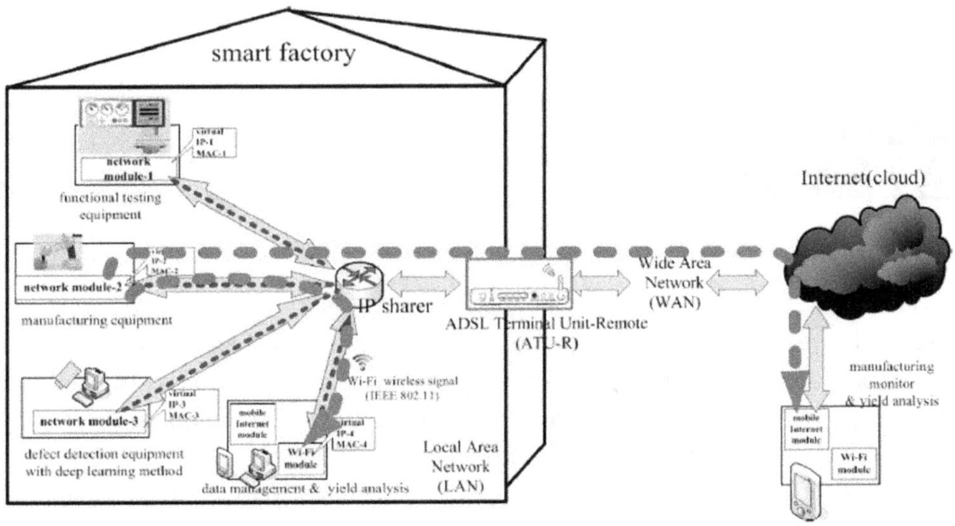

3. RESULT AND DISCUSSION

In this research, a suggested approach is employed to identify machining defects in steel semi-finished products using a convolutional neural network (CNN). As shown in Figure 4, the CNN planning used in this study comprises of several key components, including one or more convolutional layers, a terminal fully associated layer, and a combining layer. The convolutional deposit is the core component of a CNN, and it is responsible for extracting features from the input data (Ramaswamy et al., 2022). The convolutional layer applies a set of filters to the input data, and the filters are designed to detect specific features in the data, such as edges, lines, or patterns. The filters are then moved across the input data, and the output of the convolutional layer is a set of feature maps that represent the detected features.

The pooling layer, on the other hand, is used to down sample the feature maps produced by the convolutional layer. The pooling layer applies a mathematical operation, such as maximum or average pooling, to the feature maps (Balaji et al., 2022). The operation is applied to a small region of the feature maps, such as a 2x2 or 3x3 area, and it reduces the size of the feature maps by taking the maximum or average value of the pixels in the region. Pooling layers are used to reduce the computational complexity of the CNN, as well as to make the network more robust to small translations of the input data (Pragadish et al., 2022).

The terminal fully connected layer is used to make the final prediction based on the feature maps produced by the convolutional and pooling layers. The fully

Figure 4. Employed CNN architecture

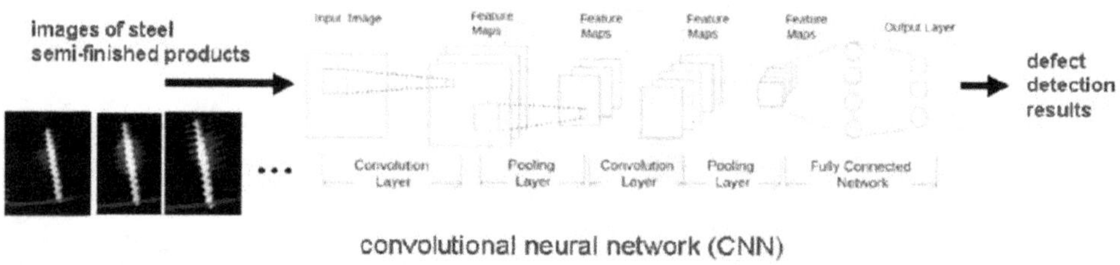

connected layer receives the feature maps as input and applies a set of weights to them to produce a final output. The output of the fully connected layer is usually a vector of probabilities that represent the likelihood of the input data belonging to each of the classes (Natrayan L et al., 2018).

In our studies, we were able to use the CNN architecture to identify machining defects larger than 1mm in steel semi-finished goods. We used a dataset of images of steel semi-finished products, which included both defective and non-defective items. We trained the CNN on this dataset and used it to classify new images as defective or non-defective. The results showed that the CNN was able to achieve high accuracy in identifying defects larger than 1mm in the steel semi-finished products (Natrayan L et al., 2021).

In addition to identifying defects, the CNN can also be used to classify the type of defects. By training the CNN on a dataset of images of different types of defects, the network can learn to recognize specific patterns associated with each type of defect (Veeman et al., 2021). This can be useful for identifying defects that are not visible to the naked eye, such as internal cracks or stress points.

Furthermore, the CNN can also be used to estimate the severity of the defects. By training the CNN on images of defects of varying severity, the network can learn to recognize patterns associated with different levels of severity. This can be useful for identifying defects that are critical and require immediate attention, versus those that can be repaired later (Natrayan L & Kumar MS. 2020).

In conclusion, the use of a CNN to identify machining defects in steel semi-finished products is a promising approach. The CNN architecture used in this study comprises of several key components, including one or more convolutional layers, a terminal fully connected layer, and a pooling layer. By training the CNN on a dataset of images of steel semi-finished products, we were able to achieve high accuracy in identifying defects larger than 1mm. The CNN can also be used to classify the type of defects and estimate the severity of the defects, making it a versatile tool for quality control in the manufacturing industry.

CONCLUSION

In general, it can improve the overall efficiency and effectiveness of the manufacturing process. This research demonstrates the potential of using deep learning techniques to identify and classify patterns in images, and it opens up new opportunities for applying these techniques in other areas of the manufacturing industry. In future research, it would be interesting to explore the use of other deep learning techniques such as recurrent neural networks (RNN) and long-short term memory neural networks (LSTM) to improve the accuracy of identifying machining defects in steel semi-finished products. Additionally, it would be interesting to investigate the potential of using these techniques for identifying other types of defects in different materials. Overall, this research highlights the potential of deep learning techniques for improving the quality control in the manufacturing industry and it opens up new opportunities for further research in this area.

REFERENCES

Alaghbari, K. A., Saad, M. H. M., Hussain, A., & Alam, M. R. (2022). Complex event processing for physical and cyber security in datacentres - recent progress, challenges and recommendations. *Journal of Cloud Computing (Heidelberg, Germany)*, *11*(1), 65. doi:10.118613677-022-00338-x

Balaji, N., Natrayan, L., Kaliappan, S., Patil, P. P., & Sivakumar, N. S. (2022). Annealed peanut shell biochar as potential reinforcement for aloe vera fiber-epoxy biocomposite: Mechanical, thermal conductivity, and dielectric properties. *Biomass Conversion and Biorefinery*, 1–9. doi:10.100713399-022-02650-7

Cui, X. (2021). Cyber-Physical System (CPS) architecture for real-time water sustainability management in manufacturing industry. *Procedia CIRP*, *99*, 543–548. doi:10.1016/j.procir.2021.03.074

Galambos, P., Rudas, I. J., Zhang, L., & Su, S.-F. (2018). Cyber-Physical Control. *Complexity*, *1–2*, 1–2. doi:10.1155/2018/9834608

Kathiravelu, P., Van Roy, P., & Veiga, L. (2019). SD-CPS: Software-defined cyber-physical systems. Taming the challenges of CPS with workflows at the edge. *Cluster Computing*, *22*(3), 661–677. doi:10.100710586-018-2874-8

Kumar, M. S. (2018). Experimental investigations on mechanical and microstructural properties of Al2O3/SiC reinforced hybrid metal matrix composite. []. IOP Publishing.]. *IOP Conference Series. Materials Science and Engineering*, *402*(1), 012123. doi:10.1088/1757-899X/402/1/012123

Kumar, M. S. (2019). Processing and characterization of AA2024/Al2O3/SiC reinforces hybrid composites using squeeze casting technique. *Iran. J. Mater. Sci. Eng*, *16*(2), 55–67.

Liu, B., Zhang, Y., Cao, X., Liu, Y., Gu, B., & Wang, T. (2020). A survey of model-driven techniques and tools for cyber-physical systems. *Frontiers of Information Technology and Electronic Engineering*, *21*(11), 1567–1590. doi:10.1631/FITEE.2000311

Liu, X. R., Meng, Y., & Chang, P. (2021). Node Importance Evaluation of Cyber-Physical System under Cyber-Attacks Spreading. *Complexity*, *2021*, 1–15. doi:10.1155/2021/6641030

Martinez, P., Al-Hussein, M., & Ahmad, R. (2022). A cyber-physical system approach to zero-defect manufacturing in light-gauge steel frame assemblies. *Procedia Computer Science*, *200*, 924–933. doi:10.1016/j.procs.2022.01.290

Meikandan, M., Karthick, M., Natrayan, L., Patil, P. P., Sekar, S., Rao, Y. S., & Bayu, M. B. (2022). Experimental investigation on tribological behaviour of various processes of anodized coated piston for engine application. *Journal of Nanomaterials*, *2022*, 1–8. doi:10.1155/2022/7983390

Mishra, A., Jha, A. V., Appasani, B., Ray, A. K., Gupta, D. K., & Ghazali, A. N. (2022). *Emerging technologies and design aspects of next generation cyber physical system with a smart city application perspective*. International Journal of Systems Assurance Engineering and Management. doi:10.100713198-021-01523-y

Nagarajan, Rajagopalan, A., Angalaeswari, S., Natrayan, L., & Mammo, W. D. (2022). Combined economic emission dispatch of microgrid with the incorporation of renewable energy sources using improved mayfly optimization algorithm. *Computational Intelligence and Neuroscience*, *2022*, 1–22. doi:10.1155/2022/6461690 PMID:35479598

Natrayan, L., & Kumar, M. S. (2020). An integrated artificial neural network and Taguchi approach to optimize the squeeze cast process parameters of AA6061/Al2O3/SiC/Gr hybrid composites prepared by novel encapsulation feeding technique. *Materials Today. Communications*, *25*, 101586. doi:10.1016/j.mtcomm.2020.101586

Natrayan, L., & Kumar, M. S. (2020). Optimization of wear behaviour on AA6061/Al2O3/SiC metal matrix composite using squeeze casting technique–Statistical analysis. *Materials Today: Proceedings*, *27*, 306–310. doi:10.1016/j.matpr.2019.11.038

Natrayan, L., Merneedi, A., Bharathiraja, G., Kaliappan, S., Veeman, D., & Murugan, P. (2021). Processing and characterization of carbon nanofibre composites for automotive applications. *Journal of Nanomaterials*, *2021*, 1–7. doi:10.1155/2021/7323885

Natrayan, L., Senthil Kumar, M., & Chaudhari, M. (2020). Optimization of squeeze casting process parameters to investigate the mechanical properties of AA6061/Al2O 3/SiC hybrid metal matrix composites by Taguchi and Anova approach. In *Advanced Engineering Optimization Through Intelligent Techniques: Select Proceedings of AEOTIT 2018* (pp. 393-406). Springer Singapore. 10.1007/978-981-13-8196-6_35

Natrayan, L., Sivaprakash, V., & Santhosh, M. S. (2018). Mechanical, microstructure and wear behavior of the material AA6061 reinforced SiC with different leaf ashes using advanced stir casting method. *International Journal of Engineering and Advanced Technology*, *8*, 366–371.

Paranthaman, V., Sundaram, K. S., & Natrayan, L. (2022). Influence of SiC particles on mechanical and microstructural properties of modified interlock friction stir weld lap joint for automotive grade aluminium alloy. *Silicon*, *14*(4), 1617–1627. doi:10.100712633-021-00944-5

Pragadish, N., Kaliappan, S., Subramanian, M., Natrayan, L., Satish Prakash, K., Subbiah, R., & Kumar, T. C. A. (2022). Optimization of cardanol oil dielectric-activated EDM process parameters in machining of silicon steel. *Biomass Conversion and Biorefinery*, 1–10. doi:10.100713399-021-02268-1

Ramaswamy, R., Gurupranes, S. V., Kaliappan, S., Natrayan, L., & Patil, P. P. (2022). Characterization of prickly pear short fiber and red onion peel biocarbon nanosheets toughened epoxy composites. *Polymer Composites*, *43*(8), 4899–4908. doi:10.1002/pc.26735

Santhosh, M. S. (2018). Investigation of mechanical and electrical properties of Kevlar/E-glass and Basalt/E-glass reinforced hybrid composites. *International Journal of Mechanical and Production Engineering Research and Development*, *8*(3), 591–598. doi:10.24247/ijmperdjun201863

Schneider, M., Lucke, D., & Adolf, T. (2019). A cyber-physical failure management system for smart factories. *Procedia CIRP*, *81*, 300–305. doi:10.1016/j.procir.2019.03.052

Streitz, N. (2019). Beyond 'smart-only' cities: Redefining the 'smart-everything' paradigm. *Journal of Ambient Intelligence and Humanized Computing*, *10*(2), 791–812. doi:10.100712652-018-0824-1

Sundaram. (2019). Smart clothes with bio-sensors for ECG monitoring. *International Journal of Innovative Technology and Exploring Engineering*, *8*(4), 298–301.

Veeman, Shree, M. V., Sureshkumar, P., Jagadeesha, T., Natrayan, L., Ravichandran, M., & Paramasivam, P. (2021). Sustainable development of carbon nanocomposites: Synthesis and classification for environmental remediation. *Journal of Nanomaterials*, *2021*, 1–21. doi:10.1155/2021/5840645

Velmurugan, G., Siva Shankar, V., Kaliappan, S., Socrates, S., Sekar, S., Patil, P. P., Raja, S., Natrayan, L., & Bobe, K. (2022). Effect of aluminium tetrahydrate nanofiller addition on the mechanical and thermal behaviour of luffa fibre-based polyester composites under cryogenic environment. *Journal of Nanomaterials*, *2022*, 1–10. doi:10.1155/2022/5970534

Xie, J., Tan, W., Fang, B., & Huang, Z. (2021). Towards a Statistical Model Checking Method for Safety-Critical Cyber-Physical System Verification. *Security and Communication Networks*, *2021*, 1–12. Advance online publication. doi:10.1155/2021/5536722

Yogeshwaran. (2015). Mechanical properties of leaf ashes reinforced aluminum alloy metal matrix composites. *International Journal of Applied Engineering Research: IJAER*, *10*(13), 11048–11052.

Chapter 22
Security for Data Communication in Cyber Physical System Limitation and Issues to Analyse the Performance Level of Networks:
A Secure Mode of Data Transmission in Networks Using Different Types of Component Layers

P. Deivendran
Velammal Institute of Technology, India

S. Soundararajan
Velammal Institute of Technology, India

G. Malathi
Vellore Institute of Technology, India

C. Geetha
https://orcid.org/0000-0002-9894-1524
R.M.K. Engineering College, India

P. Suresh Babu
Velammal College of Engineering and Technology, India

DOI: 10.4018/978-1-6684-9267-3.ch022

ABSTRACT

Network security is a secure mode of data transmission from one device to other devices in any organization that handles a huge volume of data in a fast manner. There are different levels of data and network through unauthorized person from confidentiality of the network; the authors are using some external devices like routers, hub, cable etc., for connection from one end to another end of communication. Cyber physical system (CPS) is used to build a number of devices that can be used from small wireless communication to the larger system communication locally or remotely over the networks using cloud servers. Recently, the number of users can also increases in a networked systems with processors, sensors, cloud service provider, these platforms can also be used in car, medical device, and IoT sensors for creating platforms and the network performance also increased.

INTRODUCTION

The cyber physical systems security for internet of things (IoT) gadgets and cyber physical systems (CPS). For IoT are becoming more and more significant in critical infrastructure, governance, and daily life. Example of intelligent networked systems with sensor, processors, and actuators built in that detect and communicate with the real environment while supporting real-time, assured performance in safety-critical applications (Zhu et al., 2018). As costs come down and networks, platforms (Hassoubah et al., 2015), and sensors come together more and more, a closely related filed called the internet of things is growing. These systems, whether they are used to describe a car's ability to prevent forward collision (Hong et al., 2015), a medical device's capacity to adjust to changing conditions in real-time, or the most recent internet of things innovation (Dhawan et al., 2015), are a source of competitive advantage in today's innovation economy and offer a wide range of opportunities for DHS and Homeland security enterprise missions. IoT and CPS both raise cyber security concerns and attack surfaces concurrently. Unintentional mistakes or malevolent attacks could have serious repercussions for the environment and for human lives (Singh et al., 2013). To improve security and dependability for CPS and IoT (Boppana et al., 2020), proactive and well-coordinated activities are required.

Cyber physical systems are designed systems that rely on the seamless fusion of physical components and computing algorithms (Chaganti & Boppana, 2016) . The research field of networked CPS is frequently used to describe the communications and networking difficulties connected with implementing CPS (Gao et al., 2014). More and more different kinds of technology are able to sense and affect their surroundings. These gadgets can range from tiny wireless communication devices to big systems that connect to cloud servers or communicate locally among themselves (Kim, 2014). These devices will be expected to function in a variety of situations, including those that are underwater, in the air, in production facilities (Gawanmeh & Alomari, 2018), and for personal conversations. In order to realize networked CPS, a variety of research issues in the fields like cloud computing (Bhamare et al., 2020), mobile computing, wearable computing, security, and privacy concerns, as well as challenges with indoor and outdoor localization and positioning (Lu et al., 2015), wireless communications and networking, and data analytics (Napiah et al., 2018), must all be addressed. Due to a greater emphasis on the integration of physical components into systems, the communications and networking challenges related to network CPS go beyond those of just networked system (Nazarenko & Safdar, 2019) .

In this network of embedded systems that communicate a CPS is a device with physical input and output. In other words, a CPS is a compilation of networked systems that can keep an eye on and manage real Internet of Things processes and things. The three major fundamental parts of a CPS are sensors, aggregators, and actuators (Wurm et al., 2017). Additionally, CPS system has the capacity to perceive their immediate environment and can modify and regulate the physical world (Darabseh et al., 2019). This is mostly attributable to their adaptability and capacity to alter the systems and processes runtimes by utilizing real-time computing. Indeed, CPS system are embedded in various systems (Deivendran et al., 2023), including military systems, agricultural, ecological systems, communication systems, power transmission systems are used in numerous fields (Karlof et al., 2003).

BACKGROUND

Despite its many benefits, CPS system is vulnerable to many physical or cyber safety risks, attack, and difficulties. This is because of their varied makeup, reliance on confidential and responsive information, and extensive operation. Since purposeful or unintentional exposure of these systems might have disastrous consequences, it is crucial to implement strong security measures. But this can generate excessive, mainly in terms of latency. moreover, by regularly updating software, programs, and in service systems, zero-day vulnerabilities should be reduced.

The integration of CPS systems into critical infrastructures such as the desirable target for security assaults due to its smart grid, industry, supply chain, healthcare system, and agriculture. for a variety of reasons, such as economic, criminal, military, espionage, policies, and terrorism. As a result, serious attacks against these systems can be launched using any CPS weakness as a target. Targeting can be done for a variety of security features, such as confidentiality, integrity, and availability. It is crucial to protect this system beginning several potential assaults, in order to whether internal or external, passive or active facilitate the widespread acceptance and implementation of CPS to capitalize on their betterment.

Security in Cyber Physical Systems

A cyber attack on a CPS poses a threat to both the physical world and the virtual world. This happened in 2021 when the Stuxnet virus infected a cyber physical system controlled Iranian nuclear power plant. In a traditional IT security methods and techniques are insufficient to deal with a CP's security issues. Critical dangers to the relationships between these components are missed by assessments that focus on the failure of individual components. Better methods for CPS protection are being explored by research at MIT Sloan on cyber security. This research examines threats and vulnerabilities using case studies like Stuxnet and makes recommendations for how the issues could be fixed at the most fundamental levels of device design. With the goal of identifying the major purpose of this study is to identify the main CPS security threats, vulnerabilities, and assaults and to explain the advantages and limits of the existing security solutions. This work is motivated by the requirements for a secure, accurate, dependable, efficient, and safe CPS environment. In addition to that are associated and computing complexity of the security methods is examined. Keep in mind that CPS systems demand cutting-edge security solutions that can balance system performance and security level.

CPS LAYERS AND COMPONENTS

The application layer, Perception Layer(PL), Transmission Layer(TL), Network Layer(NL), and physical layer are the five basic layer that make up the CPS architecture. These layers are shown and described in Figure 1. The work serves as the foundation for the investigation of the security concerns at the various CPS tiers. Existing IoT, CPS architectures are expanded and modified by this layered design, the below diagram shows the different types of layers are used to store the data.

Perception Layer

The recognition layer or the sensing layers are other names for the perception layer. It contains tools like sensors, actuators, aggregators, RFID tag, Global Positioning Systems (GPS), as well as numerous other devices. These devices collect real time data in order to watch, follow, and comprehend the physical world. Depending on the type of sensor, in addition to sound and light signals, examples of this data include electrical usage, hear, location, chemistry, and biology. Before being gathered and evaluated by the application layer, these IR-sensors produce real-time data within vast and local network domains. Additionally, in order to guarantee error-free and secure feedback and control commands, protecting actuators rely on approved sources, increasing the security level typically necessitates.

Figure 1. Types of layer

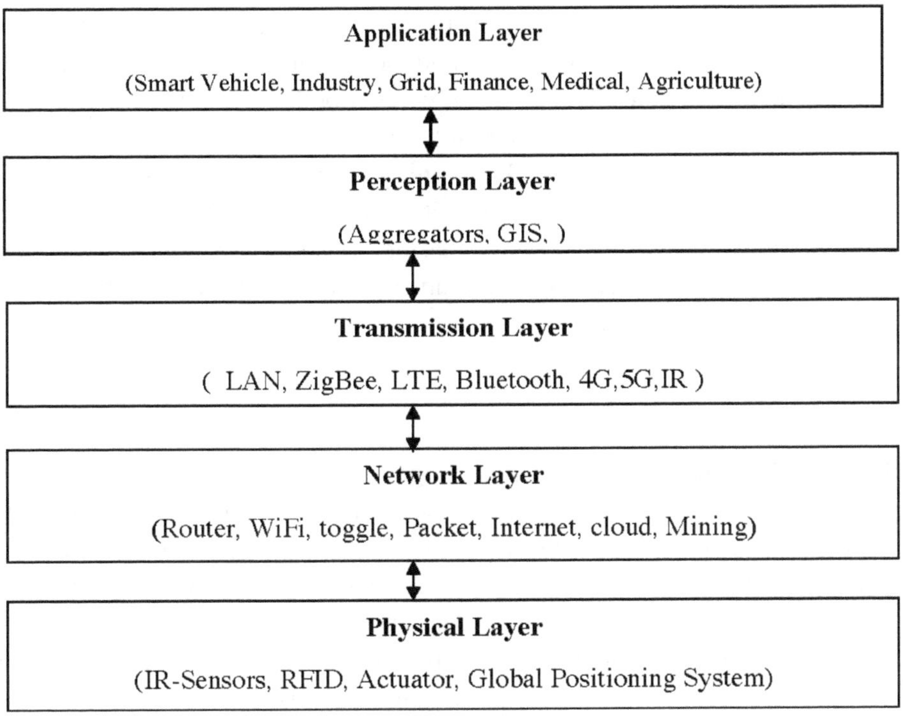

Transmission Layer

It is the second CPS layer and is also referred to as the transport layer or network layer. This layer transfers and processes information between the perception and application levels. Through the using local are networks and long-term evolution communication protocols. As an examples, which is including Bluetooth, 4G and 5G InfraRed (IR). ZigBee, and Wifi among other technologies, data transmission and interaction are made possible over the internet. The expansion the quantity of internet connected devices is addressed for this purpose using a variety of protocols, including IPV6. This layer also ensures data routing and switching utilizing cloud computing platforms, networking equipment, switching and internet gateways, firewalls, and intrusion detection and prevention systems. transmission prior to outsourcing the contents of data, it is crucial.

Application Layer

The third and most engaging layer is this one. It processes the instructions toreal world components like sensors and actuators using the data obtained from the data transfer layer. For execution. This is proficient by putting in place sophisticated algorithms for generating decisions based on the combined data. Additionally, this layer is a receiver, evaluates data from the perception layer before deciding which automatic activities should be appropriately, the data at this layer is managed via cloud computing, middleware, and data mining methods. To secure and maintain privacy, one must prevent the leak of sensitive information. Anonymization, data masking, privacy preserving and secret sharing are some of the most well known defensive measures. Additionally this layer needs a robust multi-factor authentication procedure to prevent the data.

Network Layer

Establishing Communicating with smart gadgets, edge computing nodes, cloudlets, and physical sensors other entities is the responsibility of the network communication layer. Particular application parts, in order to connect all layers and facilitate information flow in CPS. This layer typically includes both wireless and wired technology. Numerous dangers exist at the network layer, including spoofing, flooding, and routing assaults that target specific system. It takes new tactics and approaches to effectively detect ad mitigate these threats in CPS.

Application Layer

The various applications for sensors and actuators are described in this layer, including those in the smart grid, smart homes, smart transitions, and smart industries. In order to control physical devices. to compromise the application layer.

Throughout industrial systems, sensor nodes are dispersed to track physical and environmental parameters like high temperature, heaviness, or assessing the physical and as the nodes can transmit the information to neighboring nodes and so forth until the targeted base station receives the data, sensed data is aggregated at the base station. These base stations may serve as information gateways. To processing the packets of sensor data in accordance with the specified P4 series built through the P4 switches, the base station is connected to the P4 switch.

Header fields may include allocated to a bit sized user defined variable during packet header parsing so that operations can be carried out on them before leaving and exiting. The switch outside, the parse software includes match action tables as well, and each time a particular field in the table. For matches a field the package includes the related defined action is carried out on the small package fields For instance, the match key is picked as the IP address to carry out the network address translation functionality in P switches, and the matching .The IP address is set using the primitive action function IP header setting field address.. As illustrated in the match action table below, we keep track of the hostile nodes getting WELCOME packets and increase the amount of packets currently being received. Since the packet's header fields are not required to be replaced, the checksum module wasn't used in our procedure. To convert the parser programs into table, compilation is required. Dependency Graphs (TDG) and then allocating memory and other resources by mapping the TDG to particular hardware. Using minuet emulator and BMV2 the suggested workflow in the parser switch is accessed for performance and security. The below table 1.1represetn the major layers and objectives used for security purpose.

Table 1. Layer functionality

Layers	Objectives	Attack	Target	Security
Perception	Data and information using sensors, Collection	Eavesdropping, Port scan, Passive replay	Confidentiality, Privacy, Authentication	Trust management, Source authentication, Data protection
Transmission	Data and information using internet	Man in the middle, Meet in the Middle, DoS/QoS	Integrity, Availability	Strong password policy, Lightweight, dynamic encryption
Application	Data and information using vehicle	Malicious code injection, Botnets malware, Trojans, Worms	Privacy, Safety, Security	IDS, IPS, Firewalls, Strong authentication

SENSING COMPONENTS

Components of the CPS are employed to control signals or to sense information. Accordingly, the components of the CPS are divided into two groups Sensing Components, which gather and detect, information, and Controlling Components(CC), which watch over and regulate signal.

Sensors

To determine the accuracy of the data acquired, gather and collect actual data after a relationship procedure called calibration, since the judgments that will be made are dependent on the examination of this data, sensing data is crucial.

Aggregators

It is based on the choices made by the aggregators, are situated to make the information visible for the environment at the application layer. Every choice the CPS makes is based on a previous data aggregation process, because actuators heavily depend on other network nodes. Electrical processing by actuators impulses as input and produce physical performance as a result, which is another aspect of their activity, the below diagram Figure 2 shows the sending and receiving the t data from one source to designation address and based on the performance of the networks.

Figure 2. Sending the data through parse header

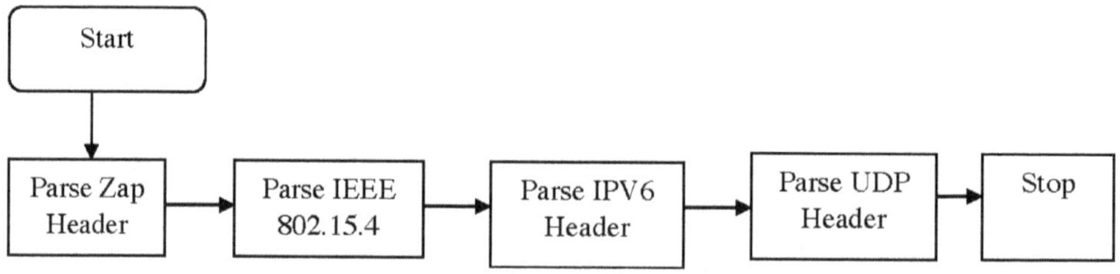

Controlling Components

These components are employed to manage and control signals. They are essential for achieving better levels of precision and defense against intentional attack and mishaps, including interference, noise, and signal jamming. The result is has become crucial to use PLCs, DCSs and all of their associated parts, including Programmable Automation Controller (PAC).

Programmable Logic Controller

These integrated industrial digital computers, for example, regulate production robotic processes, for example, the device performance and fault diagnosis processing, enabling improved elasticity and resilience. They were originally intended to replace hard wired relays. The below diagram Figure 3 gives the system architecture to analyze the system security for each and every sensor to make a good image.

Figure 3. Infracture of CPS

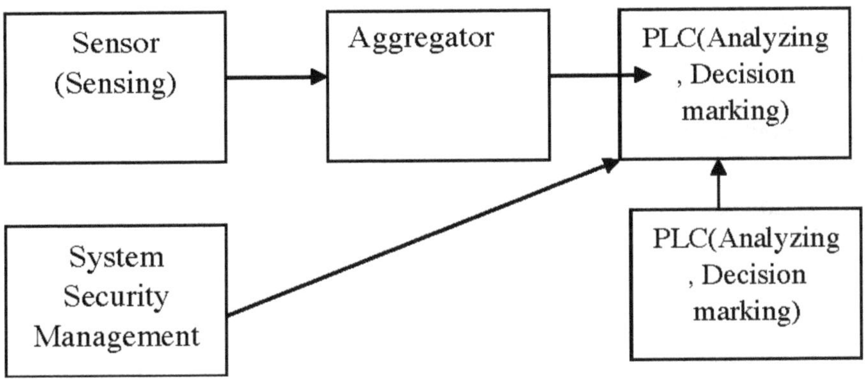

Distributed Controlled Systems

These components controlled by computer systems enable the spread of autonomous command systems across the system with the help of supervisory control from a central operator. The reliability of the DCS is raised while the cost of installation is decreased as a result of the remote supervision and monitoring procedure. In certain circumstances, DCS might resemble SCADA.

Remote Terminal Units

Electronic devices known as Remote Telemetry units are managed by microprocessors like the master terminal unit. They oppose all forms of control loops or control algorithms, in contrast to the PLC. They become better suited for wireless communications over larger geographic telemetry areas as a result. The primary function of the RTU is to connect SCADA to the physical objects through a chat platform for a supervisor that communicates telemetry data to the objects to be controlled.

Attack Detection
Step-1
Start

```
{
Transition begin Zigbee;}
```
Step-2
```
Start Zigbee
{
Packet.extract(hdr.Zigp)
0x4548: IEEE802.14.5}
```
Step-3
```
Table destnode_count{
Counter={counter_inc;
}
Tkey={hdr.lowpan.dst: exact;}
Size=512;
}
```
Step-4
```
Parser()
Verify-checjsum()
Progress()
Incresses()
Comput_checksum() Deparser()
```

CONCLUSION

At one time, it was extremely uncommon view network architecture change after an application was built because technical advancements required expensive, time consuming, and a large workforce. Complex network infrastructure is needed to handle security, latency, computing, and data processing requirements in CPSs, IoT, and machine learning. This article presented a network architecture based on a programmable network data plane for CPS applications, with an emphasis on ICS. The examination of assaults and defenses in three levels, which we suggest, is the other side of the CPS three layer architecture. Additionally, simulation and field applications. This is unacceptable, and ideally this work will advance CPS security by offering a unique CPS security architecture and a literature map of the application sector. Assists operators, researchers, and others in efficiently focusing their efforts.

REFERENCES

Bhamare, D., Zolanvari, M., Erbad, A., Jain, R., Khan, K., & Nader, M. (2020). *Cybersecurity for industrial control systems: A survey*. Washington University in St. Louis.

Boppana, R. V., Chaganti, R., & Vedula, V. (2020). *Analyzing the vulnerabilities introduced by ddos mitigation techniques for software-defined networks* (Vol. 1055). Springer. doi:10.1007/978-3-030-31239-8_14

Chaganti R. & Boppana, R. (2016). *Handling controller saturation attacks in sdn based data center network*. Research Gate.

Darabseh, A., & Freris, N. M. (2019). A software-defined architecture for control of IoT cyberphysical systems. *Cluster Computing, 22*(4), 1107–1122. doi:10.100710586-018-02889-8

Deivendran, P. (2020). Scalability Service in Data Center Persistent Storage Allocation using Virtual Machines. *International Journal of Scientific & Technology Research, 9.*

Deivendran, P. (2023, March). 2022, "Liver Infection Prediction Analysis Using Machine Learning to Evaluate Analytical Performance in Neural Networks by Optimization Techniques. *International Journal of Engineering Trends and Technology*, *71*(3), 377–384. doi:10.14445/22315381/IJETT-V71I3P240

Deivendran, P., & Naganathan, E. R. (2019, January 01). Scalability Assurance Process in replication and Migration Using Cloud Simulator. *International Journal of Networking and Virtual Organisations*, *21*(1), 112–126. doi:10.1504/IJNVO.2019.101153

Dhawan, M., Poddar, R., Mahajan, K., & Mann, V. (2015). *SPHINX: Detecting Security Attacks in Software-Defined Networks.*

Gao, Y., Peng, Y., Xie, F., Zhao, W., Wang, D., Han, X., Lu, T., & Li, Z. (2014). Analysis of security threats and vulnerability for cyber-physical systems. *Proceedings of 2013 3rd International Conference on Computer Science and Network Technology, ICCSNT 2013*. Semantic Scholar.

Gawanmeh, A., & Alomari, A. (2018). Taxonomy analysis of security aspects in cyber physical systems applications. *2018 IEEE International Conference on Communications Workshops, ICC Workshops 2018 -Proceedings,* (pp. 1–6). IEEE.

Hassoubah, R. S., Solaiman, S. M., & Abdullah, M. A. (2015). Intrusion Detection of Hello Flood Attack in WSNs Using Location Verification Scheme. *International Journal of Computer and Communication Engineering*, *4*(3), 156–165. doi:10.17706/IJCCE.2015.4.3.156-165

Hong, S., Xu, L., Wang, H., Gu, G. (2015). *Poisoning network visibility in software-defined networks: New attacks and countermeasures*. NDSS.

Karlof, C., & Wagner, D. (2003). Secure routing in wireless sensor networks: Attacks and countermeasures. *Proceedings of the 1st IEEE International Workshop on Sensor Network Protocols and Applications, SNPA 2003*, (pp. 113–127). IEEE. 10.1109/SNPA.2003.1203362

Kim, Z. (2014). An Unprecedented Look at Stuxnet, the World's First Digital Weapon. *WIRED*.

Lu, T., Lin, J., Lingling Zhao, L., Li, Y., & Peng, Y. (2015). A security architecture in cyber-physical systems: Security theories,Analysis, Simulation and application fields. *International Journal of Security and its Applications, 9*(7), 1–16.

Napiah, M. N., Idris, M. Y. I. B., Ramli, R., & Ahmedy, I. (2018). Compression Header Analyzer Intrusion Detection System (CHA - IDS) for 6LoWPAN Communication Protocol. *IEEE Access : Practical Innovations, Open Solutions*, *6*, 16623–16638. doi:10.1109/ACCESS.2018.2798626

Nazarenko, A. A., & Safdar, G. A. (2019). Survey on security and privacy issues in cyber physical systems. *AIMS Electronics and Electrical Engineering*, *3*(2), 111–143. doi:10.3934/ElectrEng.2019.2.111

Singh, V. P., Aishwarya S., Anand, U., & Sweta, J. (2013). Signal Strength based Hello Flood Attack Detection and Prevention in Wireless Sensor Networks. *Technical Report, 15.*

Wurm, J., Jin, Y., Liu, Y., Hu, S., Heffner, K., Rahman, F., & Tehranipoor, M. (2017). Introduction to Cyber-Physical System Security: A Cross-Layer Perspective. *IEEE Transactions on Multi-Scale Computing Systems*, *3*(3), 215–227. doi:10.1109/TMSCS.2016.2569446

Zhu, L., Tang, X., Shen, M., Du, X., & Guizani, M. (2018, March). Guizani. Privacy-Preserving DDoS Attack Detection Using Cross- Domain Traffic in Software Defined Networks. *IEEE Journal on Selected Areas in Communications*, *36*(3), 628–643. doi:10.1109/JSAC.2018.2815442

Compilation of References

Abid, A., Khan, M. T., & Iqbal, J. (2021). A review on fault detection and diagnosis techniques: Basics and beyond. *Artificial Intelligence Review, 54*(5), 3639–3664. doi:10.100710462-020-09934-2

Acharya, J., & Basu, A. (2020). Deep neural network for respiratory sound classification in wearable devices enabled by patient specific model tuning. *IEEE Transactions on Biomedical Circuits and Systems, 14*(3), 535–544. doi:10.1109/TBCAS.2020.2981172 PMID:32191898

Ackerman, M. J., Filart, R., Burgess, L. P., Lee, I., and Poropatich, R. K., (2010) Developing next-generation telehealth tools and technologies: patients, systems, and data perspectives. *Telemed e-Health. 16*(1), 93–95.

Acosta, M. (2018). *Skills needed in industry 4.0 and related challenges A response to Industry 4.0 challenges. Master, Aalto University School of Business*. Finlandia.

Agarwal, Y., Balaji, B., Gupta, R., Lyles, J., Wei, M., & Weng, T. (2010). *Occupancy-driven energy management for smart building automation.* Research Gate.

Agostini, L., Galati, F., & Gastaldi, L. (2020). The digitalization of the innovation process: Challenges and opportunities from a management perspective. *European Journal of Innovation Management, 23*(1), 1–12. doi:10.1108/EJIM-11-2019-0330

Agostini, L., & Nosella, A. (2020). The adoption of Industry 4.0 technologies in SMEs: Results of an international study. *Management Decision, 58*(4), 625–643. doi:10.1108/MD-09-2018-0973

Agrawal, N., & Kumar, R. (2022). Security perspective analysis of industrial cyber physical systems (I-CPS): A decade-wide survey. *ISA Transactions, 130*, 10–24. doi:10.1016/j.isatra.2022.03.018 PMID:35396141

Ahmed, A.A., Nazzal, M.A., & Darras, B.M. (2022). Cyber-Physical Systems as an Enabler of Circular Economy to Achieve Sustainable Development Goals: A Comprehensive Review. *International Journal of Precision Engineering and Manufacturing - Green Technology 9*(3), 955–975. doi:10.1007/s40684-021-00398-5

Ahuja, S. (2014). QR codes and security concerns. *International Journal of Computer Science and Information Technologies*, *5*(3), 3878–3879.

Al Breiki, H., & Rehman, M., & Salah, K., & Svetinovic, D. (2020). Trustworthy Blockchain Oracles: Review, Comparison, and Open Research Challenges. *IEEE Access*. . doi:10.1109/ACCESS.2020.2992698

Alaghbari, K. A., Saad, M. H. M., Hussain, A., & Alam, M. R. (2022). Complex event processing for physical and cyber security in datacentres - recent progress, challenges and recommendations. *Journal of Cloud Computing (Heidelberg, Germany)*, *11*(1), 65. doi:10.118613677-022-00338-x

Al-Dulaimi, A. (2021). *Cyber-Physical Systems: Recent Developments and Research Challenges*. National Institute of Standards and Technology.

Alexander, K., & Clarkson, P. J. (2000). Good design practice for medical devices and equipment, Part II: Design for validation. *Journal of Medical Engineering & Technology*, *24*(2), 53–62. doi:10.1080/030919000409311 PMID:10937360

Alguliyev, R., Imamverdiyev, Y., & Sukhostat, L. (2021). Hybrid DeepGCL model for cyber-attacks detection on cyber-physical systems. *Neural Computing & Applications*, *33*(16), 10211–10226. doi:10.100700521-021-05785-2

Ali, H.,, & Noor, S., Bashi, S., Marhaban, M.H. (2009). A Review of Pneumatic Actuators (Modeling and Control). *Australian Journal of Basic and Applied Sciences*, *3*, 440–454.

Ali, M., Reza, M. N., Shafik, K., Islam, M., Chowdhury, M., Jeong, J.-H., & Chung, S.-O. (2021). Construction and basic performance test of an ICT-based irrigation monitoring system for rice cultivation in UAE desert soil. *Korean Journal of Agricultural Science.*, *48*, 703–718. doi:10.7744/kjoas.20210059

Al-Jaroodi, J., & Mohamed, N. (2019). Blockchain in Industries: A Survey. *IEEE Access : Practical Innovations, Open Solutions*, *7*, 36500–36515. doi:10.1109/ACCESS.2019.2903554

Allen, J. (2007). Photoplethysmography and its application in clinical physiological measurement. *Physiological Measurement*, *28*(3), R1–R39. doi:10.1088/0967-3334/28/3/R01 PMID:17322588

Al-Mashud, M. (2014). PC Guided Automatic Vehicle System. *International Journal on Cybernetics & Informatics*, *3*(6), 1–10. doi:10.5121/ijci.2014.3601

Alnajjar, A. Y., Anbar, M., Manickam, S., Elejla, O., & El-Taj, H. (2016). QRphish: An Automated QR Code Phishing Detection Approach. *Journal of Engineering and Applied Sciences (Asian Research Publishing Network)*, *11*(3), 553–560.

AlZubi, A. A., Al-Maitah, M., & Alarifi, A. (2021). Cyber-attack detection in healthcare using cyber-physical system and machine learning techniques. *Soft Computing*, *25*(18), 12319–12332. doi:10.100700500-021-05926-8

Ammar, M., Haleem, A., Javaid, M., Bahl, S., & Verma, A. S. (2022). Implementing Industry 4.0 technologies in self-healing materials and digitally managing the quality of manufacturing. *Materials Today: Proceedings*, *52*, 2285–2294. doi:10.1016/j.matpr.2021.09.248

Ancarani, A., Di Mauro, C., & Mascali, F. (2019). Backshoring strategy and the adoption of Industry 4.0: Evidence from Europe. *Journal of World Business*, *54*(4), 360–371. doi:10.1016/j.jwb.2019.04.003

Annunziata, M., & Evans, P. C. (2013). *The industrial Internet @ work.* (White paper). General Electric.

Arboleda, G. Y. C., Perdomo, S. B. H., & Álvarez, E. F. E. (2020). Pymes colombianas y los Retos de la Industria 4.0. *Revista Innova ITFIP*, *7*(1), 10–17. doi:10.54198/innova07.06

Arias, L., Torres, S., Sbarbaro, D., & Farias, O. (2008). Photodiode-based sensor for flame sensing and combustion-process monitoring. *Applied Optics*, *47*(29), 5541–5549. doi:10.1364/AO.47.005541 PMID:18846197

ARTEMIS Ind. Ass. (2007). *Advanced Research and Technology for Embedded Intelligence and Systems*. ARTEMIS Ind. Assoc.

Aumasson, J.-P., Meier, W., Phan, R. C.-W., & Henzen, L. (2014). The Hash Function BLAKE (1st ed. 2014.). Springer Berlin Heidelberg: Imprint: Springer. doi:10.1007/978-3-662-44757-4

Aumasson, J.-P. (2017). *Serious cryptography: a practical introduction to modern encryption.* No Starch Press.

Avram, M., Nitu, C., Bucşan, C., & Gramescu, B. (2017). Linear pneumatic actuator. MATEC Web of Conferences. 121. 10.1051/matecconf/201712108001

Awad, A. (2020). *Cyber-Physical Systems: An Overview and Review of Research Challenges.* Research Gate. https://www.researchgate.net/publication/287233678_An_Overview_and_Some_Challenges_in_Cyber-Physical_Systems

Azhari, A. (2017). *A patch-type wireless forehead pulse oximeter for spo 2 measurement, in: 2017 IEEE Biomedical Circuits and Systems Conference (BioCAS).* IEEE. doi:10.1109/BIOCAS.2017.8325557

Baboo, P. (2015). *Principles of Mass Flow meters*. Research Gate.

Baheti, R., & Gill, H. (2011). *Cyber-physical systems.* Impact Control Technology.

Baheti, R., & Gill, H. (2011, March). Cyber-physical systems. *Impact Control Technology*, *12*(1), 161166.

Baidong, H., & Yukun, Z. (2019). Research on Quickpass Payment Terminal Application System Based on dynamic QR code. *Journal of Physics: Conference Series*, *1168*(3), 032059. doi:10.1088/1742-6596/1168/3/032059

Balaji, B., Al Faruque, M. A., Dutt, N., Gupta, R., & Agarwal, Y. (2015) Models, abstractions, and architectures. In: *Proceedings of the 52nd annual design automation conference on—DAC'15*, (pp. 1–6). ACM. 10.1145/2744769.2747936

Balaji, N., Natrayan, L., Kaliappan, S., Patil, P. P., & Sivakumar, N. S. (2022). Annealed peanut shell biochar as potential reinforcement for aloe vera fiber-epoxy biocomposite: Mechanical, thermal conductivity, and dielectric properties. *Biomass Conversion and Biorefinery*, 1–9. doi:10.100713399-022-02650-7

Balakrishnan, A., Kadiyala, R., Dhiman, G., Ashok, G., Kautish, S., Yadav, K., & Maruthi Nagendra Prasad, J. (2021). A Personalized Eccentric Cyber-Physical System Architecture for Smart Healthcare. *Security and Communication Networks*, *2021*, 1–36. doi:10.1155/2021/1747077

Bani-Hani, R., Wahsheh, Y. A., & Al-Sarhan, M. B. (2014). Secure QR code system. *2014 10th International Conference on Innovations in Information Technology* (pp. 1-6). 10.1109/INNOVATIONS.2014.6985772

Banoth, R., & Regar, R. (2023). An Introduction to Classical and Modern Cryptography. In *Classical and Modern Cryptography for Beginners*. Springer. doi:10.1007/978-3-031-32959-3_1

Barata, J., Cardoso, A., Haenisch, J., & Chaure, M. (2022). Interoperability standards for circular manufacturing in cyber-physical ecosystems: a survey. *Procedia Computer Science 207*(Kes), 3320–3329. doi:10.1016/j.procs.2022.09.390

Bartoletti, M., Chiang, J., & Lluch-Lafuente, A. (2020). *SoK: Lending Pools in Decentralized Finance*. Cornell University., doi:10.48550/arxiv.2012.13230

Basagni, S., Carosi, A., Melachrinoudis, E., Petrioli, C., & Wang, Z. M. (2008). Controlled sink mobility for prolonging wireless sensor networks lifetime. *Wireless Networks*, *14*(6), 831–858. doi:10.100711276-007-0017-x

Battaglia, M., Sampling, N., & Lavrakas, P. J. (2008). *Encyclopedia of survey research methods* (2nd ed.). Sage Publications.

Baum, M., Dukes, M., & Haman, D. (2023). Selection and Use of Water Meters for Irrigation Water Measurement. Research Gate.

Baum, M., & Kabst, R. (2014). The effectiveness of recruitment advertisements and recruitment websites: Indirect and interactive effects on applicant attraction. *Human Resource Management*, *53*(3), 353–378. doi:10.1002/hrm.21571

Beach, C., Krachunov, S., Pope, J., Fafoutis, X., Piechocki, R. J., Craddock, I., & Casson, A. J. (2018). An ultra low power personalizable wrist worn ecg monitor integrated with iot infrastructure. *IEEE Access : Practical Innovations, Open Solutions*, *6*, 44010–44021. doi:10.1109/ACCESS.2018.2864675

Belgharb, K., & Boufaida, M. (2017). Using Event B to Specify Context Awareness for Service Discovery in Pervasive Environments. *International Journal of Ambient Computing and Intelligence*, *8*(1), 1–22. doi:10.4018/IJACI.2017010101

Belogusev, V., & Egorov, A. (2019). Efficiency testing of electric rotary actuators with non-standard reduction units for lower limb exoskeletons. *Journal of Applied Engineering Science.*, *17*(4), 496–503. doi:10.5937/jaes17-20368

Beniiche, A. (2020). *A Study of Blockchain Oracles*. Cornell University. arxiv.org/pdf/2004.07140.pdf.

Berry, R. B. (2016). *Use of chest wall electromyography to detect respiratory effort during polysomnography*. National Institute of Health.

Bertoni, G., Daemen, J., Peeters, M., & Van Assche, G. (2013). Keccak. *32nd Annual international conference on the theory and applications of cryptographic techniques*. Springer.

Bertoni, G., Breveglieri, L., Koren, I., Maistri, P., & Piuri, V. (2002). A parity code based fault detection for an implementation of the advanced encryption standard. In *Proc. 17th IEEE International Symposium on Defect and Fault Tolerance in VLSI Systems* (pp. 51–59). IEEE. 10.1109/DFTVS.2002.1173501

Bertoni, G., Daemen, J., Peeters, M., & Van Assche, G. (2007). Sponge functions. *ECRYPT Hash Workshop*.

Bhamare, D., Zolanvari, M., Erbad, A., Jain, R., Khan, K., & Nader, M. (2020). *Cybersecurity for industrial control systems: A survey*. Washington University in St. Louis.

Bhatt, D. & Toit, D. & Hancke, G. (2012). *Design of a controller for a universal input/output port*. Research Gate. doi:10.1109/I2MTC.2012.6229492-https://www.researchgate.net/profile/Mdabdullah-Al-Mashud/publication/270569209/figure/fig5/AS:668301948633104@1536347009594/A-simple-block-diagram-of-I-O-interface-The-basic-input-output-device-is-a-set-of.ppm

Bhatt, H. S., Bharadwaj, S., Singh, R., & Vatsa, M. (2013, January). Recognizing surgically altered face images using multi objective evolutionary algorithm. *IEEE Transactions on Information Forensics and Security*, *8*(1), 89–100. doi:10.1109/TIFS.2012.2223684

Bibri, S. E., & Allam, Z. (2022). The Metaverse as a virtual form of data-driven smart cities: The ethics of the hyper-connectivity, datafication, algorithmization, and platformization of urban society. *Computational Urban Science*, *2*(1), 1–22. doi:10.100743762-021-00001-2 PMID:35915731

Bilal, A. (2020). *Industry 4.0 and Cyber-Physical Systems: A Comprehensive Review*. Springer. https://link.springer.com/article/10.1007/s10796-021-10153-5?error=cookies_not_supported&code=5a7bae16-af0e-4e3e-8bf3-ce9228958824

Bilir, M. O. (2020). *Qr Kod Güvenlik Farkindaliği Üzerine Ankara Ilinde Bir Araştirma*. [Unpublished Master's Thesis. Baskent University, Ankara, Türkiye].

Bilir, M. O., & Ergüner Özkoç, E. (2020). QR Kod Güvenlik Farkındalığı Üzerine Ankara İlinde Bir Araştırma. *Journal of Internet Applications and Management*, *11*(2), 113–129. doi:10.34231/iuyd.763399

Bissola, R. & Imperatori, B. (2019). HRM in the Industry 4.0 Era: Are Workers Still in The Center? *HRM 4.0 For Human-Centered Organizations*. Advancedseries in management.

Biswas, D., Simoes-Capela, N., Van Hoof, C., & Van Helleputte, N. (2019). Heart rate estimation from wrist-worn photoplethysmography: A review. *IEEE Sensors Journal, 19*(16), 6560–6570. doi:10.1109/JSEN.2019.2914166

Bleichner, M. G., & Debener, S. (2017). Concealed, unobtrusive ear-centered eeg acquisition: Ceegrids for transparent eeg. *Frontiers in Human Neuroscience, 11*, 163. doi:10.3389/fnhum.2017.00163 PMID:28439233

Bletskan, D., Kabatsii, V., & Bletskan, M. (2016). Photovoltaic Sensor for the Solar Battery Guidance on the Sun Based on GeS:Sb Layered Crystals. *Open Journal of Inorganic Non-metallic Materials., 06*(2), 7–11. doi:10.4236/ojinm.2016.62002

Blömer, J., & Seifert, J. P. (2003). *Fault based cryptanalysis of the advanced encryption standard (AES).* 7th Int. Conf. Financial Cryptography, Guadeloupe, French West Indies. 10.1007/978-3-540-45126-6_12

Bonato, P. (2010). Wearable sensors and systems. From enabling technology to clinical applications. *IEEE Engineering in Medicine and Biology Magazine, 29*(3), 25–36. doi:10.1109/MEMB.2010.936554 PMID:20659855

Boppana, R. V., Chaganti, R., & Vedula, V. (2020). *Analyzing the vulnerabilities introduced by ddos mitigation techniques for software-defined networks* (Vol. 1055). Springer. doi:10.1007/978-3-030-31239-8_14

Bordel, B., Alcarria, R., de Rivera, D. S., & Robles, T. (2018). Process execution in Cyber-Physical Systems using cloud and Cyber-Physical Internet services. *The Journal of Supercomputing, 74*(8), 4127–4169. doi:10.100711227-018-2416-4

Breidenbach, L. (2021). Chainlink 2.0: Next Steps in the Evolution of Decentralized Oracle Networks. *Chainlink Labs, 15*(April), 1–34.

Bretschneider, J., Wilde, A., Schneider, P., Hohe, H.P., & Kohler, U. (2010). *Design of multi-dimensional magnetic position sensor systems based on HallinOne® technology.* Fraunhofer EAS. . doi:10.1109/ISIE.2010.5637864

Brinda, P., Pranish, S., Pragadeesh, S., & Raju, T. (2023). Real Time Secure Clickbait and Biometric ATM User Authentication and Multiple Bank Transaction System [IJARCCE]. *International Journal of Advanced Research in Computer and Communication Engineering, 12*(5).

Broy, M., Daembkes, H., & Sztipanovits, J. (2019). Editorial to the theme section on model-based design of cyber-physical systems. *Software & Systems Modeling, 18*(3), 1575–1576. doi:10.100710270-018-0670-9

Busygin, A. G., Konoplev, A. S., & Zegzhda, D. P. (2018). Providing Stable Operation of Self-Organizing Cyber-Physical System via Adaptive Topology Management Methods Using Blockchain-Like Directed Acyclic Graph. *Automatic Control and Computer Sciences, 52*(8), 1080–1083. doi:10.3103/S0146411618080059

Caiado, R. G. G., Scavarda, L. F., Azevedo, B. D., de Mattos Nascimento, D. L., & Quelhas, O. L. G. (2022). Challenges and benefits of sustainable industry 4.0 for operations and supply chain management—A framework headed toward the 2030 agenda. *Sustainability (Basel), 14*(2), 830. doi:10.3390u14020830

Caldarelli, G., & Ellul, J. (2021). The Blockchain Oracle Problem in Decentralized Finance—A Multivocal Approach. *Applied Sciences (Basel, Switzerland), 11*(16), 7572. doi:10.3390/app11167572

Cardenas, A. A., Amin, S., & Sastry, S. S. 2008. Research Challenges for the Security of Control Systems, in: 3rd Conf. on Hot Topics in Security, USENIX. pp. 6:1- 6:6

Cardiovascular diseases (heart attack, stroke). (n.d.). WHO. https://www.who.int/westernpacific/health-topics/cardiovascular-diseases.

Castaneda, D. (2018). A review on wearable photoplethysmography sensors and their potential future applications in health care. *International Journal of Biosensors & Bioelectronics, 4*(4), 195. PMID:30906922

Castellini, L., & D'Andrea, M. & Borgarelli, N. (2014). Analysis and design of a Linear Electro-Mechanical Actuator for a High Lift System. *2014 International Symposium on Power Electronics, Electrical Drives, Automation and Motion, SPEEDAM 2014.* (pp. 243-247). IEEE. 10.1109/SPEEDAM.2014.6872035

Çataloğlu, E., & Ateşkan, A. (2014). Use of QR codes in education with examples. *Elementary Education Online, 13*(1), 5–14.

Caudill, T. S., Lofgren, R., Jennings, C. D., & Karpf, M. (2011). Commentary: Health care reform and primary care: training physicians for tomorrow's challenges. *Academic Medicine, 86*(2), 158–160. doi:10.1097/ACM.0b013e3182045f13 PMID:21270552

Chaganti R. & Boppana, R. (2016). *Handling controller saturation attacks in sdn based data center network.* Research Gate.

Chan, L., Hart, L. G., & Goodman, D. C. (2006). Geographic access to health care for rural Medicare beneficiaries. *The Journal of Rural Health, 22*(2), 140–146. doi:10.1111/j.1748-0361.2006.00022.x PMID:16606425

Charlton, P. H., Birrenkott, D. A., Bonnici, T., Pimentel, M. A. F., Johnson, A. E. W., Alastruey, J., Tarassenko, L., Watkinson, P. J., Beale, R., & Clifton, D. A. (2017). Breathing rate estimation from the electrocardiogram and photoplethysmogram: A review. *IEEE Reviews in Biomedical Engineering, 11*, 2–20. doi:10.1109/RBME.2017.2763681 PMID:29990026

Cheema, S., Malik, A., & Mahmood, T. (2021). *Cyber-Physical Systems for Industry 4.0: Challenges and Opportunities.* Science Direct. https://www.sciencedirect.com/science/article/abs/pii/S0166361520304954

Chen, H., Chen, W., Bao, S., Lu, C., Wang, L., Ma, J., Wang, P., Lu, H., Shu, F., & Bambang Oetomo, S. (2020). Design of an integrated wearable multi-sensor platform based on flexible materials for neonatal monitoring. *IEEE Access : Practical Innovations, Open Solutions*, *8*, 23732–23747. doi:10.1109/ACCESS.2020.2970469

Chen, R., Li, X., & Chen, Y. (2023). Optimal layout model of feeder automation equipment oriented to the type of fault detection and local action. *Protection and Control of Modern Power Systems*, *6*(1), 2. doi:10.118641601-022-00275-6

Chen, R., Yu, Y., Xu, X., Wang, L., Zhao, H., & Tan, H.-Z. (2019). Adaptive Binarization of QR Code Images for Fast Automatic Sorting in Warehouse Systems. *Sensors (Basel)*, *19*(24), 5466. doi:10.339019245466 PMID:31835866

Chen, T. M. (2010). Survey of cyber security issues in smart grids. In *Cyber Security, Situation Management, and Impact Assessment II; and Visual Analytics for Homeland Defense and Security II, 7709* (p. 77090D). International Society for Optics and Photonics. doi:10.1117/12.862698

Chu, H. K., Chang, C. S., Lee, R. R., & Mitra, N. J. (2013). Halftone QR codes. *ACM Transactions on Graphics*, *32*(6), 217. doi:10.1145/2508363.2508408

Chung, H. U., Kim, B. H., Lee, J. Y., Lee, J., Xie, Z., Ibler, E. M., Lee, K. H., Banks, A., Jeong, J. Y., Kim, J., Ogle, C., Grande, D., Yu, Y., Jang, H., Assem, P., Ryu, D., Kwak, J. W., Namkoong, M., Park, J. B., & Rogers, J. A. (2019). Binodal, wireless epidermal electronic systems with in-sensor analytics for neonatal intensive care. *Science*, *363*(6430), 6430. doi:10.1126cience.aau0780 PMID:30819934

Chung, H., Lee, H., & Lee, J. (2018). Finite state machine framework for instantaneous heart rate validation using wearable photoplethysmography during intensive exercise. *IEEE Journal of Biomedical and Health Informatics*, *23*(4), 1595–1606. doi:10.1109/JBHI.2018.2871177 PMID:30235152

Coma del Corral, M. J., Luquín, P. A., Cordero Guevara, J., Olea Movilla, A., Torres, G. T., & Lozano Garcia, J. (2005). Utility of a thematic network in primary health care: A controlled interventional study in a rural area. *Human Resources for Health*, *3*(1), 1–7. doi:10.1186/1478-4491-3-4 PMID:16042778

Cong S, Jianfeng M, Qingsong Y (2016). On the architecture cyber-physical. *Communications and Information Networks 11*(44), 1–21. doi:10.11959/j.issn.2096-1081.2016.041

Copaci, D., Blanco, D., Martin-Clemente, A., & Moreno, L. (2020). Flexible shape memory alloy actuators for soft robotics: Modelling and control. *International Journal of Advanced Robotic Systems*, *17*(1), 172988141988674. doi:10.1177/1729881419886747

Corò, G., & Volpe, M. (2020). Driving factors in the adoption of Industry 4.0 technologies: An investigation of SMEs. In L. De Propris & D. Bailey (Eds.), *Industry 4.0 and regional transformations* (pp. 112–132). Routledge. doi:10.4324/9780429057984-7

Coron, J.-S., Dodis, Y., Malinaud, C., & Puniya, P. (2005). Merkle-Damgård revisited: How to construct a hash function. In Shoup, V. (eds) Advances in Cryptology – CRYPTO 2005. Springer, Berlin, Heidelberg. doi:10.1007/11535218_26

Cui, X. (2021). Cyber-Physical System (CPS) architecture for real-time water sustainability management in manufacturing industry. *Procedia CIRP*, *99*, 543–548. doi:10.1016/j.procir.2021.03.074

Da He, D. (2015). An ear-worn vital signs monitor, IEEE (Inst. Electr. Electron. Eng.). *IEEE Transactions on Biomedical Engineering*, *62*(11), 2547–2552. doi:10.1109/TBME.2015.2459061

Da Xu, L., He, W., Li, S. (2014). *Internet of things in industries: A survey*. IEEE Trans. Ind. Inf.

Dafflon, B., Moalla, N., & Ouzrout, Y. (2021). The challenges, approaches, and used techniques of CPS for manufacturing in Industry 4.0: A literature review. *International Journal of Advanced Manufacturing Technology*, *113*(7-8), 2395–2412. doi:10.100700170-020-06572-4

Darabseh, A., & Freris, N. M. (2019). A software-defined architecture for control of IoT cyberphysical systems. *Cluster Computing*, *22*(4), 1107–1122. doi:10.100710586-018-02889-8

Darwish, A., & Hassanien, A. E. (2018). Cyber physical systems design, methodology, and integration: The current status and future outlook. *Journal of Ambient Intelligence and Humanized Computing*, *9*(5), 1541–1556. doi:10.100712652-017-0575-4

Das, D., Ghosal, S., & Mohanty, S. P. (2022). CoviLearn: A Machine Learning Integrated Smart X-Ray Device in Healthcare Cyber-Physical System for Automatic Initial Screening of COVID-19. *SN Computer Science*, *3*(2), 1–11. doi:10.100742979-022-01035-x PMID:35132394

daSilva, L. B. P., Soltovski, R., Pontes, J., Treinta, F. T., Leitão, P., Mosconi, E., de Resende, L. M., & Yoshino, R. T. (2022). Human Resources Management 4.0: Literature Review and Trends. *Computers & Industrial Engineering*, *108111*, 108111. doi:10.1016/j.cie.2022.108111

David, U. (2006). *OpenDNS*. Phistank. https://www.phishtank.com/

de Fine Licht K. (2020). *Artifcial intelligence, transparency, and public decision-making*. Springer.

Deivendran, P. (2020). Scalability Service in Data Center Persistent Storage Allocation using Virtual Machines. *International Journal of Scientific & Technology Research, 9.*

Deivendran, P. (2023, March). 2022, "Liver Infection Prediction Analysis Using Machine Learning to Evaluate Analytical Performance in Neural Networks by Optimization Techniques. *International Journal of Engineering Trends and Technology*, *71*(3), 377–384. doi:10.14445/22315381/IJETT-V71I3P240

Deivendran, P., & Naganathan, E. R. (2019, January 01). Scalability Assurance Process in replication and Migration Using Cloud Simulator. *International Journal of Networking and Virtual Organisations*, *21*(1), 112–126. doi:10.1504/IJNVO.2019.101153

Devarajan, M., & Ravi, L. (2019). Intelligent cyber-physical system for an efficient detection of Parkinson disease using fog computing. *Multimedia Tools and Applications*, *78*(23), 32695–32719. doi:10.100711042-018-6898-0

Dey, N., Ashour, A. S., & Fong, S. J. (2018). Medical cyber-physical systems. *Survey (London, England)*, 1–13. PMID:29525900

Dhawan, M., Poddar, R., Mahajan, K., & Mann, V. (2015). *SPHINX: Detecting Security Attacks in Software-Defined Networks.*

Dieber, B., Breiling, B., Taurer, S., Kacianka, S., Rass, S., & Schartner, P. (2017). Security for the robot operating system. *Robotics and Autonomous Systems*, *98*, 192–203. doi:10.1016/j.robot.2017.09.017

Dieffenderfer, J. P. (2015). Wearable wireless sensors for chronic respiratory disease monitoring. *2015 IEEE 12th International Conference on Wearable and implantable Body Sensor Networks (BSN)*. IEEE.

Dieffenderfer, J., Goodell, H., Mills, S., McKnight, M., Yao, S., Lin, F., Beppler, E., Bent, B., Lee, B., Misra, V., Zhu, Y., Oralkan, O., Strohmaier, J., Muth, J., Peden, D., & Bozkurt, A. (2016). Low-power wearable systems for continuous monitoring of environment and health for chronic respiratory disease. *IEEE Journal of Biomedical and Health Informatics*, *20*(5), 1251–1264. doi:10.1109/JBHI.2016.2573286 PMID:27249840

Dieste, M., Sauer, P. C., & Orzes, G. (2022). Organizational tensions in industry 4.0 implementation: A paradox theory approach. *International Journal of Production Economics*, *108532*, 108532. doi:10.1016/j.ijpe.2022.108532

Ding, C., Xu, C., & Tao, D. (2015, March). Multi-task pose-invariant face recognition. *IEEE Transactions on Image Processing*, *24*(3), 980–993. doi:10.1109/TIP.2015.2390959 PMID:25594967

Dobbertin, H., Bosselaers, A., & Preneel, B. (1996). RIPEMD-160: A strengthened version of RIPEMD. In D. Gollmann (Ed.), Lecture Notes in Computer Science: Vol. 1039. *Fast Software Encryption. FSE 1996*. Springer. doi:10.1007/3-540-60865-6_44

Dobrzynska, J., & Gijs, M. (2010). Capacitive flexible force sensor. *Procedia Engineering*, *5*, 404–407. doi:10.1016/j.proeng.2010.09.132

Dombrowski, U., & Wagner, T. (2014). Mental strain as feld of action in the 4th industrial revolution. *Procedia CIRP*, *17*, 100–105. doi:10.1016/j.procir.2014.01.077

Dong, X., Sun, S., Shi, D., Gao, F., Wang, X., & Hu, L. (2020). Quantum Collision Attacks on AES-Like Hashing with Low Quantum Random Access Memories. In *Advances in Cryptology – ASIACRYPT 2020: 26th International Conference on the Theory and Application of Cryptology and Information Security*. Springer-Verlag. 10.1007/978-3-030-64834-3_25

Dostálek, P., Vasek, V., & Dolinay, J. (2008). *Design and implementation of portable data acquisition unit in process control and supervision applications*. Research Gate.

Dostálek, P., Vasek, V., & Dolinay, J. (2008). *Design and implementation of portable data acquisition unit in process control and supervision applications.* Research Gate. https://www.researchgate.net/profile/Mdabdullah-Al-Mashud/publication/270569209/figure/fig5/AS:668301948633104@1536347009594/A-simple-block-diagram-of-I-O-interface-The-basic-input-output-device-is-a-set-of.ppm

Dunkelman, O., & Biham, E. (2006). A framework for iterative hash functions: HAIFA. *2nd NIST cryptographich hash workshop*. IEEE.

Dworkin, M. (2015). SHA-3 Standard: Permutation-Based Hash and Extendable-Output Functions. National Institute of Standards and Technology. doi:10.6028/NIST.FIPS.202

Electric Power Research Institute. (2015). *Electric sector failure scenarios and impact analyses - version 3.0. Technical report.* Nevermore Security. https://www.nevermoresecurity.com/wpcontent/uploads/2017/10/Electric-Sector-Failure-Scenarios-and-Impact-Analyses-%E2%80%93-Version-3.0.pdf

Eminagaoglu, M., Cini, E., Sert, G., & Zor, D. (2014). A two-factor authentication system with QR codes for web and mobile applications. *2014 Fifth International Conference on Emerging Security Technologies* (pp. 105-112). IEEE. 10.1109/EST.2014.19

Fang, L., Tang, L., & Shang, Q. (2016). A semi-physical model for pneumatic control valves. *Nonlinear Dynamics*, *85*(3), 1–14. doi:10.100711071-016-2790-5

Farasat, A., Gross, G., Nagi, R., & Nikolaev, A. G. (2016). Social network analysis with data fusion. *IEEE Transactions on Computational Social Systems*, *3*(2), 88–99. doi:10.1109/TCSS.2016.2613563

Fatimah, Y. A., Widianto, A., & Hanafi, M. (2020). Cyber-physical System Enabled in Sustainable Waste Management 4.0: A Smart Waste Collection System for Indonesian Semi-Urban Cities. *Procedia Manufacturing*, *43*, 535–542. doi:10.1016/j.promfg.2020.02.169

Fayad, M., Mostefaoui, A., Chouali, S., & Benbernou, S. (2022). Toward a design model-oriented methodology to ensure QoS of a cyber-physical healthcare system. *Computing*, *104*(7), 1615–1641. doi:10.100700607-022-01058-5

Fernandes, M., Corchado, J. M., & Marreiros, G. (2022). Machine learning techniques applied to mechanical fault diagnosis and fault prognosis in the context of real industrial manufacturing use-cases: A systematic literature review. *Applied Intelligence*, *52*(12), 14246–14280. doi:10.100710489-022-03344-3 PMID:35261480

Filippi, P., & Mannan, M. (2020). Blockchain as a confidence machine: The problem of trust & challenges of governance. *Technology in Society*, *62*, 101284. doi:10.1016/j.techsoc.2020.101284

Finžgar, L., & Trebar, M. (2011). Use of NFC and QR code identification in an electronic ticket system for public transport. *SoftCOM 2011, 19th International Conference on Software, Telecommunications and Computer Networks* (pp. 1-6). IEEE.

Focardi, R., Luccio, F. L., & Wahsheh, H. A. (2019). Usable security for QR code. *Journal of Information Security and Applications*, *48*, 102369. doi:10.1016/j.jisa.2019.102369

Foxley, W. (2020). *Chainlink to Provide Data for Farming Insurance Startup Arbol.* CoinDesk. www.coindesk.com/chainlink-to-provide-data-for-farming-insurance-startup-arbol

Freitas, P., Ferreira, R., Cardoso, S., & Cardoso, F. (2007). Magnetoresistive Sensors. *Journal of Physics Condensed Matter*, *19*(16), 165221–21. doi:10.1088/0953-8984/19/16/165221

Friedt, J. (2010). *Acoustic sensors*. Research Gate.

Galambos, P., Rudas, I. J., Zhang, L., & Su, S.-F. (2018). Cyber-Physical Control. *Complexity*, *1–2*, 1–2. doi:10.1155/2018/9834608

Gallasch, E., Rafolt, D., Taferner, H.P. & Fend, M. (2001). Bilateral manipulandum to synthesize ground referenced and interlimb viscoelastic loads. IEEE. . doi:10.1109/IEMBS.2001.1020394

Gao, Y., Peng, Y., Xie, F., Zhao, W., Wang, D., Han, X., Lu, T., & Li, Z. (2014). Analysis of security threats and vulnerability for cyber-physical systems. *Proceedings of 2013 3rd International Conference on Computer Science and Network Technology, ICCSNT 2013*. Semantic Scholar.

Gasperi, M. & Hurbain, P. (2009). *Potentiometer Sensors*. Springer. . doi:10.1007/978-1-4302-2454-9_6

Gauravaram, P., & Knudsen, L. R. (2010). Cryptographic hash functions. In *Handbook of Information and Communication Security* (pp. 59–79). Springer. doi:10.1007/978-3-642-04117-4_4

Gautschi, G. (2002). *Piezoelectric Sensors*. Springer. . doi:10.1007/978-3-662-04732-3_5

Gawanmeh, A., & Alomari, A. (2018). Taxonomy analysis of security aspects in cyber physical systems applications. *2018 IEEE International Conference on Communications Workshops, ICC Workshops 2018 -Proceedings,* (pp. 1–6). IEEE.

Ge, H., Yue, D., Xie, X. P., Deng, S., & Hu, S. L. (2017). Analysis of cyber physical systems security issue via uncertainty approaches. In Proceedings Advance Computing Methods Life System Model. Simulation. Nanjing, China: Springer. doi:10.1007/978-981-10-6370-1_42

Gensolen, F., Cathebras, G., Martin, L., & Robert, M. (2005). *An Integrated Image Motion Sensor for Micro Camera Module. IEEE Transactions on Very Large Scale Integration (VLSI)*. Systems.

Ghinea, O. (2021). Industry 4.0 and Cyber-Physical Systems: A Review of Emerging Research Trends. NIH.

Giesberts, R., Sluiter, V., & Verkerke, G. (2018). Design and Test of a New Inductive Force Sensor. *Sensors (Basel)*, *18*(7), 2079. doi:10.339018072079 PMID:29958474

Golby, J. (2010). Advances in inductive position sensor technology. Sensor Review -. *Sensor Review*, *30*(2), 142–147. doi:10.1108/02602281011022742

Gospel, H. (2015). Varieties of qualifications, training, and skills in long-term care: A German, Japanese, and UK Comparison. *Human Resource Management*, *54*(5), 833–850. doi:10.1002/hrm.21714

Goswami, S., Roy, P., Dey, N., & Chakraborty, S. (2016). Wireless body area networks combined with mobile cloud computing in healthcare: A survey. *Advances in Medical Technologies and Clinical Practice*, *7*, 388–402. doi:10.4018/978-1-5225-0140-4.ch016

Grabowska, S., & Saniuk, S. (2022). Business models in the industry 4.0 environment—Results of web of science bibliometric analysis. *Journal of Open Innovation*, *8*(1), 19. doi:10.3390/joitmc8010019

Guan, X., Xu, Z., Jia, Q. S., Liu, K., & Zhou, Y. (2018). Cyber-physical model for efficient and secured operation of CPES or energy Internet. *Science China. Information Sciences*, *61*(11), 3–5. doi:10.100711432-018-9621-8

Guest, G., Bunce, A., & Johnson, L. (2006). How many interviews are enough? An experiment with data saturation and variability. *Field Methods*, *18*(1), 59–82. doi:10.1177/1525822X05279903

Gu, J., Qi, X., Wang, Y., Liu, F., & Zhang, C. (2014). Advances in methods for networked and cyber-physical system. *Journal of Control Science and Engineering*, *2014*, 1–2. doi:10.1155/2014/496075

Guo, Y., Li, Y., Li, J. (2020). *Design and Implementation of Cyber-Physical Systems for Industry 4.0: A Review*. NIH. https://pubmed.ncbi.nlm.nih.gov/32838362/

Guo, H., Lan, C., Zhou, Z., Sun, P., Wei, D., & Li, C. (2017). Transparent, flexible, and stretchable ws 2 based humidity sensors for electronic skin. *Nanoscale*, *9*(19), 6246–6253. doi:10.1039/C7NR01016H PMID:28466937

Gupta, A., & Singh, A. (2022). *An Intelligent Healthcare Cyber Physical Framework for Encephalitis Diagnosis Based on Information Fusion and Soft-Computing Techniques*. Ohmsha. doi:10.100700354-022-00175-1

Gupta, P., Krishna, C., Rajesh, R., Ananthakrishnan, A., Vishnuvardhan, A., Patel, S. S., Kapruan, C., Brahmbhatt, S., Kataray, T., Narayanan, D., Chadha, U., Alam, A., Selvaraj, S. K., Karthikeyan, B., Nagalakshmi, R., & Chandramohan, V. (2022). *Industrial internet of things in brilliant manufacturing: a review, approaches, opportunities, open challenges, and future directions*. International Journal on Interactive Design and Manufacturing., doi:10.100712008-022-01075-w

Gu, Y. (2018). Comparison between scalp eeg and behind-the-ear eeg for development of a wearable seizure detection system for patients with focal epilepsy. *Sensors (Basel)*, *18*(1), 29. PMID:29295522

Guzman, J. D., Deckro, R. F., Robbins, M. J., Morris, J. F., & Ballester, N. A. (2014). An analytical comparison of social network measures. *IEEE Transactions on Computational Social Systems*, *1*(1), 35–45. doi:10.1109/TCSS.2014.2307451

Haas, J. (2001). *New Type of Hydraulic Rotary Actuator; Development and Application in a Large Telerobotic Manipulator System*. American Society of Mechanical Engineers. . doi:10.1115/IMECE2001/FPST-25015

Habraken, M., & Bondarouk, T. (2017). Smart industry research in the field of HRM: Resetting job design as an example of upcoming challenges. In *Electronic HRM in the smart era*. Emerald Publishing Limited. doi:10.1108/978-1-78714-315-920161009

Hagos, D. H. (2016). The performance of network-controlled mobile data offloading from LTE to WiFi networks. *Telecommunication Systems*, *61*(4), 675–694. doi:10.100711235-015-0061-2

Hai-Wu Lee. (2021). Design of Multi-Functional Access Control System. *IEEE Access, 9*.

Haque, SMahfuzul SAziz, SRahman, M. (2014). Review of Cyber-Physical System in Health care. *International Journal of Distributed Sensor Networks*.

Hassoubah, R. S., Solaiman, S. M., & Abdullah, M. A. (2015). Intrusion Detection of Hello Flood Attack in WSNs Using Location Verification Scheme. *International Journal of Computer and Communication Engineering*, *4*(3), 156–165. doi:10.17706/IJCCE.2015.4.3.156-165

Has, Z., Rahmat, M., Husain, A., Ishaque, K., Ghazali, R., Ahmad, M., Sam, M., & Rozali, S. M. (2014). Robust Position Tracking Control of an Electro-Hydraulic Actuator in the Presence of Friction and Internal Leakage. *Arabian Journal for Science and Engineering*, *39*(4), 2965–2978. doi:10.100713369-013-0888-3

Hatzivasilis, G. (2017). Password-Hashing Status. *Cryptography*, *1*(2), 10. doi:10.3390/cryptography1020010

He, F., Zhuang, J., & Rao, N. S. V. (2020). Discrete game-theoretic analysis of defense in correlated cyber-physical systems. *Annals of Operations Research*, *294*(1–2), 741–767. doi:10.100710479-019-03381-1

Henning, A. (2006). Comprehensive Model for Thermopneumatic Actuators and Microvalves. Microelectromechanical Systems. *Journal of Microelectromechanical Systems*, *15*(5), 1308–1318. doi:10.1109/JMEMS.2006.879665

Herdt, V., & Drechsler, R. (2022). Advanced virtual prototyping for cyber-physical systems using RISC-V: Implementation, verification and challenges. *Science China. Information Sciences*, *65*(1), 1–17. doi:10.100711432-020-3308-4

Hernández-Sampieri, R., & Torres, C. P. M. (2018). *Metodología de la investigación: las rutas cuantitativa, cualitativa y mixta* (1st ed.). McGraw-Hill Interamericana.

Hofman, W. (2015). *Federated platforms for seamless interoperability in the Physical Internet*. Research Gate.

Hofman, W. (2015). *Federated platforms for seamless interoperability in the Physical Internet*. Research Gate. https://www.researchgate.net/profile/Wout-Hofman/publication/282816170/figure/fig2/AS:284122183421953@1444751415086/The-concepts-service-protocol-and-interface.png

Hong, S., Xu, L., Wang, H., Gu, G. (2015). *Poisoning network visibility in software-defined networks: New attacks and countermeasures*. NDSS.

Huang, C., Wang, Y., Li, X., Ren, L., Zhao, J., Hu, Y., Zhang, L., Fan, G., Xu, J., Gu, X., Cheng, Z., Yu, T., Xia, J., Wei, Y., Wu, W., Xie, X., Yin, W., Li, H., Liu, M., & Cao, B. (2020). Clinical features of patients infected with 2019 novel coronavirus in wuhan, China. *Lancet*, *395*(10223), 497–506. doi:10.1016/S0140-6736(20)30183-5 PMID:31986264

Humayed, A., Lin, J., Li, F., & Luo, B. (2017). Cyber-physical systems security-a survey. *IEEE Internet of Things Journal*, *4*(6), 1802–1831. doi:10.1109/JIOT.2017.2703172

Hummel, V., Hyra, K., Ranz, F., & Schuhmacher, J. (2015). Competence development for the holistic design of collaborative work systems in the logistics learning factory. *Procedia CIRP*, *32*, 76–78. doi:10.1016/j.procir.2015.02.111

Hung, S. H., Yao, C. Y., Fang, Y. J., Tan, P., Lee, R. R., Sheffer, A., & Chu, H. K. (2020). Micrography QR Codes. *IEEE Transactions on Visualization and Computer Graphics*, *26*(9), 2834–2847. doi:10.1109/TVCG.2019.2896895 PMID:30716038

IBM. (2022). *Cost of a Data Breach Report 2022*. IBM. https://www.ibm.com/downloads/cas/3R8N1DZJ

Ibrahim, D. (2002). Thermistor Temperature Sensors. Newnes. doi:10.1016/B978-075065556-9/50005-0

Ináncsi, M. (2022). Cybersecurity Challenges of the Civilian Unmanned Aircraft Systems. *The Military Engineer*, *17*(2), 205–216. doi:10.32567/hm.2022.2.14

Inkulu, A. K., Bahubalendruni, M. R., & Dara, A. (2022). Challenges and opportunities in human robot collaboration context of Industry 4.0-a state of the art review. *Industrial Robot: the international journal of robotics research and application*, *49*(2), 226-239. doi:10.1108/IR-04-2021-0077

Isezaki, T., Kadone, H., Niijima, A., Aoki, R., Watanabe, T., Kimura, T., & Suzuki, K. (2019). Sock-type wearable sensor for estimating lower leg muscle activity using distal emg signals. *Sensors (Basel)*, *19*(8), 1954. doi:10.339019081954 PMID:31027302

Ivlev, I., Kneppo, P., & Bartak, M. (2015). Method for selecting expert groups and determining the important of experts' judgments for the purpose of managerial decision-making tasks in health system. *E+M. Ekonomie a Management*, *18*(2), 57–72. doi:10.15240/tul/001/2015-2-005

J.M. C, J. M, & J.B. R. (2019). Successful implementation of a reflective practice curriculum in an internal medicine residency training program. *Journal of General Internal Medicine*, *34*(2, Supplement), S847–S848.

Jani, M. (2017). Designing shape memory alloy linear actuators: A review. *Journal of Intelligent Material Systems and Structures*, *28*(13), 1699–1718. doi:10.1177/1045389X16679296

Jan, S. U., Abbasi, I., Algarni, F., & Khan, S. (2022). A Verifiably Secure ECC Based Authentication Scheme for Securing IoD using FANET. *IEEE Access : Practical Innovations, Open Solutions*, *10*, 23. doi:10.1109/ACCESS.2022.3204271

Javaid, M., Haleem, A., Singh, R. P., & Suman, R. (2022, September). An integrated outlook of Cyber-Physical systems for Industry 4.0: Topical practices, architecture, and applications. *Green Technologies and Sustainability*, *1*(1), 100001. doi:10.1016/j.grets.2022.100001

Javaid, M., Haleem, A., Singh, R. P., & Suman, R. (2022a). Enabling flexible manufacturing system (FMS) through the applications of industry 4.0 technologies. *Internet of Things and Cyber-Physical Systems*, *2*, 49–62. doi:10.1016/j.iotcps.2022.05.005

Javaid, M., Haleem, A., Singh, R. P., & Suman, R. (2022b). Artificial intelligence applications for industry 4.0: A literature-based study. *Journal of Industrial Integration and Management*, *7*(01), 83–111. doi:10.1142/S2424862221300040

Javaid, M., Haleem, A., Singh, R. P., Suman, R., & Gonzalez, E. S. (2022). Understanding the adoption of Industry 4.0 technologies in improving environmental sustainability. *Sustainable Operations and Computers*, *3*, 203–217. doi:10.1016/j.susoc.2022.01.008

Jayakumari, D., Rupa, C., & Nikhila, V. (2021). Pentatope Based Elliptic Curve Encryption for Privacy and Protection of Multimedia Data. In V. L. N. Komanapalli, N. Sivakumaran, & S. Hampannavar (Eds.), *Advances in Automation, Signal Processing, Instrumentation, and Control. i-CASIC 2020. Lecture Notes in Electrical Engineering* (Vol. 700). Springer. doi:10.1007/978-981-15-8221-9_48

Jayarathna, T., Gargiulo, G. D., & Breen, P. P. (2020). Continuous vital monitoring during sleep and light activity using carbon-black elastomer sensors. *Sensors (Basel)*, *20*(6), 1583. doi:10.339020061583 PMID:32178307

Jbair, M., Ahmad, B., Maple, C., & Harrison, R. (2022). Threat modelling for industrial cyber physical systems in the era of smart manufacturing. *Computers in Industry*, *137*, 103611. doi:10.1016/j.compind.2022.103611

Jha, A. V., Appasani, B., Ghazali, A. N., Pattanayak, P., Gurjar, D. S., Kabalci, E., & Mohanta, D. K. (2021). Smart grid cyber-physical systems: Communication technologies, standards and challenges. *Wireless Networks*, *27*(4), 2595–2613. doi:10.100711276-021-02579-1

Jiang, Z., Pajic, M., & Mangharam, R. (2012). Cyber–physical modeling of implantable cardiac medical devices. *Proceedings of the IEEE*, *100*(1), 122–137. doi:10.1109/JPROC.2011.2161241

Jin, A. (2009). Performance evaluation of a tri-axial accelerometry-based respiration monitoring for ambient assisted living. *2009 Annual International Conference of the IEEE Engineering in Medicine and Biology Society*. IEEE. 10.1109/IEMBS.2009.5333116

Johnson, R. B., & Christensen, L. B. (2013). *Educational research: Quantitative, qualitative, and mixed approaches* (5th ed.). SAGE Publications.

Ju, A., & Wang, Z. (2022). A novel fully convolutional network based on marker-controlled watershed segmentation algorithm for industrial soot robot target segmentation. *Evolutionary Intelligence*. doi:10.100712065-022-00708-z

Junejo, A. Z., Hashmani, M. A., & Memon, M. M. (2021). Empirical Evaluation of Privacy Efficiency in Blockchain Networks: Review and Open Challenges. *Applied Sciences (Basel, Switzerland)*, *11*(15), 7013. doi:10.3390/app11157013

Kagermann, H. (2015). Change through digitization-Value creation in the age of Industry 4.0. Management of Permanent Change, National Academy of Science and Engineering. Springer. doi:10.1007/978-3-658-05014-6_2

Kagermann, H., Lukas, W. D., & Wahlster, W. (2011). Industrie 4.0: Mit dem Internet der Dinge auf dem Weg zur 4. industriellen Revolution. *VDI nachrichten, 13*(1), 2-3.

Kagermann, H., & Wahlster, W. (2022). Ten years of Industrie 4.0. *Sci*, *4*(3), 26. doi:10.3390ci4030026

Kagermann, H., Wahlster, W., & Helbig, J. (2013). *Recommendations for implementing the strategic initiative INDUSTRIE 4.0: Final report of the Industrie 4.0 Working Group. Final Report of the Industrie 4.0*. National Academy of Science and Engineering.

Kaiser, F. K., Wiens, M., & Schultmann, F. (2021). Use of digital healthcare solutions for care delivery during a pandemic-chances and (cyber) risks referring to the example of the COVID-19 pandemic. *Health and Technology*, *11*(5), 1125–1137. doi:10.100712553-021-00541-x PMID:33875933

Kandasamy K, Srinivas S, Achuthan K, Rangan VP (2020) IoT cyber risk: a holistic analysis of cyber risk assessment frameworks, risk vectors, and risk ranking process. *Eurasip Journal on Information Security 2020*(1). doi:10.1186/s13635-020-00111-0

Kapsalis, I. (2013). *Security of QR Codes*. [Unpublished master's thesis, Norwegian University of Science and Technology, Trondheim, Norway].

Karasar, N. (2012). *Scientific Research Method* (15th ed.). Nobel Publishing.

Karlof, C., & Wagner, D. (2003). Secure routing in wireless sensor networks:Attacks and countermeasures. *Proceedings of the 1st IEEE International Workshop on Sensor Network Protocols and Applications, SNPA 2003*, (pp. 113–127). IEEE. 10.1109/SNPA.2003.1203362

Kathiravelu, P., Van Roy, P., & Veiga, L. (2019). SD-CPS: Software-defined cyber-physical systems. Taming the challenges of CPS with workflows at the edge. *Cluster Computing*, *22*(3), 661–677. doi:10.100710586-018-2874-8

Kerry, C. F., & Gallagher, P. D. (2013). *Digital signature standard (DSS)*. FIPS 186-4. doi:10.6028/NIST.FIPS.186-4

Khan, I. H., & Javaid, M. (2022). Role of Internet of Things (IoT) in adoption of Industry 4.0. *Journal of Industrial Integration and Management*, *7*(04), 515–533. doi:10.1142/S2424862221500068

Khedr, A. (2021). Smart Manufacturing and Industry 4.0: A Comprehensive Review of Emerging Trends and Technologies. NIH.

Kieseberg, P., Leithner, M., Mulazzani, M., Munroe, L., Schrittwieser, S., Sinha, M., & Weippl, E. (2010). QR code security. *8th International Conference on Advances in Mobile Computing and Multimedia* (pp. 430-435). ACM. 10.1145/1971519.1971593

Kim, Z. (2014). An Unprecedented Look at Stuxnet, the World's First Digital Weapon. *WIRED*.

Kim, B. S., Nam, S., Jin, Y., & Seo, K. M. (2020). Simulation framework for cyber-physical production system: Applying concept of LVC interoperation. *Complexity*, *2020*, 1–11. Advance online publication. doi:10.1155/2020/4321873

Kim, J., Park, C., Ahn, J., Ko, Y., Park, J., & Gallagher, J. C. (2017). Real-time UAV sound detection and analysis system. *2017 IEEE Sensors Applications Symposium (SAS),* (pp. 1-5). IEEE. 10.1109/SAS.2017.7894058

Kim, S.-Y., Kim, J.-D., Kim, Y.-S., Song, H.-J., & Park, C. (2015). Resistance Temperature Detector Sensor with a Copper Pattern on the Printed Circuit Board. *International Journal of Control and Automation.*, *8*(8), 67–74. doi:10.14257/ijca.2015.8.8.08

Klitou, D., Conrads, J., Rasmussen, M., Probst, L., & Pedersen, B. (2017). Digital Transformation Monitor Germany: Industrie 4.0. *European Union*.

Kluemper, D. H., Mitra, A., & Wang, S. (2016). Social Media use in HRM. *Research in Personnel and Human Resources Management*, *34*, 153–207. doi:10.1108/S0742-730120160000034011

Kohnke, O. (2017). *It's not just about technology: The people side of digitization. Shaping the digital enterprise, 69-91*. Springer.

Kondruss, B. (2023). Cyber attacks news today 2023. *Kon Briefing*. https://konbriefing.com/en-topics/cyber-attacks.html

Korte, C., & Quodbach, J. (2018). 3D-Printed Network Structures as Controlled-Release Drug Delivery Systems: Dose Adjustment, API Release Analysis and Prediction. *AAPS PharmSciTech*, *19*(8), 3333–3342. doi:10.120812249-018-1017-0 PMID:29855799

Krombholz, K., Frühwirt, P., Kieseberg, P., Kapsalis, I., Huber, M., & Weippl, E. (2014). QR code security: A survey of attacks and challenges for usable security. In *International Conference on Human Aspects of Information Security, Privacy, and Trust* (pp. 79-90). Springer, Cham. 10.1007/978-3-319-07620-1_8

Kumar, M. S. (2018). Experimental investigations on mechanical and microstructural properties of Al2O3/SiC reinforced hybrid metal matrix composite. []. IOP Publishing.]. *IOP Conference Series. Materials Science and Engineering*, *402*(1), 012123. doi:10.1088/1757-899X/402/1/012123

Kumar, M. S. (2019). Processing and characterization of AA2024/Al2O3/SiC reinforces hybrid composites using squeeze casting technique. *Iran. J. Mater. Sci. Eng, 16*(2), 55–67.

Kumar, M., Nikhil, N., & Singh, R. (2020). Decentralising Finance using Decentralised Blockchain Oracles. *2020 International Conference for Emerging Technology (INCET)*, Belgaum, India. 10.1109/INCET49848.2020.9154123

Lacava, G., Mercaldo, F., Martinelli, F., Santone, A., & Pizzi, M. (2022). Drone Audio recognition based on Machine Learning Techniques. *Procedia Computer Science, 207*, 848-857. doi:10.1016/j.procs.2022.09.140

Lee, E. A., & Seshia, S. A. (2017). *Introduction to embedded systems - A cyber-physical systems approach* (2nd ed.). MIT Press.

Lee, I., Sokolsky, O., Chen, S., Hatcliff, J., Jee, E., Kim, B., King, A., Mullen-Fortino, M., Park, S., Roederer, A., & Venkatasubramanian, K. K. (2012). Challenges and research directions in medical cyber–physical systems. *Proceedings of the IEEE, 100*(1), 75–90. doi:10.1109/JPROC.2011.2165270

Lee, J., Ardakani, H. D., Yang, S., & Bagheri, B. (2015). Industrial big data analytics and cyberphysical systems for future maintenance & service innovation. *Procedia CIRP*, *38*, 3–7. doi:10.1016/j.procir.2015.08.026

Lee, K. (2016). *Artificial intelligence, automation, and the economy*. The White. House Blog.

Leitao, P., Karnouskos, S., Ribeiro, L., Lee, J., Strasser, T., & Colombo, A. W. (2016). Smart agents in industrial cyber–physical systems. *Proceedings of the IEEE, 104*(5), 1086–1010. doi:10.1109/JPROC.2016.2521931

Leurent, G., & Peyrin, T. (2020). *SHA-1 is a Shambles*. International Association for Cryptologic Research.

Li, L., Fan, M., & Wang, G. (2018). LWSQR: Lightweight Secure QR Code. *International Conference on Frontiers in Cyber Security* (pp. 241-255). Springer. 10.1007/978-981-13-3095-7_19

Li, A., Shan, S., & Gao, W. (2012, January). Coupled bias-variance tradeoff for cross-pose face recognition. *IEEE Transactions on Image Processing*, *21*(1), 305–315. doi:10.1109/TIP.2011.2160957 PMID:21724510

Liang, J., Zhao, H., Li, X., & Zhao, H. (2017). Face recognition system based on deep residual network. *Proc. 3rd Workshop Adv. Res. Technol. Ind. (WARTIA)*. Atlantis Press. 10.2991/wartia-17.2017.69

Liboni, L. B., Cezarino, L. O., Jabbour, C. J. C., Oliveira, B. G., & Stefanelli, N. O. (2019). Smart industry and the pathways to HRM 4.0: Implications for SCM. *Supply Chain Management*, *24*(1), 124–146. doi:10.1108/SCM-03-2018-0150

Li, C., Guo, X., & Wang, X. (2021). An Autonomous Cyber-Physical Anomaly Detection System Based on Unsupervised Disentangled Representation Learning. *Security and Communication Networks, 2021*, 1–17. doi:10.1155/2021/1626025

Li, H.-T. (2016). Lower-limb motion classification for hemiparetic patients through imu and emg signal processing. *2016 International Conference on Biomedical Engineering (BME-HUST)*. IEEE. 10.1109/BME-HUST.2016.7782096

Lin, X., Zhang, T., Cao, J., Wen, H., Fei, T., Liu, S., Wang, R., Ren, H., & Zhao, H. (2020). Flexible piezoresistive sensors based on conducting polymer-coated fabric applied to human physiological signals monitoring. *Journal of Bionics Engineering, 17*(1), 55–63. doi:10.100742235-020-0004-9

Liu, B., Szalachowski, P., & Zhou, J. (2020). A first look into defi oracles. Computer Science, Cryptography and Security. https://arxiv.org/abs/2005.04377

Liu, B., Zhang, Y., Cao, X., Liu, Y., Gu, B., & Wang, T. (2020). A survey of model-driven techniques and tools for cyber-physical systems. *Frontiers of Information Technology and Electronic Engineering, 21*(11), 1567–1590. doi:10.1631/FITEE.2000311

Liu, C., Zhang, X., Zhao, L., Liu, F., Chen, X., Yao, Y., & Li, J. (2018). Signal quality assessment and lightweight qrs detection for wearable ecg smartvest system. *IEEE Internet of Things Journal, 6*(2), 1363–1374. doi:10.1109/JIOT.2018.2844090

Liu, G., Guo, Y.-W., Zhu, Q.-S., Huang, B.-Y., & Wang, L. (2011). Estimation of respiration rate from three-dimensional acceleration data based on body sensor network. *Telemedicine Journal and e-Health, 17*(9), 705–711. doi:10.1089/tmj.2011.0022 PMID:22035321

Liu, X. R., Meng, Y., & Chang, P. (2021). Node Importance Evaluation of Cyber-Physical System under Cyber-Attacks Spreading. *Complexity, 2021*, 1–15. doi:10.1155/2021/6641030

Liu, Z., & Wang, J. (2020). Human-cyber-physical systems: Concepts, challenges, and research opportunities. *Frontiers of Information Technology and Electronic Engineering, 21*(11), 1535–1553. doi:10.1631/FITEE.2000537

Li, X., Wang, X., & Tian, M. (2020). A new type of pneumatic linear reciprocating actuator. *Journal of Physics: Conference Series, 1650*(2), 022085. doi:10.1088/1742-6596/1650/2/022085

Long, C., & Loveday, P. (2007). A thermo-hydraulic wax actuation system for high force and large displacement applications. *Proceedings of SPIE - The International Society for Optical Engineering*. Spie Digital Library. 10.1117/12.714737

Lorenz, M., Rüßmann, M., Strack, R., Lueth, K. L., & Bolle, M. (2015). *Man, and machine in industry 4.0: How will technology transform the industrial workforce through 2025.* The Boston Consulting Group, 2. http://hdl.voced.edu.au/10707/405644

Lu, K., Zhang, H., & Liu, W. (2020). *Cyber-Physical Systems for Industry 4.0: A Comprehensive Survey.* Semantic Scholar. https://www.semanticscholar.org/paper/Cyber-Physical-System-(CPS)-Based-Industry-4.0:-A-Lu/84008fe9c7558924b49cfd0772ec20ce86df354f

Lu, T., Lin, J., Lingling Zhao, L., Li, Y., & Peng, Y. (2015). A security architecture in cyber-physical systems: Security theories,Analysis, Simulation and application fields. *International Journal of Security and its Applications, 9*(7), 1–16.

Maini, R. (2021). *Industrial Internet of Things and Industry 4.0: A Comprehensive Survey*. IEEE.

Ma, M., Lin, W., Pan, D., Lin, Y., Wang, P., Zhou, Y., & Liang, X. (2018). Data and Decision Intelligence for Human-in-the-Loop Cyber-Physical Systems: Reference Model, Recent Progresses and Challenges. *Journal of Signal Processing Systems for Signal, Image, and Video Technology, 90*(8–9), 1167–1178. doi:10.100711265-017-1304-0

Marnewick, A. L., & Marnewick, C. (2019). The ability of project managers to implement industry 4.0-related projects. *IEEE Access : Practical Innovations, Open Solutions*, *8*, 314–324. doi:10.1109/ACCESS.2019.2961678

Martinez, P., Al-Hussein, M., & Ahmad, R. (2022). A cyber-physical system approach to zero-defect manufacturing in light-gauge steel frame assemblies. *Procedia Computer Science*, *200*, 924–933. doi:10.1016/j.procs.2022.01.290

Martin, K. (2020). *Cryptography: The key to digital security, how it works, and why it matters*. WW Norton & Company.

Matos, J., & Laine, J. (1998). A carbon macro-network from the controlled pyrolysis of saccharose. *Journal of Materials Science Letters*, *17*(8), 649–651. doi:10.1023/A:1006664122346

MAXQDA. (2023) *Home*. Maxqda. https://www.maxqda.com

Meikandan, M., Karthick, M., Natrayan, L., Patil, P. P., Sekar, S., Rao, Y. S., & Bayu, M. B. (2022). Experimental investigation on tribological behaviour of various processes of anodized coated piston for engine application. *Journal of Nanomaterials*, *2022*, 1–8. doi:10.1155/2022/7983390

Melia, T., De La Oliva, A., Soto, I., Serrano, P., & Aguiar, R. (2007). Network controlled handovers: Challenges and possibilities. *Wireless Personal Communications*, *43*(3), 959–974. doi:10.100711277-007-9256-5

Melnyk, M., Kernytskyy, A., Lobur, M., Szermer, M., Zajac, P., & Zabierowski, W. (2014). *Study of Characteristics of MEMS Thermo-Electric Actuators*. Research Gate.

Mendoza-Vázquez, J. R., Tlelo-Cuautle, E., Vázquez-Gonzalez, J. L., & Escudero-Uribe, A. Z. (2009). Simulation of a parallel mechanical elbow with 3 DOF. *Journal of Applied Research and Technology*, *7*(2), 113–123. doi:10.22201/icat.16656423.2009.7.02.494

Mihailescu, M. I., & Nita, S. L. (2023). Getting Started in Cryptography and Cryptanalysis. In *Pro Cryptography and Cryptanalysis with C++23*. Apress. doi:10.1007/978-1-4842-9450-5_1

Mikhailov, A. N., Rodin, A. B., & Smirnova, M. I. (2018). Humanization of Engineering Education in Conditions of the Process of Industry 4.0 Forming. *4th International Conference on Information*. IEEE. 10.1109/INFORINO.2018.8581805

Milenković, A., Otto, C., & Jovanov, E. (2006). Wireless sensor networks for personal health monitoring: Issues and an implementation. *Computer Communications, 29*(13), 2521–2533. doi:10.1016/j.comcom.2006.02.011

Miles, M. B., & Huberman, A. M. (1994). *Qualitative data analysis: An Expanded Sourcebook* (2nd ed.). SAGE.

Miric, S., Schuck, M., Tuysuz, A., & Kolar, J. (2018). *Double Stator Linear-Rotary Actuator with a Single Set of Mover Magnets*, (pp. 750-757). IEEE. . doi:10.1109/ECCE.2018.8557575

Mishra, A., Jha, A. V., Appasani, B., Ray, A. K., Gupta, D. K., & Ghazali, A. N. (2022). *Emerging technologies and design aspects of next generation cyber physical system with a smart city applicationperspective*. International Journal of Systems Assurance Engineering and Management. doi:10.100713198-021-01523-y

Mishra, A., Kumar, V., Singh, V. K., & Sharma, A. (2020). An overview of Industry 4.0: Definition, components, and applications. *Journal of Advanced Research, 24*, 381–392.

Mishra, D. P., & Ray, P. (2018). *Fault detection, location and classification of a transmission line*. Springer London. doi:10.100700521-017-3295-y

Mitnick, K. D., & Simon, W. L. (2011). *The art of deception controlling the human element of security*. John Wiley & Sons.

Mlynczak, M. (2017). *Decomposition of the cardiac and respiratory components from impedance pneumography signals*. BIOSIGNALS.

Mohamad, N., Salleh, A., Hashim, N., Abd Aziz, M., & Zakaria, Z. (2016). Modeling of Wireless Energy Transfer Circuit and Implementation of Wi-tricity System Control using Infrared and PIC Microcontroller. *IACSIT International Journal of Engineering and Technology, 8*, 1282–1294.

Mosterman, P. J., & Zander, J. (2016). Cyber-physical systems challenges: A needs analysis for collaborating embedded software systems. *Software & Systems Modeling, 15*(1), 5–16. doi:10.100710270-015-0469-x

Mueller, E., Chen, X. L., & Riedel, R. (2017). Challenges and Requirements for the Application of Industry 4.0: A Special Insight with the Usage of Cyber-Physical System. [English Edition]. *Chinese Journal of Mechanical Engineering, 30*(5), 1050–1057. doi:10.100710033-017-0164-7

Nagarajan. (2022). Combined economic emission dispatch of microgrid incorporating renewable energy sources using improved mayfly optimization algorithm. *Computational Intelligence and Neuroscience*, 2022.

Nagarajan, Rajagopalan, A., Angalaeswari, S., Natrayan, L., & Mammo, W. D. (2022). Combined economic emission dispatch of microgrid with the incorporation of renewable energy sources using improved mayfly optimization algorithm. *Computational Intelligence and Neuroscience, 2022*, 1–22. doi:10.1155/2022/6461690 PMID:35479598

Napiah, M. N., Idris, M. Y. I. B., Ramli, R., & Ahmedy, I. (2018). Compression Header Analyzer Intrusion Detection System (CHA - IDS) for 6LoWPAN Communication Protocol. *IEEE Access : Practical Innovations, Open Solutions*, *6*, 16623–16638. doi:10.1109/ACCESS.2018.2798626

Narayanan, A. S. (2012). QR codes and security solutions. *International Journal of Computer Science and Telecommunications*, *3*(7), 69–72.

Nasim, M., Charbey, R., Prieur, C., & Brandes, U. (2016). Investigating Link Inference in Partially Observable Networks: Friendship Ties and Interaction. IEEE Trans. Computing. *Soc. Syst.*, *3*(3), 113–119.

Natrayan, L., Senthil Kumar, M., & Chaudhari, M. (2020). Optimization of squeeze casting process parameters to investigate the mechanical properties of AA6061/Al 2 O 3/SiC hybrid metal matrix composites by Taguchi and Anova approach. In *Advanced Engineering Optimization Through Intelligent Techniques: Select Proceedings of AEOTIT 2018* (pp. 393-406). Springer Singapore. 10.1007/978-981-13-8196-6_35

Natrayan, L., & Kumar, M. S. (2020). An integrated artificial neural network and Taguchi approach to optimize the squeeze cast process parameters of AA6061/Al2O3/SiC/Gr hybrid composites prepared by novel encapsulation feeding technique. *Materials Today. Communications*, *25*, 101586. doi:10.1016/j.mtcomm.2020.101586

Natrayan, L., & Kumar, M. S. (2020). Optimization of wear behaviour on AA6061/Al2O3/SiC metal matrix composite using squeeze casting technique–Statistical analysis. *Materials Today: Proceedings*, *27*, 306–310. doi:10.1016/j.matpr.2019.11.038

Natrayan, L., Merneedi, A., Bharathiraja, G., Kaliappan, S., Veeman, D., & Murugan, P. (2021). Processing and characterization of carbon nanofibre composites for automotive applications. *Journal of Nanomaterials*, *2021*, 1–7. doi:10.1155/2021/7323885

Natrayan, L., Sivaprakash, V., & Santhosh, M. S. (2018). Mechanical, microstructure and wear behavior of the material AA6061 reinforced SiC with different leaf ashes using advanced stir casting method. *International Journal of Engineering and Advanced Technology*, *8*, 366–371.

Nazarenko, A. A., & Safdar, G. A. (2019). Survey on security and privacy issues in cyber physical systems. *AIMS Electronics and Electrical Engineering*, *3*(2), 111–143. doi:10.3934/ElectrEng.2019.2.111

Neuman, C. (2009). Challenges in security for cyber-physical systems. In *DHS Workshop on Future Directions in Cyber-Physical Systems Security*, (pp. 22–24). IEEE.

Ng, F. (2020). *Ultrasonic Sensors*. Research Gate., doi:10.13140/RG.2.2.33638.78404

Nguyen, X. T., & Luu, Q. K. (2020). Factors affecting adoption of industry 4.0 by small-and medium-sized enterprises: A case in Ho Chi Minh city, Vietnam. *The Journal of Asian Finance. Economics and Business*, *7*(6), 255–264. doi:10.13106/jafeb.2020

Nicolescu, R., Huth, M., Radanliev, P., & De Roure, D. (2018). Mapping the values of IoT. J Inf. *Technoloy*, *33*(4), 345–360.

Noriega, M., Ramírez, R., López, R., Vaca, M., Morales, J., Terres, H., Lizardi, A., & Chávez, S. (2015). Thermocouples calibration and analysis of the influence of the length of the sensor coating. *Journal of Physics: Conference Series*, *582*(1), 012029. doi:10.1088/1742-6596/582/1/012029

O'Connor, J., Aumasson, J.-P., Neves, S., & Wilcox-O'Hearn, Z. (2020). BLAKE3: one function, fast everywhere. GitHub. https://github.com/BLAKE3-team/BLAKE3

Oh, S. W., Bang, H., & Hwang, J. (2010). *Light-weight RFID device interface for controlling RFID tag memory access*. Research Gate.

Oh, S. W., Bang, H., & Hwang, J. (2010). *Light-weight RFID device interface for controlling RFID tag memory access*. Research Gate. https://www.researchgate.net/profile/Se-Won-Oh-2/publication/224128273/figure/fig1/AS:302750387720197@1449192725712/RFID-air-interface-and-RFID-reader-interface.png

Okkonen, J., Vuori, V., & Palvalin, M. (2019). Digitalization changing work: Employees' view on the benefits and hindrances. *International Conference on Information Technology & Systems,* (pp. 165-176). Springer, Cham. 10.1007/978-3-030-11890-7_17

Ollington, N., Gibb, J., & Harcourt, M. (2013). Online social networks: An emergent recruiter tool for attracting and screening. *Personnel Review*, *42*(3), 248–265. doi:10.1108/00483481311320390

Otto, C., Milenkovic, A., Sanders, C., & Jovanov, E. (2006). System architecture of a wireless body area sensor network for ubiquitous health monitoring. *J. Mob Mult.*, *1*(4), 307–326.

Owasp. (2019). *Index.* Owasp. https://www.owasp.org/index.php/Qrljacking

Ozkan, H., Ozhan, O., Karadana, Y., Gulcu, M., Macit, S., & Husain, F. (2019). A portable wearable tele-ecg monitoring system. *IEEE Transactions on Instrumentation and Measurement*, *69*(1), 173–182. doi:10.1109/TIM.2019.2895484

Oztemel, E., & Gursev, S. (2020). Literature review of Industry 4.0 and related technologies. *Journal of Intelligent Manufacturing*, *31*(1), 127–182. doi:10.100710845-018-1433-8

Paar, C., & Pelzl, J. (2009). *Understanding cryptography: a textbook for students and practitioners*. Springer Science & Business Media.

Pang, Y., Jian, J., Tu, T., Yang, Z., Ling, J., Li, Y., Wang, X., Qiao, Y., Tian, H., Yang, Y., & Ren, T.-L. (2018). Wearable humidity sensor based on porous graphene network for respiration monitoring. *Biosensors & Bioelectronics*, *116*, 123–129. doi:10.1016/j.bios.2018.05.038 PMID:29879538

Pantho, J. (2020). A Comprehensive Survey of Cyber-Physical Systems: From Theory to Applications. IEEE. https://ieeexplore.ieee.org/document/6096958

Pan, X. (2012). Research and implementation of access control system based on RFID and FNN-face recognition. *Proc. 2nd Int. Conf. Intell. Syst. Design Eng. Appl*, (pp. 716-719). IEEE. 10.1109/ISdea.2012.400

Pan, Y., & Zhang, L. (2022). Integrating BIM and AI for Smart Construction Management: Current Status and Future Directions. *Archives of Computational Methods in Engineering*. doi:10.100711831-022-09830-8

Paranthaman, V., Sundaram, K. S., & Natrayan, L. (2022). Influence of SiC particles on mechanical and microstructural properties of modified interlock friction stir weld lap joint for automotive grade aluminium alloy. *Silicon*, *14*(4), 1617–1627. doi:10.100712633-021-00944-5

Park, S. O., Xiao, B., Leung, V., & Jeong, Y. S. (2013). Advanced technologies and applications for Highly-Reliable Cyber Physical System (HRCPS). *The Journal of Supercomputing*, *66*(1), 1–3. doi:10.100711227-013-1027-3

Parvin, S., Hussain, F. K., Hussain, O. K., Thein, T., & Park, J. S. (2013). Multi-cyber framework for availability enhancement of cyber physical systems. *Computing*, *95*(10–11), 927–948. doi:10.100700607-012-0227-7

Pattepu, S., Mukherjee, A., Routray, S., Mukherjee, P., Qi, Y., & Datta, A. (2022). Multi-antenna relay based cyber-physical systems in smart-healthcare NTNs: An explainable AI approach. *Cluster Computing*, *0*. doi:10.100710586-022-03632-0

Peng, L. H., Cao, X. H., Sun, C. Y., Cheng, Y., & Jin, S. (2018). Energy efficient jamming attack schedule against remote state estimation in wireless cyber-physical systems. *Neurocomputing*, *272*, 571–583. doi:10.1016/j.neucom.2017.07.036

Pérez-Rojas, J. G. (2020). Retos de las instituciones de educación superior para su articulación en la Industria 4.0. *Revista CEA*, *6*(11), 9–11. doi:10.22430/24223182.1584

Pierre, A., Gaikwad, A., & Arias, A. (2017). Charge-integrating organic heterojunction phototransistors for wide-dynamic-range image sensors. *Nature Photonics*, *11*(3), 193–199. doi:10.1038/nphoton.2017.15

Pittalia, P. P. (2019). A comparative study of hash algorithms in cryptography. *International Journal of Computer Science and Mobile Computing*, *8*(6), 147–152.

Piuzzi, E., Pisa, S., Pittella, E., Podesta, L., & Sangiovanni, S. (2018). Low-cost and portable impedance plethysmography system for the simultaneous detection of respiratory and heart activities. *IEEE Sensors Journal*, *19*(7), 2735–2746. doi:10.1109/JSEN.2018.2887303

Pivoto, D. G., de Almeida, L. F., da Rosa Righi, R., Rodrigues, J., Lugli, A. B., & Alberti, A. M. (2021). Cyber-physical systems architectures for industrial internet of things applications in Industry 4.0: A literature review. *Journal of Manufacturing Systems*, *58*, 176–192. doi:10.1016/j.jmsy.2020.11.017

Polenghi, A., Cattaneo, L., & Macchi, M. (2023). *A framework for fault detection and diagnostics of articulated collaborative robots based on hybrid series modelling of Artificial Intelligence algorithms*. Springer.

Ponce Romero, J. M., Hallett, S. H., & Jude, S. (2017). Leveraging big data tools and technologies: Addressing the challenges of the water quality sector. *Sustainability (Basel)*, *9*(12), 2160. doi:10.3390u9122160

Poomvichid, T., Patirupanusara, P., & Ketcham, M. (2012). The QR code for audio watermarking using Genetic algorithm. *International Conference on Machine Learning and Computer Science* (pp. 171-174). IEEE.

Poovendran, R. (2010). Cyber-Physical Systems: Close Encounters Between Two Parallel Worlds [Point of View]. *Proceedings of the IEEE*, *98*(8), 1363–1366. doi:10.1109/JPROC.2010.2050377

Poriņš, R. & Apse-Apsitis, P. (2019). *Capacitive proximity sensing hardware comparison*. IEEE. . doi:10.1109/RTUCON48111.2019.8982317

Pozzi, R., Rossi, T., & Secchi, R. (2023). Industry 4.0 technologies: Critical success factors for implementation and improvements in manufacturing companies. *Production Planning and Control*, *34*(2), 139–158. doi:10.1080/09537287.2021.1891481

Pragadish, N., Kaliappan, S., Subramanian, M., Natrayan, L., Satish Prakash, K., Subbiah, R., & Kumar, T. C. A. (2022). Optimization of cardanol oil dielectric-activated EDM process parameters in machining of silicon steel. *Biomass Conversion and Biorefinery*, 1–10. doi:10.100713399-021-02268-1

Preneel, B., Govaerts, R., & Vandewalle, J. (1994). Hash functions based on block ciphers: a synthetic approach. In D. R. Stinson (Ed.), Lecture Notes in Computer Science: Vol. 773. *Advances in Cryptology — CRYPTO' 93*. Springer. . doi:10.1007/3-540-48329-2_31

President's Council of Advisors on Science and Technology. (2007) *Leadership under challenge: Information technology R&D in a competitive world*. President's council of advisors on science and technology (PCAST).

Priem, R. (2020). Distributed Ledger Technology for Securities Clearing and Settlement: Benefits, Risks, and Regulatory Implications. *Financial Innovation*, *6*(1), 11. doi:10.118640854-019-0169-6

Priyadarshini, I., Kumar, R., Tuan, L. M., Son, L. H., Long, H. V., Sharma, R., & Rai, S. (2021). A new enhanced cyber security framework for medical cyber physical systems. *Software-Intensive Cyber-Physical Systems*, *35*(3–4), 159–183. doi:10.100700450-021-00427-3

Priya, P. I., Muthurajkumar, S., & Daisy, S. S. (2022). Data Fault Detection in Wireless Sensor Networks Using Machine Learning Techniques. *Wireless Personal Communications*, *122*(3), 2441–2462. doi:10.100711277-021-09001-1

Qian, L., Luo, Z., Du, Y., & Guo, L. (2009). *Cloud computing: An overview*. Springer.

Qian, Y., Ye, F., & Chen, H.-H. (2022). Cryptographic Techniques. In Y. Qian, F. Ye, & H.-H. Chen (Eds.), *Security in Wireless Communication Networks*., doi:10.1002/9781119244400.ch4

Qiao, G., Liu, G., Shi, Z., Wang, Y., Ma, S., & Lim, T. (2017). A review of electromechanical actuators for More/All Electric aircraft systems. *Proceedings of the Institution of Mechanical Engineers. Part C, Journal of Mechanical Engineering Science*, *232*(22), 095440621774986. doi:10.1177/0954406217749869

Rachim, V. P., & Chung, W.-Y. (2016). Wearable noncontact armband for mobile ecg monitoring system. *IEEE Transactions on Biomedical Circuits and Systems*, *10*(6), 1112–1118. doi:10.1109/TBCAS.2016.2519523 PMID:27214910

Radanliev, P., De Roure, D., Nicolescu, R., Huth, M., Santos, O. (2021). Artificial intelligence and the Internet of Things in Industry 4.0. *Interact*, 1–10.

Radanliev, P., De Roure, D., Nicolescu, R., Huth, M., & Santos, O. (2022). Digital twins: Artificial intelligence and the IoT cyber-physical systems in Industry 4.0. *International Journal of Intelligent Robotics and Applications*, *6*(1), 171–185. doi:10.100741315-021-00180-5

Raj, A., Dwivedi, G., Sharma, A., de Sousa Jabbour, A. B. L., & Rajak, S. (2020). Barriers to the adoption of industry 4.0 technologies in the manufacturing sector: An inter-country comparative perspective. *International Journal of Production Economics*, *224*, 107546. doi:10.1016/j.ijpe.2019.107546

Rajkumar, S.C, Jegatha, L., Vijayakumar, P., & Karthick, K.R. (2023). Secure session key pairing and a lightweight key authentication scheme for liable drone services. *Cyber Security and Applications*. doi:10.1016/j.csa.2022.100012

Ramaswamy, R., Gurupranes, S. V., Kaliappan, S., Natrayan, L., & Patil, P. P. (2022). Characterization of prickly pear short fiber and red onion peel biocarbon nanosheets toughened epoxy composites. *Polymer Composites*, *43*(8), 4899–4908. doi:10.1002/pc.26735

Ramsden, E. (2006). *Hall-Effect Sensors*. Hall-Effect Sensors.

Randazzo, V. (2019). Ecg watch: a real time wireless wearable ecg. *IEEE International Symposium on Medical Measurements and Applications (MeMeA)*. IEEE 10.1109/MeMeA.2019.8802210

Rashid, A., & Ali, A. (2018). Performance Analysis of Low-Cost Infrared Sensors for Multi-Robot Localization and Communication. *International Journal of Computer Applications*, *182*(26), 23–29. doi:10.5120/ijca2018918139

Rezaeibagha, F., Mu, Y., Huang, X., Yang, W., & Huang, K. (2019). *Fully Secure Lightweight Certificateless Signature Scheme for IIoT* (pp. 1). IEEE Access. . doi:10.1109/ACCESS.2019.2944631

Ribeiro, L., & Björkman, M. (2017). Transitioning from standard automation solutions to cyber physical production systems: An assessment of critical conceptual and technical challenges. *IEEE Systems Journal*, *12*(4), 3816–3827. doi:10.1109/JSYST.2017.2771139

Rivest, R. (1992). RFC1321: The MD5 message-digest algorithm. In RFC Editor.

Rose, J., Lukic, V., Milon, T., & Cappuzzo, A. (2016). *Sprinting to value in Industry 4.0.* BCG Perspectives. https://www.bcg.com/publications/2016/lean-manufacturing-technology-digital-sprinting-to-value-industry-40

Rupa, C., Srivastava, G., Gadekallu, T., Maddikunta, P., & Bhattacharya, S. (2020). Security and privacy of UAV data using blockchain technology. *Journal of Information Security and Applications, 55.*

Rusconi, A., Costantini, S., & Prelini, C. (2022). *Micro Speakers*. Springer. . doi:10.1007/978-3-030-80135-9_20

Sachdeva, S., Malhotra, J., & Kumar, M. (2022). L-band PON (NG-PON2) fault detection/monitoring and PWR using C-band ASEN and FBGs. *Optical and Quantum Electronics*, *54*(7), 1–17. doi:10.100711082-022-03854-x

Sadeghi-Nasab, A., & Rafe, V. (2023). A comprehensive review of the security flaws of hashing algorithms. *J Comput Virol Hack Tech*, *19*(2), 287–302. doi:10.100711416-022-00447-w

Sadun, A., Jalani, J., & Sukor, J.A. (2016). *Force Sensing Resistor (FSR): a brief overview and the low-cost sensor for active compliance control.* SPIE. . doi:10.1117/12.2242950

Saeed, T., Khan, M. A., Khan, M. A., & Siddiqui, J. A. (2021). Cyber-physical systems in manufacturing: A survey. *Journal of Manufacturing Systems*, *59*, 384–400.

Sandborn, M., Wright, C., White, J. (2021). CaSTL: Towards cyber-physical authentication of additively manufactured components. *Procedia Computer Science 198*(2018), 764–769. doi:10.1016/j.procs.2021.12.319

Saniuk, S., Grabowska, S., & Straka, M. (2022). Identification of Social and Economic Expectations: Contextual Reasons for the Transformation Process of Industry 4.0 into the Industry 5.0 Concept. *Sustainability (Basel)*, *14*(3), 1391. doi:10.3390u14031391

Santhosh, M. S. (2018). Investigation of mechanical and electrical properties of Kevlar/E-glass and Basalt/E-glass reinforced hybrid composites. *International Journal of Mechanical and Production Engineering Research and Development*, *8*(3), 591–598. doi:10.24247/ijmperdjun201863

Sawlikar, A. P., Khan, Z. J., & Akojwar, S. G. (2016). Efficient energy saving cryptographic techniques with software solution in wireless network. [IJSE]. *International Journal of Synthetic Emotions*, *7*(2), 78–96. doi:10.4018/IJSE.2016070105

Schneider, M., Lucke, D., & Adolf, T. (2019). A cyber-physical failure management system for smart factories. *Procedia CIRP*, *81*, 300–305. doi:10.1016/j.procir.2019.03.052

Schneider, P., & Sting, F. J. (2020). Employees' perspectives on digitalization-induced change: Exploring frames of industry 4.0. *Academy of Management Discoveries*, *6*(3), 406–435. doi:10.5465/amd.2019.0012

Schneier, B. (2007). *Applied cryptography: protocols, algorithms, and source code in C.* John Wiley & Sons.

Schuh, G., Gartzen, T., Rodenhauser, T., & Marks, A. (2015). Promoting work-based learning through industry 4.0. *Procedia CIRP*, *32*, 82–87. doi:10.1016/j.procir.2015.02.213

Senna, P. P., Ferreira, L. M. D., Barros, A. C., Roca, J. B., & Magalhaes, V. (2022). Prioritizing barriers for the adoption of Industry 4.0 technologies. *Computers & Industrial Engineering*, *171*, 108428. doi:10.1016/j.cie.2022.108428

Sethuraman, S. C., Vijayakumar, V., & Walczak, S. (2020). Cyber Attacks on Healthcare Devices Using Unmanned Aerial Vehicles. *Journal of Medical Systems*, *44*(1), 29. doi:10.100710916-019-1489-9 PMID:31838588

Shadwani, M., Sachan, S., & Sachan, P. (2018). Capacitive Sensing & Its Applications. doi:10.13140/RG.2.2.15752.32001

Shahriar, M. R., Al Sunny, S. M. N., Liu, X., Leu, M. C., Hu, L., & Nguyen, N. T. (2018) MTComm based virtualization and integration of physical machine operations with digital-twins in cyber-physical manufacturing cloud. *Proceedings - 5th IEEE International Conference on Cyber Security and Cloud Computing 2018* (pp. 46–51). IEEE. 10.1109/CSCloud/EdgeCom.2018.00018

Sharma, T., & Aarthy, S. L. (2016). An automatic attendance monitoring system using RFID and IOT using cloud. *Proc. Online Int. Conf. Green Eng. Technol. (IC-GET)*, (pp. 1-4). IEEE. 10.1109/GET.2016.7916851

Sharma, V. (2012). A study of malicious QR codes. *International Journal of Computational Intelligence and Information Security*, *3*(5), 21–26.

Sherrit, S., Bao, X., & Jones, C.M., Aldrich, J., Blodget, C.J., Moore, J., & Carson, J.W., Goullioud, R. & Donotsendmail, B. (2011). Piezoelectric stack actuator life test. *IEEE Aerospace Conference*, (pp. 1-8). IEEE.

Shewale, M., Razban, A., Deshmukh, S., Mulik, S., & Zambare, H. (2018). *Design and Implementation of Position Estimator Algorithm on Voice Coil Motor.*, *1-5*, 1–5. doi:10.1109/I2CT.2018.8529546

Shields, J. & Konefat, E. (2021). *Modeling of Piezoceramic Actuators for Control.* Intech Open. . doi:10.5772/intechopen.96727

Shin, D., & Yao, H. (2013). A User Study of Security Warnings for Detecting QR Code Based Attacks on Android Phone. *Journal of Wireless Mobile Networks, Ubiquitous Computing and Dependable Applications*, *4*(4), 49–64.

Shmueli, E., Singh, V. K., Lepri, B., & Pentland, A. (2014). Sensing, understanding, and shaping social behavior. *IEEE Transactions on Computational Social Systems*, *1*(1), 22–34. doi:10.1109/TCSS.2014.2307438

Sigov, A., Ratkin, L., Ivanov, L. A., & Xu, L. D. (2022). Emerging enabling technologies for industry 4.0 and beyond. *Information Systems Frontiers*, 1–11. doi:10.100710796-021-10213-w

Silva, J. E. (2003). *An overview of cryptographic hash functions and their uses.* (White paper). SANS. https://www.sans.org/white-papers/879/

Silva, V. L., Kovaleski, J. L., Pagani, R. N., Corsi, A., & Gomes, M. A. S. (2020). Human factor in smart industry: A literature review. *Future Studies Research Journal: Trends and Strategies*, *12*(1), 87–111. doi:10.24023/FutureJournal/2175-5825/2020.v12i1.473

Singh, V. P., Aishwarya S., Anand, U., & Sweta, J. (2013). Signal Strength based Hello Flood Attack Detection and Prevention in Wireless Sensor Networks. *Technical Report, 15*.

Singh, M., & Khilar, P. M. (2017). A Range Free Geometric Technique for Localization of Wireless Sensor Network (WSN) Based on Controlled Communication Range. *Wireless Personal Communications*, *94*(3), 1359–1385. doi:10.100711277-016-3686-x

Sleurs, K., Seys, S. F., Bousquet, J., Fokkens, W. J., Gorris, S., Pugin, B., & Hellings, P. W. (2019). Mobile health tools for the management of chronic respiratory diseases. *Allergy*, *74*(7), 1292–1306. doi:10.1111/all.13720 PMID:30644567

Song, S., Hu, J., Chen, D., Chen, W., & Wu, Z. (2020). An Event-Triggered Approach to Robust Fault Detection for Nonlinear Uncertain Markovian Jump Systems with Time-Varying Delays. *Circuits, Systems, and Signal Processing*, *39*(7), 3445–3469. doi:10.100700034-019-01327-3

Sony, M., Antony, J., Mc Dermott, O., & Garza-Reyes, J. A. (2021). An empirical examination of benefits, challenges, and critical success factors of industry 4.0 in manufacturing and service sector. *Technology in Society*, *67*, 101754. doi:10.1016/j.techsoc.2021.101754

Sood, S. K., & Mahajan, I. (2018). A Fog Assisted Cyber-Physical Framework for Identifying and Preventing Coronary Heart Disease. *Wireless Personal Communications*, *101*(1), 143–165. doi:10.100711277-018-5680-y

Soon, T. J. (2008). QR code. *Synthesis Journal*, *3*, 59–78.

Sotoodeh, K. (2019). Actuator selection and sizing for valves. *SN Applied Sciences*, *1*(10), 1207. doi:10.100742452-019-1248-z

Span'o, E., Di Pascoli, S., & Iannaccone, G. (2016). Low-power wearable ecg monitoring system for multiple-patient remote monitoring. *IEEE Sensors Journal*, *16*(13), 5452–5462. doi:10.1109/JSEN.2016.2564995

Srilatha, S., Pallavi, A. R., & Uma, R. (2017). Srinivas Koppu, Quantitative Performance Analysis of Face Recognition System. International Conference on Trends in Electronics and Informatics ICEI. IEEE.

Stevens, M., Bursztein, E., Karpman, P., Albertini, A., & Markov, Y. (2017). The First Collision for Full SHA-1. In J. Katz & H. Shacham (Eds.), Lecture Notes in Computer Science: Vol. 10401. *Advances in Cryptology – CRYPTO 2017. CRYPTO 2017*. Springer., doi:10.1007/978-3-319-63688-7_19

Streitz, N. (2019). Beyond 'smart-only' cities: Redefining the 'smart-everything' paradigm. *Journal of Ambient Intelligence and Humanized Computing*, *10*(2), 791–812. doi:10.100712652-018-0824-1

Sundaram. (2019). Smart clothes with bio-sensors for ECG monitoring. *International Journal of Innovative Technology and Exploring Engineering*, *8*(4), 298–301.

Sun, H., Oh, B.-S., Guo, X., & Lin, Z. (2019, December). Improving the Doppler Resolution of Ground-Based Surveillance Radar for Drone Detection. *IEEE Transactions on Aerospace and Electronic Systems*, *55*(6), 3667–3673. doi:10.1109/TAES.2019.2895585

Sun, R. (2020). Potential of full human–machine symbiosis through truly intelligent cognitive systems. *AI & Society*, *35*(1), 17–28. doi:10.100700146-017-0775-7

Suresh Babu, C. V. (2022). *IoT and its Applications*. Anniyappa Publications.

Suresh Babu, C. V. (2023a). *Principles of Information Security*. Anniyappa Publications.

Suresh Babu, C. V. (2023b). *Network Security*. Anniyappa Publications.

Swain, R. R., & Khilar, P. M. (2017). Composite Fault Diagnosis in Wireless Sensor Networks Using Neural Networks. *Wireless Personal Communications*, *95*(3), 2507–2548. doi:10.100711277-016-3931-3

Taleb, I., Amine Ouis, M. E., & Mammar, M. O. (2014). Access control using automated face recognition: Based on the PCA & LDA algorithms. *Proc. 4th Int. Symp. ISKO-Maghreb Concepts Tools Knowl. Manage. (ISKO-Maghreb)*, (pp. 1-5). IEEE. 10.1109/ISKO-Maghreb.2014.7033455

Tang, J., Zhang, J., Yao, L., Li, J., Zhang, L., & Su, Z. (2008). Arnetminer: extraction and mining of academic social networks. In *Proceedings of the 14th ACM SIGKDD international conference on Knowledge discovery and data mining*. ACM. 10.1145/1401890.1402008

Tao, F., Qi, Q., Wang, L., & Nee, A. Y. C. (2019). Digital Twins and Cyber–Physical Systems toward Smart Manufacturing and Industry 4.0: Correlation and Comparison. *Engineering (Beijing)*, *5*(4), 653–661. doi:10.1016/j.eng.2019.01.014

Teja, P. R., & Kumaar, A. N. (2018). QR Code based Path Planning for Warehouse Management Robot. *2018 International Conference on Advances in Computing, Communications and Informatics* (pp. 1239-1244). IEEE. 10.1109/ICACCI.2018.8554760

Teng, X.-F., Zhang, Y.-T., Poon, C. C. Y., & Bonato, P. (2008). Wearable medical systems for p-Health. *IEEE Reviews in Biomedical Engineering*, *1*, 62–74. doi:10.1109/RBME.2008.2008248 PMID:22274900

Terán, A. (2013). *El capital humano, la estrategia y el desarrollo y acumulación de capacidades, competencias organizacionales: su impacto en innovación en el sector biofarmacéutico en México.*

Thakker, M. (2013). *Microphone Sensor for Sound Analysis and Breakdown Detection Using LabVIEW*. Research Gate.

Tikani, R., Vahdati, N., & Ziaei-Rad, S. (2012). Two-Mode Operation Engine Mount Design for Automotive Applications. *Shock and Vibration, 19*(6), 1267–1280. doi:10.1155/2012/651591

Tiwari, S. (2016). An introduction to QR code technology. *2016 International Conference on Information Technology* (pp. 39-44). IEEE. 10.1109/ICIT.2016.021

Tolmach, P., Li, Y., Lin, S.-W., & Liu, Y. (2021). *Formal Analysis of Composable DeFi Protocols*. Cornell University., doi:10.1007/978-3-662-63958-0_13/arxiv.2103.00540

Tong, G. A., Li, S., Wu, W., & Du, D. Z. (2016). Effector Detection in Social Networks. *IEEE Transactions on Computational Social Systems, 3*(4), 151–163. doi:10.1109/TCSS.2016.2627811

Traganos, K., Grefen, P., Vanderfeesten, I., Erasmus, J., Boultadakis, G., & Bouklis, P. (2021). The HORSE framework: A reference architecture for cyber-physical systems in hybrid smart manufacturing. *Journal of Manufacturing Systems 61*(November 2020), 461–494. doi:10.1016/j.jmsy.2021.09.003

Tripathi, D., Biswas, A., Tripathi, A. K., Singh, L. K., & Chaturvedi, A. (2022). *An integrated approach of designing functionality with security for distributed cyber-physical systems*. Springer US. doi:10.100711227-022-04481-9

Tsao, K.-Y., Girdler, T., & Vassilakis, V. G. (2022). A survey of cyber security threats and solutions for UAV communications and flying ad-hoc networks. *Ad Hoc Networks, 133*, 1570-8705

Tsigkanos, C., Kehrer, T., & Ghezzi, C. (2016). Architecting dynamic cyber-physical spaces. *Computing, 98*(10), 1011–1040. doi:10.100700607-016-0509-6

Tsukima, M., Kai, T., Takahashi, K., Takeuchi, T., & Koyama, K. (2017). Development of Tri-Stable Rotary Solenoid Actuator. *IEEJ Transactions on Industry Applications., 137*(8), 681–686. doi:10.1541/ieejias.137.681

Tutty, L. M., Rothery, R. M., & Grinnell, J. R. (1996). *Analyzing Your Own Data, Qualitative Research for Social Workers; Publisher*. Allyn and Bacon.

Tyszberowicz, S., & Faitelson, D. (2020). Emergence in cyber-physical systems: Potential and risk. *Frontiers of Information Technology and Electronic Engineering, 21*(11), 1554–1566. doi:10.1631/FITEE.2000279

Ural, N., & Örenç, Ö. (2019). *Uygulamalı Şifreleme ve Şifre Çözme Yöntemleri*. Pusula.

Van-Dijck, J., & Poell, T. (2013). Understanding social media logic. *Media and Communication, 1*(1), 2–14. doi:10.17645/mac.v1i1.70

Vangipuram, S. L. T., Mohanty, S. P., & Kougianos, E. (2021). CoviChain: A Blockchain Based Framework for Nonrepudiable Contact Tracing in Healthcare Cyber-Physical Systems During Pandemic Outbreaks. *SN Computer Science, 2*(5), 1–16. doi:10.100742979-021-00746-x PMID:34179827

Veeman, Shree, M. V., Sureshkumar, P., Jagadeesha, T., Natrayan, L., Ravichandran, M., & Paramasivam, P. (2021). Sustainable development of carbon nanocomposites: Synthesis and classification for environmental remediation. *Journal of Nanomaterials*, *2021*, 1–21. doi:10.1155/2021/5840645

Velmurugan, G., Siva Shankar, V., Kaliappan, S., Socrates, S., Sekar, S., Patil, P. P., Raja, S., Natrayan, L., & Bobe, K. (2022). Effect of aluminium tetrahydrate nanofiller addition on the mechanical and thermal behaviour of luffa fibre-based polyester composites under cryogenic environment. *Journal of Nanomaterials*, *2022*, 1–10. doi:10.1155/2022/5970534

Venedikian, G. (2022). *Message Digest vs Secure Hash Algorithm vs RIPEMD*. CUNY New York City College of Technology. https://www.coursehero.com/file/56888501/hash-comparisondocx/

Verma, R. (2022). Smart City Healthcare Cyber Physical System: Characteristics, Technologies and Challenges. *Wireless Personal Communications*, *122*(2), 1413–1433. doi:10.100711277-021-08955-6 PMID:34462622

Von Rosenberg, W., Chanwimalueang, T., Goverdovsky, V., Looney, D., Sharp, D., & Mandic, D. P. (2016). Smart helmet: Wearable multichannel ecg and eeg. *IEEE Journal of Translational Engineering in Health and Medicine*, *4*, 4. doi:10.1109/JTEHM.2016.2609927 PMID:27957405

Wan, B., Xu, C., Mahapatra, R. P., & Selvaraj, P. (2021). Understanding the Cyber-Physical System in International Stadiums for Security in the Network from Cyber-Attacks and Adversaries using AI. *Wireless Personal Communications*, *127*(2), 1207–1224. doi:10.100711277-021-08573-2

Wane, A. R., Jamankar, S. P., & Chandure, O. V. (2013). An effective mechanism for ensuring security of QR code. *International Journal of Advanced Research in Computer Science*, *4*(6), 175–179.

Wang, D., Wu, S., Lin, Z., Wu, L., Yuan, X., Zhou, Y., Wang, H., & Ren, K. (2021). Towards a First Step to Understand Flash Loan and Its Applications in DeFi Ecosystem. *Proceedings of the Ninth International Workshop on Security in Blockchain and Cloud Computing*. ACM. 10.1145/3457977.3460301

Wang, X., Xu, L., & Li, X. (2020). *Smart Manufacturing for Industry 4.0: A Review*. Research Gate. https://www.researchgate.net/publication/321280578_Intelligent_Manufacturing_in_the_Context_of_Industry_40_A_Review

Wang, X., Yin, Y. L., & Yu, H. (2005). Finding Collisions in the Full SHA-1. In Shoup, V. (eds) Advances in Cryptology – CRYPTO 2005. Springer, Berlin, Heidelberg. doi:10.1007/11535218_2

Wang, Y., Lin, Z., Liang, X., Xu, W., Yang, Q., & Yan, G. (2016). On modeling of electrical cyber-physical systems considering cyber security. *Frontiers of Information Technology and Electronic Engineering* *17*(5), 465–478. doi:10.1631/FITEE.1500446

Wang, B., Liu, C., Xiao, Y., Zhong, J., Li, W., Cheng, Y., Hu, B., Huang, L., & Zhou, J. (2017). Ultrasensitive cellular fluorocarbon piezoelectret pressure sensor for self-powered human physiological monitoring. *Nano Energy*, *32*, 42–49. doi:10.1016/j.nanoen.2016.12.025

Wang, B., Zhou, H., Yang, G., Li, X., & Yang, H. (2022). Human Digital Twin (HDT) Driven Human-Cyber-Physical Systems: Key Technologies and Applications. [English Edition]. *Chinese Journal of Mechanical Engineering, 35*(1), 11. doi:10.118610033-022-00680-w

Wang, X., Chen, Y., & Wu, D. (2020). Cyber-physical systems and Industry 4.0: A comprehensive review. *IEEE Access : Practical Innovations, Open Solutions, 8*, 135948–135960.

Wang, Y., Fagiani, F. E., Ho, K. E., & Matson, E. T. (2022). A Feature Engineering Focused System for Acoustic UAV Payload Detection. *International Conference on Agents and Artificial Intelligence.* Scite Press. 10.5220/0010843800003116

Wan, J., Cai, H., & Zhou, K. (2015). Industrie 4.0: enabling technologies. In: *Proceedings of 2015 international conference on intelligent computing and internet of things,* (pp. 135–140). IEEE.

Wan, J., Yan, H., Li, D., Zhou, K., & Zeng, L. (2013). Cyber-physical systems for optimal energy management scheme of autonomous electric vehicle. *The Computer Journal, 56*(8), 947–956. doi:10.1093/comjnl/bxt043

Wan, K., & Alagar, V. (2014). Context-aware security solutions for cyber-physical systems. *Mobile Networks and Applications, 19*(2), 212–226. doi:10.100711036-014-0495-x

Warasart, M., & Kuacharoen, P. (2012). Paper-based document authentication using digital signature and QR code. *4th International Conference on Computer Engineering and Technology*. Springer.

Wazwaz, A. A., Herbawi, A. O., Teeti, M. J., & Hmeed, S. Y. (2018). Raspberry Pi and computers-based face detection and recognition system. *Proc. 4th Int. Conf. Comput. Technol. Appl. (ICCTA)*, (pp. 171-174). IEEE. 10.1109/CATA.2018.8398677

Wei, C., Wu, C., Fu, X., Dong, X., He, K., Hong, J., & Wang, X. (2022). Preimage attacks on 4-round keccak by solving multivariate quadratic systems. *International Conference on Information Security and Cryptology*. Springer. 10.1007/978-3-031-08896-4_10

Wei, S., Liao, X., Pang, J., & Zhou, Y. (2021). Recent Progress of Fluxgate Magnetic Sensors: Basic Research and Application. *Sensors (Basel), 21*(4), 1500. doi:10.339021041500 PMID:33671507

Wei-Wei, C. (2017). China cardiovascular diseases report 2015: A summary. *Journal of Geriatric Cardiology : JGC, 14*(1), 1. PMID:28270835

White, J., Clarke, S., Groba, C., Dougherty, B., Thompson, C., & Schmidt, D. C. (2010). R&D challenges and solutions for mobile cyber-physical applications and supporting Internet services. *Journal of Internet Services and Applications, 1*(1), 45–56. doi:10.100713174-010-0004-9

Wijesundera, R. P. & Abeyratne, M. & Kalupahana, I. & Buddhika, W. & Kumara, S., Jayatissa, N., Harambearachchi, J., & Kalingamudali, S. (2012). *An Inexpensive Universal Serial Bus Interfaced Home Automation System.* Research Gate.

Wolf, M., & Serpanos, D. (2017). Safety and security in cyber-physical systems and internet-of-things systems. *Proceedings of the IEEE, 106*(1), 9–20. doi:10.1109/JPROC.2017.2781198

World Economic Forum. (2020). *The Future of Jobs Report 2020*. WEF. https://www.weforum.org/reports/the-future-of-jobs-report-2020

Wu, Q., Shi, Y., & Wang, Q. (2020). *Industry 4.0 and Cyber-Physical Systems: A Comprehensive Review*. Springer. https://link.springer.com/article/10.1007/s10796-021-10153-5?

Wu, C., Guo, S., Wu, Y., Ai, J., & Xiong, N. N. (2020). Networked Fault Detection of Field Equipment from Monitoring System Based on Fusing of Motion Sensing and Appearance Information. *Multimedia Tools and Applications*, *79*(23–24), 16319–16348. doi:10.100711042-020-08885-8

Wu, D., Greer, M. J., Rosen, D. W., & Schaefer, D. (2013). Cloud manufacturing: Strategic vision and state-of-the-art. *Journal of Manufacturing Systems*, *32*(4), 564–579. doi:10.1016/j.jmsy.2013.04.008

Wu, G., & Li, Z. S. (2021). Cyber—Physical Power System (CPPS): A review on measures and optimization methods of system resilience. *Frontiers of Engineering Management*, *8*(4), 503–518. doi:10.100742524-021-0163-3

Wurm, J., Jin, Y., Liu, Y., Hu, S., Heffner, K., Rahman, F., & Tehranipoor, M. (2017). Introduction to Cyber-Physical System Security: A Cross-Layer Perspective. *IEEE Transactions on Multi-Scale Computing Systems*, *3*(3), 215–227. doi:10.1109/TMSCS.2016.2569446

Xie, J., Tan, W., Fang, B., & Huang, Z. (2021). Towards a Statistical Model Checking Method for Safety-Critical Cyber-Physical System Verification. *Security and Communication Networks*, *2021*, 1–12. doi:10.1155/2021/5536722

Xu, K., Jiang, X., & Chen, W. (2019). Photoplethysmography motion artifacts removal based on signal noise interaction modeling utilizing envelope filtering and time-delay neural network. *IEEE Sensors Journal*, *20*(7), 3732–3744. doi:10.1109/JSEN.2019.2960370

Xu, K., Jiang, X., Lin, S., Dai, C., & Chen, W. (2020). Stochastic modeling based nonlinear bayesian filtering for photoplethysmography de-noising in wearable devices. *IEEE Transactions on Industrial Informatics*, *16*(11), 7219–7230. doi:10.1109/TII.2020.2988097

Xu, L., Liu, Z., Zhai, H., Chen, X., Sun, R., Lyu, S., Yangyang, F., Yi, Y., Chen, Z., Jin, L., Zhang, J., Li, Y., & Ye, T. (2020). Moisture-Resilient Graphene-Dyed Wool Fabric for Strain Sensing. *ACS Applied Materials & Interfaces*, (11), 13265–13274. doi:10.1021/acsami.9b20964 PMID:32105063

Xu, X., Lu, Y., Vogel-Heuser, B., & Wang, L. (2021). Industry 4.0 and Industry 5.0—Inception, conception and perception. *Journal of Manufacturing Systems*, *61*, 530–535. doi:10.1016/j.jmsy.2021.10.006

Yamin, M., & Sen, A. A. (2018). Improving Privacy and Security of User Data in Location Based Services. International Journal of Ambient Computing. *Intel.*, *9*(1), 19–42.

Yang, F., & Gu, S. (2021). Industry 4.0, a revolution that requires technology and national strategies. *Complex & Intelligent Systems*, *7*(3), 1311–1325. doi:10.100740747-020-00267-9

Yang, G., Pang, G., Pang, Z., Gu, Y., Mantysalo, M., & Yang, H. (2019). Non-invasive flexible and stretchable wearable sensors with nanobased enhancement for chronic disease care. *IEEE Reviews in Biomedical Engineering, 12*, 34–71. doi:10.1109/RBME.2018.2887301 PMID:30571646

Yang, J., Lei, Z., Yi, D., & Li, S. (2015, April). Person-specific face anti spoofing with subject domain adaptation. *IEEE Transactions on Information Forensics and Security, 10*(4), 797–809. doi:10.1109/TIFS.2015.2403306

Yao, X., Zhou, J., Lin, Y., Li, Y., Hongnian, Y., & Liu, Y. (2019). Smart manufacturing based on cyber-physical systems and beyond. *Journal of Intelligent Manufacturing, 30*(8), 2805–2817. doi:10.100710845-017-1384-5

Yildiz, S., Dorman, L., Kuperman, W. A., Sabra, K., Roux, P., Green, D., Fried, S., & Schmidt, H. (2014). Using hydrophones as vector sensors. *The Journal of the Acoustical Society of America, 135*(4_Supplement), 2361. doi:10.1121/1.4877778

Yin and X. & Liu, X. (2018). Multi-Task Convolutional Neural Network for Pose-Invariant Face Recognition. *IEEE Transactions On Image Processing, 27*(2).

Yogeshwaran. (2015). Mechanical properties of leaf ashes reinforced aluminum alloy metal matrix composites. *International Journal of Applied Engineering Research: IJAER, 10*(13), 11048–11052.

Yoo, H., & Shon, T. (2016). Challenges and research directions for heterogeneous cyber—physical system based on IEC 61850: Vulnerabilities, security requirements, and security architecture. *Future Generation Computer Systems, 61*, 128–136. doi:10.1016/j.future.2015.09.026

Young, S., Hung, J., & Li, Z. (2020). Industry 4.0 and Cyber-Physical Systems: A Review of Recent Advances. *World Scientific*.

You, S. (2020). *Unsupervised automatic seizure detection for focal-onset seizures recorded with behind-the-ear eeg using an anomaly-detecting generative adversarial network, Comput*. Methods Progr. Biomed. doi:10.1016/j.cmpb.2020.105472

Yu, Y., & Zichichi, A. (2021). Enhancing Trust in Trust Services: Towards an Intelligent Human-input-based Blockchain Oracle (IHiBO). *Proceedings of the 54th Hawaii International Conference on System Sciences,* (pp. 1-10). Scholar Space. 10.24251/HICSS.2021.001

Yuan, H., Fauroux, J.-C., Chapelle, F., & Balandraud, X. (2017). A review of rotary actuators based on shape memory alloys. *Journal of Intelligent Material Systems and Structures*. . doi:10.1177/1045389X16682848

Zangiacomi, A., Pessot, E., Fornasiero, R., Bertetti, M., & Sacco, M. (2020). Moving towards digitalization: A multiple case study in manufacturing. *Production Planning and Control, 31*(2-3), 143–157. doi:10.1080/09537287.2019.1631468

Zennaro, M., Bezzateev, S., Giunchiglia, F., & Zennaro, M. (2020). Cyber-Physical Systems in Industry 4.0: A Survey. *IEEE Access : Practical Innovations, Open Solutions, 8*, 187093–187116.

Zetzsche, D. A., Arner, D. W., & Buckley, R. P. (2020). Decentralized Finance. *Journal of Financial Regulation*, *6*(2), 172–203. doi:10.1093/jfr/fjaa010

Zhang, H., Li, K., & Zhao, Y. (2020). *Cyber-Physical Systems and Industry 4.0: A Review of Recent Advances*. RSC. https://pubs.rsc.org/en/content/articlelanding/2020/qm/d0qm00025f

Zhang, H., Zhang, J., Hu, Z., Quan, L., Shi, L., Chen, J., Xuan, W., Zhang, Z., Dong, S., & Luo, J. (2019). Waist-wearable wireless respiration sensor based on triboelectric effect. *Nano Energy*, *59*, 75–83. doi:10.1016/j.nanoen.2019.01.063

Zhang, Han, Ge, Ding, Ding, & Yue, & Duo. (2020). A Survey Of Cyber Attacks On Cps: Recent Advances And Challenges. IEEE.

Zhang, J., Tang, J., & Li, J. (2007). *Expert finding in a social network*. In *12th International Conference on Database Systems for Advanced Applications*, Bangkok, Thailand.

Zhang, Q., Zeng, X., Hu, W., & Zhou, D. (2017). A machine learning-empowered system for long-term motion tolerant wearable monitoring of blood pressure and heart rate with ear-ecg/ppg. *IEEE Access : Practical Innovations, Open Solutions*, *5*, 10547–10561. doi:10.1109/ACCESS.2017.2707472

Zhang, S., & Yoshino, K. (2008). DWT-Based Watermarking Using QR Code. *Science Journal of Kanagawa University*, *19*, 3–6.

Zhang, X., & Yan, X. (2011). Continuous Rotary Motor Actuated by Multiple Segments of Shape Memory Alloy Wires. *Journal of Materials Engineering and Performance*, *21*(12), 2643–2649. doi:10.100711665-012-0390-1

Zhao, Y. (2022). Towards Trustworthy DeFi Oracles: Past, Present and Future. arXiv preprint arXiv:2201.02358

Zhao, J., Li, N., Yu, H., Wei, Z., Liao, M., Chen, P., Wang, S., Shi, D., Sun, Q., & Zhang, G. (2017). Highly sensitive mos2 humidity sensors array for noncontact sensation. *Advanced Materials*, *29*(34), 1702076. doi:10.1002/adma.201702076 PMID:28692765

Zhou, Y., Yao, W., He, X., Li, T., Wang, S., & Han, Y. (2022). Flow velocity sensors arrangement for vegetated channels. *Frontiers in Plant Science*, *13*, 960103. doi:10.3389/fpls.2022.960103 PMID:36035729

Zhu, X.-G & Wang, Y. (2006). *Research and design of permanent magnet actuator for higher opening capability*. Research Gate.

Zhu, L., Tang, X., Shen, M., Du, X., & Guizani, M. (2018, March). Guizani. Privacy-Preserving DDoS Attack Detection Using Cross- Domain Traffic in Software Defined Networks. *IEEE Journal on Selected Areas in Communications*, *36*(3), 628–643. doi:10.1109/JSAC.2018.2815442

Zuo, J.-Y & Yan, G.-Z & Gao, Z.-J & Lu, Q.-H. (2005). *New piezoelectric linear actuator*. Research Gate.

About the Contributors

Thanigaivelan Rajasekaran currently works at A.K.T.Memorial College of Engineering and Technology does research in Mechanical Engineering and Manufacturing Engineering. Their current project is 'Natural convective heat transfer' and surface modification of stainless steel

C. Jegadheesan serves as an Associate Professor at the Kongu Engineering College. He specializes in Mechatronics System Design, Electric Mobility and Hydrogen Fuel.

Kaliappan S. is Head of the Mechanical and Mechatronics Engineering Department. at Velammal Institute of Technology, faculty member since 2011. He completed his Ph.D. and M.E at Anna University, Chennai and his B.E at R.V.S College of Engineering & Technology, Dindigul. He is having more than 23 years experience . His research interests lie in the area of Computational Fluid Dynamics, Thermal Sciences, Internal Combustion Engine, Automotive system, Research Methodologies, Composites, etc. He has served as technical committee head in conferences and workshop and worked as the reviewers and Editorial Board Member for several National and International Journals. He published almost 53 research articles in various national and international journals and conferences, organized STTPs, FDPs, Conferences and other technical events. He holds international Australian Patent Grants and published Indian Patents .He co-authored 15 books . He is Life member in ISTE, International Association of Engineers, member in Universal Association of Mechanical and Aeronautical Engineers, Institute Of Research Engineers And Doctors. He got Young Researcher Award in InSc awards-2021

* * *

Backiyaraj A. has completed bachelor degree in mechatronics and master degree in Automobile engineering.

Chinnahajisagari Mohammad Akram is currently an Assistant Professor at Department of Mechanical Engineering, S.R.K. Institute of Technology, Vijayawada, Andhra Pradesh. He received his M.Tech. in Thermal Engg. from K.L. (Deemed-to-be) University, Andhra Pradesh and his B.E. in Mechanical from Osmania University, Hyderabad. Mr. Akram, jointly published five Indian patents including two grants, several research articles in peer-reviewed international journals / conferences and co-authored two books as well as a book chapter. He is a life member in several professional bodies like ISDS, SESD, I2OR, IFERP, SDIWC, IAEng, etc. and is a reviewer to refereed international journals such as Biofuels, JERR. Mr. Akram has been an active participant in a range of learning and development events like FDPs, STTPs, workshops, webinars, conferences. His research areas of interest cover Solar Desalination, Alternate Fuels, Heat Transfer, Renewable Energy, etc.

Aykut Aydın holds an MSc degree in cyber security. His current research interests focus on QR code related threats.

Atchaya B. works in the Department of ECE, Velammal Institute of Technology, Chennai.

C.V. Suresh Babu is a pioneer in content development. A true entrepreneur, he founded Anniyappa Publications, a company that is highly active in publishing books related to Computer Science and Management. Dr. C.V. Suresh Babu has also ventured into SB Institute, a center for knowledge transfer. He holds a Ph.D. in Engineering Education from the National Institute of Technical Teachers Training & Research in Chennai, along with seven master's degrees in various disciplines such as Engineering, Computer Applications, Management, Commerce, Economics, Psychology, Law, and Education. Additionally, he has UGC-NET/SET qualifications in the fields of Computer Science, Management, Commerce, and Education. Currently, Dr. C.V. Suresh Babu is a Professor in the Department of Information Technology at the School of Computing Science, Hindustan Institute of Technology and Science (Hindustan University) in Padur, Chennai, Tamil Nadu, India.

Kirubadurai Balaraman received the B.E. degree in Aeronautical Engineering from Veltech Engineering College, Chennai. He got master's degree in Thermal Engineering from Government College Of Technology, Coimbatore. He has eight years working experience in teaching. Currently working as an Assistant Professor in Aeronautical department in Vel Tech Rangarajan Dr.Sagunthala R&D Institute of Science and Technology, Chennai, India. His research interests are in jet engine component design, heat transfer, nanofluid heat transfer.

Geetha C. is working as a Professor in the Department of Computer Science and Engineering. She graduated in Computer Science at Madurai Kamaraj University, Tamilnadu, India. She completed Master Degree in Computer Science and Engineering at Sathyabama University, Chennai, India. She is awarded with Ph.D. in the field of Wireless Sensor Networks at Manonmaniam Sundaranar University, Tirunelveli, India. She has teaching experience of 25 years. She has around 25 papers in National and International journal and conference publications. Her main area of interest includes Artificial Intelligence, Sensor Networks and IoT.

Eduardo Cornejo-Velazquez is an Investigate Professor at the Autonomous University of the State of Hidalgo. His research interests are related to Digital Transformation and Strategic Innovation. He holds a Ph.D. in Strategic Planning and Technology Management (Popular Autonomous University of the State of Puebla.), a Master's Degree in Computer Ciences (Autonomous University of the State of Hidalgo). He has publications in several national and international journals and publishers. He is a member of the National System of Researchers in Mexico.

Swethi E. completed her B.E in the year 2022 with first Class. Currently working in software company.

Kaviha G., Department of ECE, Velammal Institute of Technology, Chennai

Malathi G. is working as a Professor in the School of Computer Science and Engineering, Vellore Institute of Technology, Chennai, Tamil Nadu, India. She has over 20 years of experience in teaching and research. Her area of specialization is Image Processing and Healthcare Analytics. She has been a resource person in FDPs, workshops, and Seminars. Currently, she is guiding 3 Ph.D. scholars and has several publications in reputed International Journals. She has authored a few book chapters. She has filed and published two patents. She received Best Outstanding Faculty Award in Computer Science 'by VENUS Foundation in July 2018.

Sangeetha Ganesan received her B.E degree in Computer Science and Engineering from Periyar University Salem, M.E degree in Computer Science and Engineering from Anna University, Chennai, and Ph.D., in Faculty of Information and Communication Engineering, Anna University, Chennai. She is currently working as an Associate Professor in R.M.K. College of Engineering and Technology,

Puduvoyal. Her research interests are Distributed Computing, Cloud Computing, Security, Data Science and Machine Learning.

Sethuram Rao G. is working as an Assistant Professor in Velammal Institute of Technology, Chennai. He has 18 years of teaching experience. He has published many papers in international journals and conference proceedings. His research interest includes image processing, IOT, deep learning, machine learning and embedded systems.

Shanmugaraj G. received his Ph.D. in Anna University, Chennai. He has received the M. E. degree in Applied Electronics from University of Madras, Chennai. He has published several papers in International Journals and Conferences. He also published two Engineering books and two patent publications. He has a teaching experience of 20 years. He is currently working in Velammal Institute of Technology as Associate Professor in Electronics and Communication Engineering Department. His present research interests include VLSI signal processing algorithms Machine learning, Networking with IoT.

Arunkumar K. is working as Assistant professor in Aeronautical Department, MLR Institute of Technology. I am specialized in Aerodynamics (Morphing Analysis of wings and propellers).

Mohanambal K. is working as an Assistant Professor in Velammal Engineering College. He has received the M. E. degree in Computer Science from Anna University, India in May 2007, and B. E. degree in Information Technology from Bharathiyar University in April 2004. He has published papers in various International Conferences and International Journals in the domain of Machine Learning and IoT.

Natrayan L. is currently working as an Assistant Professor in the Department of Mechanical Engineering, SIMATS deemed University – Chennai, India. He completed PhD from Vellore Institute of Technology - India. He received B.E (Mechanical Engineering), M.E (Engineering Design) degrees from Anna University - Chennai and MBA (Operational Management) degree from RVD University-Rajasthan, India. His research work involves the study of Non-homogenous materials and advanced optimization.

Jeyamani Latha, Associate professor, Department of Electronics and Communication Engineering, Velammal Institute of Technology, Chennai. She received the B.E degree in Electronics and Communication Engineering from Government College of Engineering, Tirunelveli, India in 1989 and M.E degree in Applied

Electronics from Karunya University, Coimbatore, India in 2007. She works currently as an Associate professor, Department of Electronics and Communication Engineering, Velammal Institute of Technology, Chennai; and she has 18 years of teaching experience. She is a lifetime member of ISTE with number LM60834 from 2009 onwards.

Ramya Maranan is an accomplished researcher working in the Department of Research and Innovation at Saveetha School of Engineering, SIMATS in Chennai, India. With a passion for pushing the boundaries of knowledge and driving innovation, Ramya plays a vital role in advancing the research activities of the institution. Ramya's work primarily revolves around conducting research and development activities within their area of specialization. They are involved in designing and executing experiments, collecting and analyzing data, and disseminating their findings through scholarly publications. Ramya's dedication to research demonstrates its commitment to advancing scientific understanding and promoting technological advancements. Their work has the potential to create a positive impact on society and contribute to the overall academic and scientific community.

Clavel-Maqueda Mireya has a PhD in Strategic Planning and Technology Management from the Universidad Popular Autónoma del Estado de Puebla. She is currently a professor at the Universidad Autónoma del Estado de Hidalgo. Scientific interests: Digitalization and Technological Innovation Strategies. She is a member of the National System of Researchers in Mexico.

Susmitha N. works at the Department of ECE, Velammal Institute of Technology., Chennai.

Öner Özer is a versatile network engineer with a background involving complex and large telecommunications projects for over 10+ years. He holds an MSc degree in cyber security.

Deivendran P. is working as an Associate Professor in the Department of Information Technology at Velammal Institute of Technology. He graduated in Mathematics at Madurai Kamaraj University, Chennai,Tamilnadu, India. He Master degree Engineering in Information Technology at Sathyabama University, Chennai, India. He is awarded Ph.D., in the field of Cloud Computing at Manonmaniam Sundaranar University,Tirunelveli, India. He is in teaching profession for more than 21 years. He has presented and published number of papers in National and International Journals, Conference, Patents and Books. His main area of interest includes Cloud Computing, Computer Networks and Web Service.

Suresh Babu P. is working as an Assistant Professor in the Department of Information Technology at Velammal College of Engineering and Technology . He is awarded Master in the field of Image Processing at Anna University,Chennai India. He is in teaching profession for more than 10 years. He has presented and published number of papers in National and International Journals, Conference, Patents and Books. His main area of interest includes Computer Networks and Image Processing.

Jaganraj R. is Head / Aeronautical Engineering at the Vel Tech University, Chennai. He is DGCA certified drone pilot and mentor for drone based startups. His research interests are in product development in Unmanned Systems, Drones, and Defense systems & mathematical modeling. He developed various products in drones & their technologies as part of funded & consultancy projects in collaboration with national & international industries & institutes.

Kavitha Rajarathinam is working as a Professor in Velammal College of Engineering and Technology, Madurai. She is having nearly 24 years of experience in teaching. She completed B.E in R.V.S College of Engineering and Technology affiliated to Madurai Kamaraj University, Tamilnadu, India, in the year 1997 and M.E in Computer Science and Engineering in Annamalai University, Chidambaram., Tamilnadu, India, in the year 2004. She obtained her Ph.D in software testing in Anna University, Tamilnadu, India, in the year 2014. Her area of interest includes software engineering, Software testing and Data Mining and analytics. She has published few books and also she has published many research papers in international and national journals, international and national conferences.

Kaliappan S. is an accomplished researcher working in the Division of Research and Development at Lovely Professional University. With a keen interest in advancing knowledge and innovation, Kaliappan plays a vital role in contributing to the university's research endeavours. Kaliappan`s work primarily focuses on conducting research and development activities within their area of specialization. They are likely involved in designing and executing experiments, collecting and analyzing data, and publishing their findings in reputable scientific journals. Through their research efforts, Kaliappan aims to contribute to the existing body of knowledge and drive innovation in their field.

Radhika S. is a post graduate in Computer Science and Engineering and has completed her Ph.D. in the field of Image Processing and Neural Networks. She has around 27 years of experience in teaching and 18 years in research. She has around 21 publications in International Journals and Conferences and participated in various Faculty Development Programmes that enhanced her knowledge in technical arenas. Her research areas include Data Deduplication and Machine learning. She has bagged the Best Teacher award twice in her career and was also awarded with the IBM TGMC Mentor award. She has delivered lectures in various Faculty Development Programmes and has been invited as Chair/keynote speaker in Conferences.

Socrates S. is an accomplished professional currently working at Velammal Institute of Technology in Chennai. With a strong educational background, he holds a Master's degree in Plastic Technology from CIPET Chennai. Socrates has developed a keen interest in the field of polymer composites and is recognized as an active and dedicated researcher in this area. Throughout his career, Socrates has demonstrated a passion for exploring the applications and advancements in polymer composites. His research efforts have focused on enhancing the mechanical properties, durability, and sustainability of polymer-based materials. Socrates has actively contributed to the development of innovative solutions in the field, aiming to improve the performance and versatility of these materials in various industries.

Soundararajan S. is working as a Professor in the Department of COMPUTER SCIENCE and ENGINEERING at Velammal Institute of Technology. He secured Master of Engineering in Computer Science and Engineering at Sathyabama Institute of Science and Technology, Chennai, Tamilnadu, India. He secured Ph.D. in Computer Science and Engineering at Anna University, Chennai, Tamilnadu, India. He is in the field of Computer Science and Engineering at Velammal Institute of Technology, Chennai, TamilNadu, India.. He is in teaching profession for more than 24 years. He has presented 36 number of papers in National and International Journals, Conference and Symposiums. His main area of interest includes Mobile Computing, Computer Networks and Artificial Intelligence.

S. Selvakanmani received M.E degree in Digital Communication and Network Engineering from Arulmigu Kalasalingam College of Engineering, in 2006 and completed her Doctoral degree from Anna University, Chennai in 2019. She is currently working as Associate Professor in the department of Information Technology in R.M.K Engineering College, Chennai, India. She is having more than 15 years of experience in academic, administration, research and innovations. She has published more than 23 research papers in various SCI and Scopus indexed journals. She is a recipient of Women Researcher Award from ScienceFather 2020. She acted as an expert member of various world level International conferences organized by IEEE, Springer and etc. Her research interest includes Wireless communication, Machine learning, Deep learning, Data Science and Computer algorithms. She is a member of IEEE, ACM and CSI.

Raju Shanmugam is a Teacher, Researcher and Administrator having 30 years of higher education institutions

G. Vanya Sree is working as an Assistant professor in CVR college of engineering and has 8 years of teaching experience.

Gurkan Tuna is a Professor at the Department of Computer Programming of Trakya University (2006–. . ..). He has authored several papers in international conference proceedings and refereed journals, and has been actively serving as a reviewer for international journals and conferences. His current research interests include smart grid, ad hoc and sensor networks, machine learning, and cyber security.

Yogalakshmi V. completed her B.E in the year 2022 with First class.currently working in software compnany

Index

P

Q

R

S

T

V

W

Premier Reference Source
Tax Audit and Taxation in the Paradigm of Sustainable Development
TAX

Premier Reference Source
Controlling Epidemics With Mathematical and Machine Learning Models

Premier Reference Source
School-Museum Relationships and Teaching Social Sciences in Formal Education

Premier Reference Source
Sociological Perspectives on Sustainable Development and Poverty Reduction in Rural Populations

www.ingramcontent.com/pod-product-compliance
Lightning Source LLC
LaVergne TN
LVHW080611300926
849162LV00005B/1119